Drug Delivery to the Oral Cavity

Molecules to Market

DRUGS AND THE PHARMACEUTICAL SCIENCES

DRUGS AND THE PHARMACEUTICAL SCIENCES

A Series of Textbooks and Monographs

1. Pharmacokinetics, *Milo Gibaldi and Donald Perrier*
2. Good Manufacturing Practices for Pharmaceuticals: A Plan for Total Quality Control, *Sidney H. Willig, Murray M. Tuckerman, and William S. Hitchings IV*
3. Microencapsulation, *edited by J. R. Nixon*
4. Drug Metabolism: Chemical and Biochemical Aspects, *Bernard Testa and Peter Jenner*
5. New Drugs: Discovery and Development, *edited by Alan A. Rubin*
6. Sustained and Controlled Release Drug Delivery Systems, *edited by Joseph R. Robinson*
7. Modern Pharmaceutics, *edited by Gilbert S. Banker and Christopher T. Rhodes*
8. Prescription Drugs in Short Supply: Case Histories, *Michael A. Schwartz*
9. Activated Charcoal: Antidotal and Other Medical Uses, *David O. Cooney*
10. Concepts in Drug Metabolism (in two parts), *edited by Peter Jenner and Bernard Testa*
11. Pharmaceutical Analysis: Modern Methods (in two parts), *edited by James W. Munson*
12. Techniques of Solubilization of Drugs, *edited by Samuel H. Yalkowsky*
13. Orphan Drugs, *edited by Fred E. Karch*
14. Novel Drug Delivery Systems: Fundamentals, Developmental Concepts, Biomedical Assessments, *Yie W. Chien*
15. Pharmacokinetics: Second Edition, Revised and Expanded, *Milo Gibaldi and Donald Perrier*
16. Good Manufacturing Practices for Pharmaceuticals: A Plan for Total Quality Control, Second Edition, Revised and Expanded, *Sidney H. Willig, Murray M. Tuckerman, and William S. Hitchings IV*
17. Formulation of Veterinary Dosage Forms, *edited by Jack Blodinger*
18. Dermatological Formulations: Percutaneous Absorption, *Brian W. Barry*
19. The Clinical Research Process in the Pharmaceutical Industry, *edited by Gary M. Matoren*
20. Microencapsulation and Related Drug Processes, *Patrick B. Deasy*
21. Drugs and Nutrients: The Interactive Effects, *edited by Daphne A. Roe and T. Colin Campbell*
22. Biotechnology of Industrial Antibiotics, *Erick J. Vandamme*

23. Pharmaceutical Process Validation, *edited by Bernard T. Loftus and Robert A. Nash*
24. Anticancer and Interferon Agents: Synthesis and Properties, *edited by Raphael M. Ottenbrite and George B. Butler*
25. Pharmaceutical Statistics: Practical and Clinical Applications, *Sanford Bolton*
26. Drug Dynamics for Analytical, Clinical, and Biological Chemists, *Benjamin J. Gudzinowicz, Burrows T. Younkin, Jr., and Michael J. Gudzinowicz*
27. Modern Analysis of Antibiotics, *edited by Adjoran Aszalos*
28. Solubility and Related Properties, *Kenneth C. James*
29. Controlled Drug Delivery: Fundamentals and Applications, Second Edition, Revised and Expanded, *edited by Joseph R. Robinson and Vincent H. Lee*
30. New Drug Approval Process: Clinical and Regulatory Management, *edited by Richard A. Guarino*
31. Transdermal Controlled Systemic Medications, *edited by Yie W. Chien*
32. Drug Delivery Devices: Fundamentals and Applications, *edited by Praveen Tyle*
33. Pharmacokinetics: Regulatory • Industrial • Academic Perspectives, *edited by Peter G. Welling and Francis L. S. Tse*
34. Clinical Drug Trials and Tribulations, *edited by Allen E. Cato*
35. Transdermal Drug Delivery: Developmental Issues and Research Initiatives, *edited by Jonathan Hadgraft and Richard H. Guy*
36. Aqueous Polymeric Coatings for Pharmaceutical Dosage Forms, *edited by James W. McGinity*
37. Pharmaceutical Pelletization Technology, *edited by Isaac Ghebre-Sellassie*
38. Good Laboratory Practice Regulations, *edited by Allen F. Hirsch*
39. Nasal Systemic Drug Delivery, *Yie W. Chien, Kenneth S. E. Su, and Shyi-Feu Chang*
40. Modern Pharmaceutics: Second Edition, Revised and Expanded, *edited by Gilbert S. Banker and Christopher T. Rhodes*
41. Specialized Drug Delivery Systems: Manufacturing and Production Technology, *edited by Praveen Tyle*
42. Topical Drug Delivery Formulations, *edited by David W. Osborne and Anton H. Amann*
43. Drug Stability: Principles and Practices, *Jens T. Carstensen*

44. Pharmaceutical Statistics: Practical and Clinical Applications, Second Edition, Revised and Expanded, *Sanford Bolton*
45. Biodegradable Polymers as Drug Delivery Systems, *edited by Mark Chasin and Robert Langer*
46. Preclinical Drug Disposition: A Laboratory Handbook, *Francis L. S. Tse and James J. Jaffe*
47. HPLC in the Pharmaceutical Industry, *edited by Godwin W. Fong and Stanley K. Lam*
48. Pharmaceutical Bioequivalence, *edited by Peter G. Welling, Francis L. S. Tse, and Shrikant V. Dinghe*
49. Pharmaceutical Dissolution Testing, *Umesh V. Banakar*
50. Novel Drug Delivery Systems: Second Edition, Revised and Expanded, *Yie W. Chien*
51. Managing the Clinical Drug Development Process, *David M. Cocchetto and Ronald V. Nardi*
52. Good Manufacturing Practices for Pharmaceuticals: A Plan for Total Quality Control, Third Edition, *edited by Sidney H. Willig and James R. Stoker*
53. Prodrugs: Topical and Ocular Drug Delivery, *edited by Kenneth B. Sloan*
54. Pharmaceutical Inhalation Aerosol Technology, *edited by Anthony J. Hickey*
55. Radiopharmaceuticals: Chemistry and Pharmacology, *edited by Adrian D. Nunn*
56. New Drug Approval Process: Second Edition, Revised and Expanded, *edited by Richard A. Guarino*
57. Pharmaceutical Process Validation: Second Edition, Revised and Expanded, *edited by Ira R. Berry and Robert A. Nash*
58. Ophthalmic Drug Delivery Systems, *edited by Ashim K. Mitra*
59. Pharmaceutical Skin Penetration Enhancement, *edited by Kenneth A. Walters and Jonathan Hadgraft*
60. Colonic Drug Absorption and Metabolism, *edited by Peter R. Bieck*
61. Pharmaceutical Particulate Carriers: Therapeutic Applications, *edited by Alain Rolland*
62. Drug Permeation Enhancement: Theory and Applications, *edited by Dean S. Hsieh*
63. Glycopeptide Antibiotics, *edited by Ramakrishnan Nagarajan*
64. Achieving Sterility in Medical and Pharmaceutical Products, *Nigel A. Halls*
65. Multiparticulate Oral Drug Delivery, *edited by Isaac Ghebre-Sellassie*
66. Colloidal Drug Delivery Systems, *edited by Jörg Kreuter*

67. Pharmacokinetics: Regulatory • Industrial • Academic Perspectives, Second Edition, *edited by Peter G. Welling and Francis L. S. Tse*
68. Drug Stability: Principles and Practices, Second Edition, Revised and Expanded, *Jens T. Carstensen*
69. Good Laboratory Practice Regulations: Second Edition, Revised and Expanded, *edited by Sandy Weinberg*
70. Physical Characterization of Pharmaceutical Solids, *edited by Harry G. Brittain*
71. Pharmaceutical Powder Compaction Technology, *edited by Göran Alderborn and Christer Nyström*
72. Modern Pharmaceutics: Third Edition, Revised and Expanded, *edited by Gilbert S. Banker and Christopher T. Rhodes*
73. Microencapsulation: Methods and Industrial Applications, *edited by Simon Benita*
74. Oral Mucosal Drug Delivery, *edited by Michael J. Rathbone*
75. Clinical Research in Pharmaceutical Development, *edited by Barry Bleidt and Michael Montagne*
76. The Drug Development Process: Increasing Efficiency and Cost Effectiveness, *edited by Peter G. Welling, Louis Lasagna, and Umesh V. Banakar*
77. Microparticulate Systems for the Delivery of Proteins and Vaccines, *edited by Smadar Cohen and Howard Bernstein*
78. Good Manufacturing Practices for Pharmaceuticals: A Plan for Total Quality Control, Fourth Edition, Revised and Expanded, *Sidney H. Willig and James R. Stoker*
79. Aqueous Polymeric Coatings for Pharmaceutical Dosage Forms: Second Edition, Revised and Expanded, *edited by James W. McGinity*
80. Pharmaceutical Statistics: Practical and Clinical Applications, Third Edition, *Sanford Bolton*
81. Handbook of Pharmaceutical Granulation Technology, *edited by Dilip M. Parikh*
82. Biotechnology of Antibiotics: Second Edition, Revised and Expanded, *edited by William R. Strohl*
83. Mechanisms of Transdermal Drug Delivery, *edited by Russell O. Potts and Richard H. Guy*
84. Pharmaceutical Enzymes, *edited by Albert Lauwers and Simon Scharpé*
85. Development of Biopharmaceutical Parenteral Dosage Forms, *edited by John A. Bontempo*
86. Pharmaceutical Project Management, *edited by Tony Kennedy*

87. Drug Products for Clinical Trials: An International Guide to Formulation • Production • Quality Control, *edited by Donald C. Monkhouse and Christopher T. Rhodes*

88. Development and Formulation of Veterinary Dosage Forms: Second Edition, Revised and Expanded, *edited by Gregory E. Hardee and J. Desmond Baggot*

89. Receptor-Based Drug Design, *edited by Paul Leff*

90. Automation and Validation of Information in Pharmaceutical Processing, *edited by Joseph F. deSpautz*

91. Dermal Absorption and Toxicity Assessment, *edited by Michael S. Roberts and Kenneth A. Walters*

92. Pharmaceutical Experimental Design, *Gareth A. Lewis, Didier Mathieu, and Roger Phan-Tan-Luu*

93. Preparing for FDA Pre-Approval Inspections, *edited by Martin D. Hynes III*

94. Pharmaceutical Excipients: Characterization by IR, Raman, and NMR Spectroscopy, *David E. Bugay and W. Paul Findlay*

95. Polymorphism in Pharmaceutical Solids, *edited by Harry G. Brittain*

96. Freeze-Drying/Lyophilization of Pharmaceutical and Biological Products, *edited by Louis Rey and Joan C. May*

97. Percutaneous Absorption: Drugs–Cosmetics–Mechanisms–Methodology, Third Edition, Revised and Expanded, *edited by Robert L. Bronaugh and Howard I. Maibach*

98. Bioadhesive Drug Delivery Systems: Fundamentals, Novel Approaches, and Development, *edited by Edith Mathiowitz, Donald E. Chickering III, and Claus-Michael Lehr*

99. Protein Formulation and Delivery, *edited by Eugene J. McNally*

100. New Drug Approval Process: Third Edition, The Global Challenge, *edited by Richard A. Guarino*

101. Peptide and Protein Drug Analysis, *edited by Ronald E. Reid*

102. Transport Processes in Pharmaceutical Systems, *edited by Gordon L. Amidon, Ping I. Lee, and Elizabeth M. Topp*

103. Excipient Toxicity and Safety, *edited by Myra L. Weiner and Lois A. Kotkoskie*

104. The Clinical Audit in Pharmaceutical Development, *edited by Michael R. Hamrell*

105. Pharmaceutical Emulsions and Suspensions, *edited by Francoise Nielloud and Gilberte Marti-Mestres*

106. Oral Drug Absorption: Prediction and Assessment, *edited by Jennifer B. Dressman and Hans Lennernäs*

107. Drug Stability: Principles and Practices, Third Edition, Revised and Expanded, *edited by Jens T. Carstensen and C. T. Rhodes*

108. Containment in the Pharmaceutical Industry, *edited by James P. Wood*

109. Good Manufacturing Practices for Pharmaceuticals: A Plan for Total Quality Control from Manufacturer to Consumer, Fifth Edition, Revised and Expanded, *Sidney H. Willig*

110. Advanced Pharmaceutical Solids, *Jens T. Carstensen*

111. Endotoxins: Pyrogens, LAL Testing, and Depyrogenation, Second Edition, Revised and Expanded, *Kevin L. Williams*

112. Pharmaceutical Process Engineering, *Anthony J. Hickey and David Ganderton*

113. Pharmacogenomics, *edited by Werner Kalow, Urs A. Meyer, and Rachel F. Tyndale*

114. Handbook of Drug Screening, *edited by Ramakrishna Seethala and Prabhavathi B. Fernandes*

115. Drug Targeting Technology: Physical • Chemical • Biological Methods, *edited by Hans Schreier*

116. Drug–Drug Interactions, *edited by A. David Rodrigues*

117. Handbook of Pharmaceutical Analysis, *edited by Lena Ohannesian and Anthony J. Streeter*

118. Pharmaceutical Process Scale-Up, *edited by Michael Levin*

119. Dermatological and Transdermal Formulations, *edited by Kenneth A. Walters*

120. Clinical Drug Trials and Tribulations: Second Edition, Revised and Expanded, *edited by Allen Cato, Lynda Sutton, and Allen Cato III*

121. Modern Pharmaceutics: Fourth Edition, Revised and Expanded, *edited by Gilbert S. Banker and Christopher T. Rhodes*

122. Surfactants and Polymers in Drug Delivery, *Martin Malmsten*

123. Transdermal Drug Delivery: Second Edition, Revised and Expanded, *edited by Richard H. Guy and Jonathan Hadgraft*

124. Good Laboratory Practice Regulations: Second Edition, Revised and Expanded, *edited by Sandy Weinberg*

125. Parenteral Quality Control: Sterility, Pyrogen, Particulate, and Package Integrity Testing: Third Edition, Revised and Expanded, *Michael J. Akers, Daniel S. Larrimore, and Dana Morton Guazzo*

126. Modified-Release Drug Delivery Technology, *edited by Michael J. Rathbone, Jonathan Hadgraft, and Michael S. Roberts*

127. Simulation for Designing Clinical Trials: A Pharmacokinetic-Pharmacodynamic Modeling Perspective, *edited by Hui C. Kimko and Stephen B. Duffull*

128. Affinity Capillary Electrophoresis in Pharmaceutics and Biopharmaceutics, *edited by Reinhard H. H. Neubert and Hans-Hermann Rüttinger*

129. Pharmaceutical Process Validation: An International Third Edition, Revised and Expanded, *edited by Robert A. Nash and Alfred H. Wachter*

130. Ophthalmic Drug Delivery Systems: Second Edition, Revised and Expanded, *edited by Ashim K. Mitra*

131. Pharmaceutical Gene Delivery Systems, *edited by Alain Rolland and Sean M. Sullivan*

132. Biomarkers in Clinical Drug Development, *edited by John C. Bloom and Robert A. Dean*

133. Pharmaceutical Extrusion Technology, *edited by Isaac Ghebre-Sellassie and Charles Martin*

134. Pharmaceutical Inhalation Aerosol Technology: Second Edition, Revised and Expanded, *edited by Anthony J. Hickey*

135. Pharmaceutical Statistics: Practical and Clinical Applications, Fourth Edition, *Sanford Bolton and Charles Bon*

136. Compliance Handbook for Pharmaceuticals, Medical Devices, and Biologics, *edited by Carmen Medina*

137. Freeze-Drying/Lyophilization of Pharmaceutical and Biological Products: Second Edition, Revised and Expanded, *edited by Louis Rey and Joan C. May*

138. Supercritical Fluid Technology for Drug Product Development, *edited by Peter York, Uday B. Kompella, and Boris Y. Shekunov*

139. New Drug Approval Process: Fourth Edition, Accelerating Global Registrations, *edited by Richard A. Guarino*

140. Microbial Contamination Control in Parenteral Manufacturing, *edited by Kevin L. Williams*

141. New Drug Development: Regulatory Paradigms for Clinical Pharmacology and Biopharmaceutics, *edited by Chandrahas G. Sahajwalla*

142. Microbial Contamination Control in the Pharmaceutical Industry, *edited by Luis Jimenez*

143. Generic Drug Product Development: Solid Oral Dosage Forms, *edited by Leon Shargel and Izzy Kanfer*

144. Introduction to the Pharmaceutical Regulatory Process, *edited by Ira R. Berry*

145. Drug Delivery to the Oral Cavity: Molecules to Market, *edited by Tapash K. Ghosh and William R. Pfister*

Drug Delivery to the Oral Cavity

Molecules to Market

Tapash K. Ghosh
U.S. Food and Drug Administration
Rockville, Maryland, U.S.A.

William R. Pfister
NexMed, (U.S.A.) Inc.
Robbinsville, New Jersey, U.S.A.

Taylor & Francis
Taylor & Francis Group

Boca Raton London New York Singapore

FIRST INDIAN REPRINT, 2014

Published in 2005 by
CRC Press
Taylor & Francis Group
6000 Broken Sound Parkway NW, Suite 300
Boca Raton, FL 33487-2742

Printed and bound in India by Bhavish Graphics.

International Standard Book Number-10: 0-8247-8293-3 (Hardcover)
International Standard Book Number-13: 978-0-8247-8293-1 (Hardcover)
Library of Congress Card Number 2004062072

Library of Congress Cataloging-in-Publication Data

Drug delivery to the oral cavity: molecules to market / Tapash K. Ghosh, William R. Pfister, editors,
p. cm. -- (Drugs and the pharmaceutical sciences)
Includes bibliographical references and index.
ISBN 0-8247-8293-3 (alk. paper)
1. Oral medication. 2. Oral mucosa. I. Ghosh, Tapash K. II. Pfister, William R. III. Series.

RM162.I56 2005
615'.6--dc22 2004062072

T&F informa

Taylor & Francis Group
is the Academic Division of T&F Informa plc.

Visit the Taylor & Francis Web site at
http://www.taylorandfrancis.com

and the CRC Press Web site at
http://www.crcpress.com

FOR SALE IN SOUTH ASIA ONLY

Contents

Preface

The oral cavity (OC) and its highly permeable mucosal tissues have been taken advantage of for decades as a site of absorption for delivery of drugs to the systemic circulation (oral transmucosal delivery, OTD), and for local delivery to the subjacent tissues (oral mucosal delivery, OMD). Administration of an active agent in a dosage form intended to release the drug in the oral cavity is referred to herein as an intraoral delivery system or intraoral dosage form (IOD). Intraoral delivery may provide for a local effect (i.e., breath freshening), absorption via the oral mucosal tissues for either a local drug effect (i.e., analgesia) or systemic drug effect (i.e., smoking cessation), or, with most drugs, provide for systemic absorption along the segments of the gastrointestinal tract (GIT).

Myriad IODs and drug products are now widely prescribed for systemic diseases, as are OMD products for local treatment of halitosis, bacterial infections, periodontal disease, and other conditions of the mouth. Mouthwash and dentifrice active ingredients are now widely used in a variety of personal care products for local delivery to the oral cavity, periodontal pocket, and the overlying mucosal tissue. The first commercially successful oral transmucosal drug delivery system (OTDS) was a sublingual tablet introduced in the early 1960s containing nitroglycerin for the symptomatic treatment of angina pectoris. Over the past 30 years, many other OTDS have been commercialized for systemic drug delivery and treatment of angina pectoris (i.e., nitroglycerin), moderate to severe pain (i.e., fentanyl), smoking addiction (i.e., nicotine), and so on.

More recently, fast-dissolving or mouth-dissolving tablets and novel film delivery systems have been developed. Dozens of oral mucosal delivery systems (OMDS) targeted at the personal care products market for the local delivery of medicinal agents to the oral cavity have been developed. These IODs include quick-dissolving films, mouthwash, and dentifrice products containing antibacterial agents, fluoride, flavoring agents, and the like. The innovative approaches in OTD and OMD delivery system designs, fundamental understanding of the mechanisms of mucoadhesion, and development of oral transmucosal permeation enhancer technology have paved the way in the development of sophisticated controlled-release delivery strategies for systemic and local therapy, respectively. More recently, interest has focused on the oral transmucosal route for delivery of biotechnology products (i.e., proteins, peptides, antisense compounds, etc.). Due to the high permeability and lower metabolic activity of the oral mucosal tissue compared to skin and the gastrointestinal tract, there are high hopes for the success of the oral transmucosal route for delivery of second-generation biotechnology products, and perhaps for delivery of insulin as well.

The early product successes focused on the "easy-to-deliver" drugs (i.e., nitroglycerin sublingual tablets, nicotine gum, and lozenges) paved the way in the development of new technologies, which may be effective in overcoming the future challenges of controlled delivery of the "difficult-to-deliver" drugs having higher molecular weight (greater than 400 Daltons) such as polypeptides and protein-based therapeutics. These new OTDS arising from our fundamental understanding of transport processes, pharmacology, and biochemistry of the mucosal tissue, and development of predictive modeling, advances in material science, and developments of transmucosal permeation enhancers will allow us to meet the challenges of the future with new and improved IODs for treatment of many other diseases.

The basic science and recent research relating to the oral cavity and oral mucosal drug delivery has been published extensively in the literature, which has emphasized the basic structure, function, biochemistry, and permeability of the oral cavity. Similarly, the scientific foundation is now well established for development of oral mucosal delivery systems, and an understanding of the influence of saliva, mucin, cellular models for predicting mucosal drug transport, and metabolism have also been discussed in numerous review articles and several books, which sets the stage for the topics covered in this book.

The purpose of this book is to bring into perspective the practical and applied aspects of pharmaceutical development of new solid-state dosage forms for OTD and OMD systems, and the strategies that have been employed for effective systemic and local drug delivery to the oral mucosa. The many dental and pharmaceutical prescriptions and OTC products under development and those that have been successfully commercialized are the main focus of this book. It is not the intent to review the numerous nutriceuticals, dietary supplements, vitamins, and consumer products (i.e., gums, breath fresheners, mouthwashes, lozenges, etc.) because these intraoral formulations (i.e., chewable tablets, liquids, sprays, etc.) are too numerous to adequately address in this volume. Furthermore, the traditional pharmaceutical products marketed as syrups, drops, or suspensions are not reviewed. Rather, the main focus of this book is to provide an overview of the new and emerging IODs and technologies that address the unmet patient and market needs for improved local and systemic drug therapy.

This book consists of 14 chapters that provide the reader with a comprehensive, succinct, state-of-the-art review of the science and innovative technologies currently available for the development of novel delivery systems targeted to the oral cavity for systemic and local delivery. Chapter 1 provides an overview of the fundamentals such as the underlying physiology, biochemistry, and anatomy of the oral mucosa; the various sites of oral mucosal absorption (i.e., gingival, buccal, palatal, sublingual, periodontal, etc.) for local and systemic drug delivery; the local environment of the oral cavity and its impact on drug absorption, distribution, metabolism, and excretion (ADME); and the metabolic barrier to oral mucosal drug absorption. The advantages of systemic drug delivery by the oral transmucosal route and various IODs are discussed. A summary of the IODs in various stages of research as well as those on the market, and a snapshot of the future trends and the next generation products and technologies are also presented, which sets the stage for the chapters to follow. Preformulation considerations are an important first step in the development of IODs.

The key elements in preclinical assessment necessary to predict mucosal permeation, selection of drug candidates and formulation components, as well as delivery system designs are highlighted in Chapter 2. The preclinical assessment of oral mucosal drug delivery and delivery systems is the next important step in the development process. The in vitro and in vivo models used to assess oral transmucosal drug absorption, the fundamental techniques for the in vitro assessment of mucobioadhesive polymers, the ideal characteristics of mucoadhesive pharmaceutical excipients and criteria for their selection, the effects of saliva and mucin on drug permeation, and finally the importance of selection of appropriate animal models for safety assessment of local irritation and sensitization are presented in Chapter 2. The advantages of systemic drug delivery by the oral transmucosal route are discussed and examples of products for the delivery of analgesics, CNS active agents, antiemetics, and so on are also presented in Chapter 2.

The challenges to the formulation scientist in overcoming the barrier properties of the oral mucosa and effective use of various chemical classes of mucosal permeation enhancers (MPE) to upregulate drug absorption are reviewed and summarized in Chapter 3.

A review of recent work in the field of OTM delivery of proteins and peptides and the challenges that need to be overcome to minimize the local enzymatic metabolism of four labile compounds and strategies to maximize their absorption are discussed in Chapter 4.

The next steps in the development cycle consist of the design, clinical development, manufacture, testing, and marketing of oral mucosal delivery systems.

There are now dozens of commercial OTD and OMD products on the market including: quick-dissolving (QD), slow-dissolving (SD), and nondissolving (ND) IODs such as sublingual tablets, quick-dissolving tablets, chewing gums, patches, devices, liposomes, microparticle delivery systems, and so on. Chapters 5 through 10 focus on some of the more recently developed and commercialized SD and ND IODs. Sublingual delivery of nitroglycerin was one of the first life-saving drugs introduced for the prevention of angina pectoris and heart attacks. The products (sublingual tablets) for delivery of organic nitrates including nitroglycerin and isosorbide dinitrate for the treatment of angina pectoris are reviewed in Chapter 5.

Development of a novel oral mucosal technology for delivery of melatonin for normalization of circadian rhythms is discussed in Chapter 6. The many innovative IODs for delivery of nicotine (i.e., gums, lozenges, sublingual tablets, inhalators, sprays, etc.) for smoking cessation therapy, including products on the market and those in development, are reviewed in detail in Chapter 7. The formulation development and preclinical, and clinical assessment of a novel dry powder needleless injection device for the delivery of conventional drugs, polypeptides, and proteins to the mucosal tissue of the oral cavity for local and systemic drug delivery are presented in Chapter 8. Oral TMDS (i.e., liposomes, fibers, microparticles, sustained release depots, etc.) for treatment of periodontal disease are reviewed and contrasted, and the general classes of products for oral hygiene including mouthwash and dentifrice products are reviewed in Chapter 9. The use of local anesthetics in dental and oral surgical applications and the clinical development of a novel mucoadhesive local anesthetic patch are highlighted in Chapter 10.

Novel QD dosage forms that dissolve in the mouth without water, offer patient convenience, and patent life extension for many drugs are reviewed in Chapters 11 through 14. The various QD tablet delivery system designs, considerations in material/excipient selection, manufacturing technologies, testing procedures, and points to consider in scale-up from laboratory to commercial production of quick-dispersing oral drug delivery systems including the chemistry, manufacturing, and controls (CMC) are discussed in Chapter 11. A new innovative QD intraoral drug delivery technology, which does not require water and offers the patient convenience for delivery of a wide variety of drugs, is highlighted in Chapter 12. The considerations in process development and scale-up of QD IOD tablets, and the principals in optimizing process scale-up conditions and product quality issues, are reviewed in Chapter 13. Finally, the scientific and regulatory considerations in developing QD IOD tablets from a clinical pharmacology and biopharmaceutics perspective are critically reviewed in Chapter 14.

In addition, a useful reference guide is provided on worldwide companies developing intraoral drug delivery technologies and products (Appendix 1), as well as selected books and market research reports dealing with intraoral drug delivery (Appendix 2). The reader may find the list of abbreviations at the end of this book useful to better understanding the unique acronyms used in the IOD literature.

We believe that this book offers a wealth of up-to-date information organized in a logical sequence corresponding to the various stages of research, development, and commercialization of IOD products. Our authors were selected from industry, academia, and government for their expertise and reputation in their selected areas to objectively present a balanced view of the state-of-the-art in IOD product development. Their insights will prove useful to the pharmaceutical scientists in industry and academia who are involved in the development of the next generation of IOD products. This book was written especially for the pharmaceutical development scientists, but should be instructive to R&D managers and those involved in the various stages of laboratory testing, manufacturing, clinical evaluation, marketing, and regulatory affairs. The importance of the book to medical professionals involved in the prescription and use of these emerging dosage forms also should be recognized.

Tapash K. Ghosh

William R. Pfister

The Editors

Tapash K. Ghosh, Ph.D. is employed by the Office of Clinical Pharmacology and Biopharmaceutics, CDER, FDA, Rockville, MD. He received his B.S. in pharmacy and M.S. in pharmaceutical chemistry from Jadavpur University, India and a Ph.D. in pharmaceutics from the University of Louisiana at Monroe, Louisiana. Before joining FDA, Dr. Ghosh held faculty positions at the College of Pharmacy of Howard University, Washington, D.C., and Massachusetts College of Pharmacy at Boston, MA. He also held scientific and management positions at Lavipharm Laboratories, Inc. and G & W Laboratories, Inc. Dr. Ghosh has published and presented several scientific papers in the areas of controlled drug delivery and is editor of the books *Transdermal and Topical Drug Delivery Systems* (1997) and *Theory and Practice of Contemporary Pharmaceutics* (2004).

William R. Pfister, Ph.D. is Senior Director Preclinical Affairs at NexMed (U.S.A.), Inc., Robbinsville, NJ where he is responsible for preclinical safety testing for all R&D product development activities. Dr. Pfister received his B.S. degree in biochemistry from the University of New Hampshire, M.S. and Ph.D. degrees in pharmacology and toxicology from Purdue University School of Pharmacy and Pharmacal Sciences. He is a Diplomate of the American Board of Toxicology (DABT), and is a Certified Regulatory Affairs Professional (RAC). Dr. Pfister has over 20 years of experience in the areas of pharmaceutical research and development, technical services, regulatory affairs, quality assurance, quality control, preclinical safety testing, and project management. He has held management positions at Lavipharm Laboratories, Inc., Pharmetrix Division of TCPI, Pharma Patch PLC, Pharmetrix Corporation, Dow Corning Corporation, American Critical Care, and Hoffmann-La Roche. Dr. Pfister is a member of several scientific organizations including the Controlled Release Society, American Society of Pharmaceutical Scientists, American Chemical Society, Society of Toxicology, American Society for Quality, and Regulatory Affairs Professionals Society. Dr. Pfister has published and presented over 95 scientific papers in the areas of pharmacology, toxicology, controlled drug delivery, and transdermal drug delivery, and is editor of the book *Transdermal and Topical Drug Delivery Systems* (1997). He is an inventor of several patents on novel intraoral, and transdermal, controlled-release, and drug delivery systems. Dr. Pfister has focused his interests over the past 15 years on the research and development of drug delivery systems.

Contributors

William Abraham, Ph.D. Monsanto, St. Louis, Missouri, E-mail: WilliamAbraham@monsanto.com

John Anthony Bauer, Ph.D. Center for Cardiovascular Medicine, Columbus Children's Research Institute, Columbus, Ohio, E-mail: bauerj@ccri.net

Brian J. Bellhouse Medical Engineering Unit, University of Oxford, Oxford, United Kingdom

Luce Bénès Laboratories 3M Sante, Pithiviers, France, E-mail: lbenes1@mmm.com

Terry L. Burkoth, Ph.D. PaceLine Consulting, Palo Alto, California, E-mail: terry@burkoth.com

Dhruba J. Chatterjee, Ph.D. Food and Drug Administration, Clinical Pharmacology and Biopharmaceutics, Rockville, Maryland, E-mail: chatterjee@cder.fda.gov

Emmanuel O. Fadiran, Ph.D. Food and Drug Administration, Clinical Pharmacology and Biopharmaceutics, Rockville, Maryland, E-mail: fadirane@cder.fda.gov

Gina Fernandez Noven Pharmaceuticals, Inc., Miami, Florida

Tapash K. Ghosh, Ph.D. Food and Drug Administration, Clinical Pharmacology and Biopharmaceutics, Rockville, Maryland, E-mail: fadriane@cder.fda.gov

Donald B. Hare Food and Drug Administration, Clinical Pharmacology and Biopharmaceutics, Rockville, Maryland, E-mail: hare@cder.fda.gov

Françoise Horrière Laboratories 3M Sante, Pithiviers, France, E-mail: fhorriere@mmm.com

John P. Hunt, Ph.D. Food and Drug Administration, Clinical Pharmacology and Biopharmaceutics, Rockville, Maryland, E-mail: huntj@cder.fda.gov

Venkateswar R. Jarugula, Ph.D. Food and Drug Administration, Clinical Pharmacology and Biopharmaceutics, Rockville, Maryland, E-mail: jarugulav@cder.fda.gov

Bhaskara R. Jasti, Ph.D. Department of Pharmaceutics & Medicinal Chemistry, TJ Long School of Pharmacy and Health Sciences, University of the Pacific, Stockton, California, E-mail: bjasti@pacific.edu

David P. Kanios Noven Pharmaceuticals, Inc., Miami, Florida

Rajendra Khankari CIMA Labs, Inc., Eden Prairie, Minnesota

Sung Yun Kwon, Ph.D. TheraJect, Fremont, California, E-Mail: sykwon @TheraJect.Com

Lawrence J. Lesko, Ph.D. Food and Drug Administration, Clinical Pharmacology and Biopharmaceutics, Rockville, Maryland, E-mail: leskol@cder.fda.gov

Bing Li, Ph.D. University of Wisconsin, Madison, Wisconsin

Xiaoling Li, Ph.D. Department of Pharmaceutics & Medicinal Chemistry, TJ Long School of Pharmacy and Health Sciences, University of the Pacific, Stockton, California

David Longridge, Ph.D. David Longridge Associates Ltd., Blyth, Northumberland, United Kingdom, E-mail: davidlongridge@aol.com

David L. Lucking Noven Pharmaceuticals, Inc., Miami, Florida

Juan A. Mantelle, Ph.D. Noven Pharmaceuticals, Inc., Miami, Florida, E-mail: jmantelle@noven.com

Venugopal Marasanapalle Department of Pharmaceutics & Medicinal Chemistry, TJ Long School of Pharmacy and Health Sciences, University of the Pacific, Stockton, California

Roy L. McQuinn 3M Pharmaceuticals, St. Paul, Minnesota, E-mail: rlmcquinn@mmm.com

Huw Millward, Ph.D. The National Centre for Product Design & Development Research, UWIC, United Kingdom, E-mail: hmillward-pdr@uwic.ac.uk

Janakiram Nyshadham CV Therapeutics, Inc., Palo Alto, California

S. Indiran Pather, D. Pharm TransOral Pharmaceuticals, Inc., Corte Madera, California, E-mail: ipather@transoral.com

William R. Pfister, Ph.D. NexMed (U.S.A.), Inc., Robbinsville, New Jersey, E-mail: wpfister@nexmed.com

Joseph R. Robinson, Ph.D. University of Wisconsin, Madison, Wisconsin, E-mail: jrrobinson@pharmacywisc.edu

Srikonda V. Sastry, Ph.D. Xenoport, Inc., Santa Clara, California, E-mail: sastry@xenoport.com

Jennifer L. Seifert Division of Pharmacology, College of Pharmacy, The Ohio State University, Columbus, Ohio

Kuldeepak Sharma, Ph.D. Fremont, California, E-mail: kuuldeepak_sharma@hotmail.com

John Siebert, Ph.D. CIMA Labs, Inc., Eden Prairie, Minnesota

Vijaya K. Tammara, Ph.D. Merck & Co., Inc., Blue Bell, Pennsylvania E-mail: Vijay_tammara@merck.com

Su Il Yum, Ph.D. Durect Corporation, Cupertino, California

Eric Yum, D.D.S. Camarillo Dental Practice, Camarillo, California

Dedication

The book is dedicated to my parents, the late Bejoy Ghosh and Promila Ghosh for their blessings, and to my wife Subhra, my daughter Mimi, and my son Ricky for their patience, support, and understanding.

Tapash K. Ghosh

I thank my wife Gail, daughters Dianna and Nicole, and son William, who were patient and forgiving of the long hours spent on editing this book and the time spent away from home required to prepare many of its chapters.

William R. Pfister, Sr.

1

Intraoral Delivery Systems: An Overview, Current Status, and Future Trends*

William R. Pfister and Tapash K. Ghosh

CONTENTS

*Disclaimer: No official support or endorsement of this chapter by the Food and Drug Administration is intended or should be inferred.

I. INTRODUCTION

Since time immemorial, oral drug administration has been one of the most convenient and widely accepted routes of delivery for most therapeutic agents. Traditionally, oral dosage forms refer to tablets, capsules, and liquid preparations taken orally, swallowed, and transiting the gastrointestinal tract (GIT) for post-buccal absorption. However, some undesirable physiological properties of the gastrointestinal (GI) tract limit the feasibility of administration of some molecules by this route (i.e., proteins, polypeptides, etc.). The relatively poor absorption, presence of abundant digestive enzymes in the GI lumen and epithelium, post-absorption efflux (i.e., by P-glycoprotein, etc.), and first-pass metabolism by the hepatic enzymes and subsequent elimination, limit the ability of many drugs to reach therapeutic levels by this route. For the last few decades, researchers have been developing intraoral delivery systems (IDS) that can produce desirable drug exposure for optimum therapeutic effect. As a result, as evident from the abundance of scientific and patent literature over the last twenty years, nontraditional oral dosage forms (e.g., buccal, sublingual, etc.) have been or are being developed with emphasis on pregastric absorption by the various tissues of the oral cavity with the intention to avoid first-pass and gut-wall metabolism, to enhance bioavailability, or improve convenience of dosing. The target sites for local drug delivery in the oral cavity include the following: buccal, sublingual, periodontal, periodontal pocket, peribuccal, perilingual, tongue (i.e., lingual), and gum (i.e., gingival). Other adjacent postintraoral sites where drug targeting may be desirable include the pharynx, larynx, adenoids, and tonsils. However, there are also a number of disadvantages to intraoral drug delivery where the objective is local absorption by the oral mucosal route. The relatively small surface area, as compared to the skin, and GIT and significant loss of the applied dose due to swallowing and salivary flow are two major limitations.

There is a plethora of literature (i.e., books, journal articles, and patents) available on traditional oral dosage forms (i.e., immediate, modified, and controlled-release tablets and capsules) for delivery of drugs to the GIT.[1] The objective of this book is to provide a comprehensive overview of the various nontraditional intraoral dosage forms (IODs) whose site of delivery is predominantly to the oral cavity as opposed to the GIT. These IODs are formulated for local as well as systemic drug delivery. However, for the most part the IODs are intended to disintegrate, dissolve, or release the drug in the oral cavity, where it then has the opportunity to either be locally absorbed, in part or whole, and/or swallowed and subsequently absorbed along the GIT. The various types of intraoral dosage forms for local and systemic drug delivery are highlighted in Figure 1.1, and illustrate the breadth of technologies that have evolved over the past several decades used to deliver drugs locally or systemically. These include liquids (solutions, sprays, syrups, injections, etc.), semisolids (i.e., ointments, pastes, etc.), and solid dosage forms (i.e., quick-dissolve and slow-dissolve tablets, sublingual tablets, lozenges, films, filaments, gums, patches, "lollipops," microparticles, drug delivery devices, etc.). Many of these dosage forms are discussed in detail in subsequent chapters.

This chapter provides an overview of intraoral drug delivery systems, the current status of technologies and products on the market and in development, and future trends of various intraoral drug delivery systems with a view to set the stage for the subsequent chapters. A review of certain intraoral dosage forms such as patches and tablets is presented in Chapter 2. The advantages and disadvantages of intraoral drug delivery, the local anatomical sites within the oral cavity for targeted localized as well as systemic drug delivery, and methods to enhance drug permeation from these sites are discussed in Chapter 3. Delivery of peptide drugs via the buccal route is reviewed in detail in Chapter 4. Oral transmucosal delivery of nicotine and melatonin, and sublingual transmucosal delivery of organic nitrates are described in Chapters 5, 6, and 7, respectively. Chapter 8 reviews a novel technology (the Oral PowderJect® System) being developed to provide needle-free and pain-free minimally invasive injection of

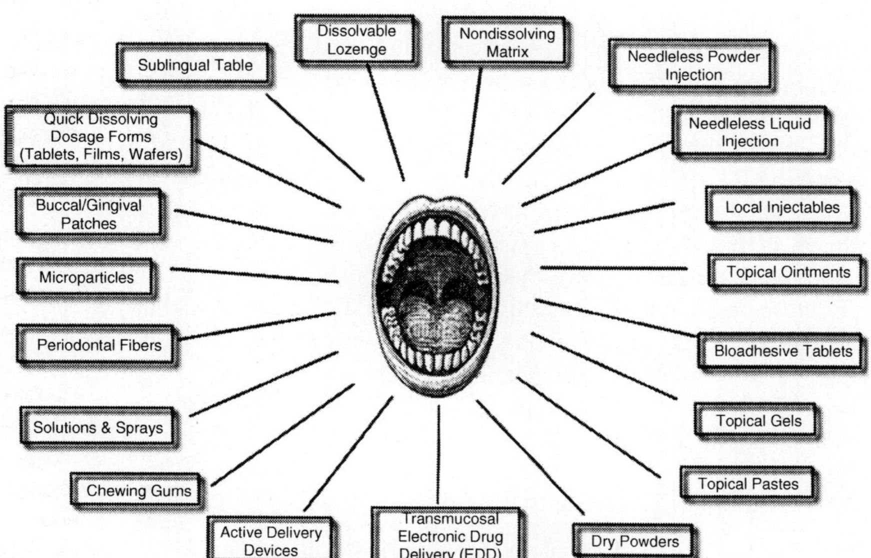

Figure 1.1 Types of intraoral dosage forms.

solid powder particles of traditional drugs. Drugs from biotechnology such as proteins, peptides, and oligonucleotides, genetic vaccines, and other therapeutic compounds for local dental applications as well as systemic drug delivery are also discussed. Chapters 9 and 10 are devoted to drug delivery systems for the treatment of periodontal and dental diseases for oral transmucosal delivery of antibiotics, and local anesthetics for dental and oral surgical premedication, respectively. Because of the significant current interest in quick-dissolving/dispersing oral drug delivery systems, the balance of the chapters (Chapters 11 through 14) in this book have been devoted to these dosage forms. These chapters provide comprehensive reviews of different quick-dispersing oral delivery systems, process development, and scale-up design approaches used for development and optimization of such products. Finally, developmental considerations for intraoral dosage forms from a clinical pharmacology and biopharmaceutics perspective are provided in Chapter 14. The main focus of this chapter is to provide a general overview of the IODs for the prescription and over-the-counter (OTC) drug market as well as a brief overview of the many new intraoral consumer products now evolving into new market segments (i.e., breath fresheners, antiseptics, such as lozenges, gums, quick-dissolve strips, and teeth whitening strips).[2]

II. INTRAORAL DRUG DELIVERY

A. Product Market

The market for intraoral drug delivery products has undergone a period of innovation and significant expansion over the past decade. The worldwide market for IOD products is in excess of five billion annually and is expected to continue to grow at a rapid pace. The main driver of this market growth is the fast-dissolve prescription and OTC market segments. The worldwide market for fast-dissolve dosage forms was in excess of one billion in 2002 and the market segment allocated by therapeutic class is shown in Figure 1.2. The OTC market segment represents 33 percent of sales for the fast-dissolve dosage forms with the balance representing prescription drug products. The fast-dissolve market by region is shown in Figure 1.3. The United States clearly provides the greatest market opportunity for fast-dissolve dosage forms followed by Europe and Japan as a distant second and third, respectively. With multiple new product launches in the consumer care, OTC, and prescription areas, the total fast-dissolve market could reach \$1.4 billion by 2006. The Zydis® fast-dissolve dosage forms (FDDFs) represented approximately 74 percent and 76 percent of the fast-dissolve worldwide market in 2002, and 2003, respectively.[3]

B. Anatomy of Oral Cavity

The anatomy of the oral cavity is discussed in detail in Chapters 2, 3, and 8. The oral cavity contains four anatomically distinct sites with unique tissue types,

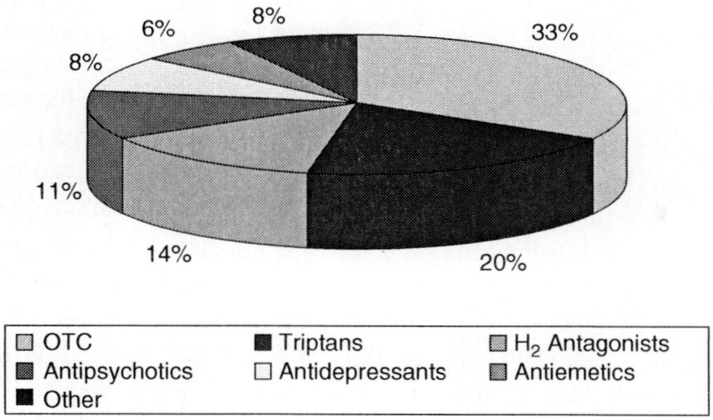

Figure 1.2 Worldwide fast-dissolving dosage form market by therapeutic class.

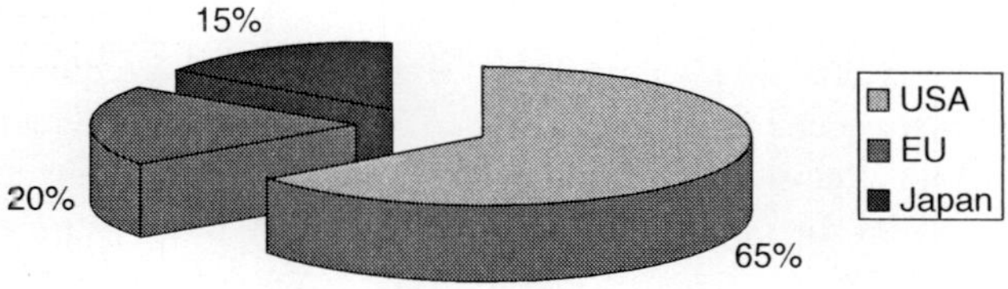

Figure 1.3 Fast-dissolve market share by region.

which include the palatal, gingival, sublingual, and buccal mucosa. The gingival tissue surrounds the teeth and holds them in place. The palatal tissue refers to the anatomical sites associated with the roof of the mouth. The sublingual tissue (i.e., under the tongue) is composed of the floor of the mouth and ventral tongue, and the buccal mucosa refers to the inside of the cheek, and inside of the upper and lower lip.

In general, oral epithelia are classified into three categories:

1. *Specialized mucosa*: This is located on the dorsum of the tongue.
2. *Masticatory mucosa*: This includes the gingival and the hard palate.
3. *Lining mucosa*: This covers the rest of the oral cavity, such as lips and cheeks, floor of the mouth, sublingual area, soft palate and alveolar mucosa.

C. Intraoral Drug Delivery

There are several generally well-accepted conventional routes of drug delivery that are less invasive compared to parenteral injectable dosage forms and they are listed in decreasing order of their available drug absorptive surface area including: oral via the GIT (301 m^2), pulmonary (100 m^2), transdermal (1 to 2 m^2), rectal (0.04 m^2), nasal (0.015 m^2), and buccal (0.01 m^2).[4] For transdermal drug delivery, the rate-limiting barrier for a compound to be absorbed into the systemic circulation is the stratum corneum, which is the keratinized layer of the skin. Because the lining mucosa of the oral cavity is not keratinized, is thinner (100 to 500 μm), and more permeable than skin, the degree of resistance for a drug to penetrate the lining mucosal membrane is much less than that of skin. Similar to the epithelium of the GIT and the skin, the epithelial lining mucosa is composed of a constantly renewing cell population, which is produced by cell division in the basal region.

The first evidence of drug absorption via the buccal mucosa was noted over 100 years ago.[5] Subsequently, in 1879, sublingual administration of nitroglycerin was reported to successfully alleviate the symptoms of classic angina pectoris.[6] Since then, oral mucosal drug delivery has drawn more and more attention because of its potential advantages over other routes of delivery. The intraoral route is one of the more preferred routes of drug administration as it is convenient and, with certain drugs, may provide a more rapid onset of action. Intraoral drug delivery overcomes hepatic first-pass metabolism and promotes rapid systemic delivery with improved bioavailability with selected drugs having the required physiochemical and biopharmaceutical characteristics. The oral mucosa provides accessibility to allow for the precise localization of the dosage form for targeted drug delivery. It also provides the opportunity to directly modify tissue permeability, inhibit protease activity, or decrease immunogenic responses to drugs. Therefore, the oral mucosa has emerged as one of the target sites for administration of drugs in a wide variety of dosage forms, particularly for those drugs targeted for local delivery in the oral cavity and those that required systemic absorption as well.[7-9]

The mouth represents the initial portal of entry to the GIT and, thus, most dosage forms placed in the mouth are expected to be swallowed, and transit the GIT either intact or when dissolved in saliva. Due to the relatively short cellular turnover time of the buccal mucosa, a properly designed buccal adhesive delivery system may remain in place, under the best circumstances, for 1 to 24 hours. However, the mouth is continuously flushed with salivary fluid to aid in swallowing, digestion of food, and to cleanse the oral cavity of pathogens such as bacteria. The daily output of saliva in humans is between 750 to 1000 ml and the pH may range from 5.8 to 7.6 when salivary fluid production is stimulated.[10] Due to the continuous production of saliva and outflow from the mouth into the GIT, drug solutions and dispersions are easily cleared from the mouth and thus provide a challenge for designing intraoral mucosal drug delivery systems intended for prolonged drug administration. This limitation, in general, results in the need for frequent dosing for those drugs that have a short half-life ($T_{1/2}$), are rapidly cleared from the oral cavity, and have poor oral bioavailability (i.e., nitroglycerin, nicotine, etc.). On this basis it is not surprising that maintaining drugs in the oral cavity through the use of an appropriate drug delivery system is a significant challenge. Historically, lozenges and troches, which are sucked on by the patient, and sublingual tablets were the traditional means for delivering drugs to the oral cavity. More recently, quick-dissolve dosage forms have become more popular, and bioadhesive tablets and gels have been used to provide a longer contact time with the absorbing tissue and thus minimize drug loss through clearance by salivary outflow.

Drug delivery via the membranes of the oral cavity can be subdivided as follows: (1) sublingual delivery, which is the administration of the drug via the sublingual mucosa (the membrane of the ventral surface of the tongue and the floor of the mouth) to the systemic circulation; (2) buccal delivery, which is the administration of the drug via the buccal mucosa (the lining of the cheek and area between the gums and upper and lower lips) to the systemic circulation; and (3) periodontal, gingival, and odontal delivery, for the local treatment of conditions of the oral cavity, principally aphthous ulcers, bacterial and fungal infections, and periodontal disease. These oral mucosal sites differ greatly from one another in terms of anatomy, permeability to an applied drug, and their ability to retain the delivery system for the desired length of time.

The sublingual route of drug absorption is by far the most widely studied of these sites of drug delivery. The sublingual mucosa is relatively permeable, allowing rapid absorption and acceptable systemic bioavailability of many drugs, and is convenient, accessible, and generally a well-accepted target for drug delivery. The buccal mucosa is somewhat less permeable than the sublingual mucosa tissue, and is generally not able to provide the rapid absorption and good bioavailability compared to sublingual administration. Both of these routes have been investigated clinically for the delivery of a wide variety of drugs. Local delivery to tissues of the oral cavity has a number of applications, including the treatment of toothache, periodontal disease, bacterial and fungal infections, aphthous ulcers, and dental stomatitis, and in facilitating tooth movement with prostaglandins.

The buccal route of drug delivery provides the opportunity for drug absorption through the buccal epithelial lining of the oral cavity (mucosa of the cheek) for local or systemic action. The noninvasive nature of administration, ease and convenience of dosing, precise localization, and, lastly, increased permeability of the buccal mucosa compared to other transepithelial routes make this a promising route of delivery. Also, the rich supply of blood vessels and lymphatics in the buccal mucosa results in rapid onset of drug action (i.e., within minutes) for those that have the requisite physicochemical profile (i.e., nicotine for smoking cessation therapy and nitroglycerin for treatment of angina pectoris). Drugs absorbed from the buccal mucosa may directly enter the systemic circulation by way of the jugular vein, minimizing the first-pass liver metabolism, and gastric acid- or enzyme-mediated degradation (salivary fluid has lower enzymatic activity than gut).[11] The presence of food or variations in the gastric emptying rate has little or no influence on drug delivery by the buccal route.[12] The continuous exposure of the oral mucosal tissues to a multitude of substances and its high cellular turnover rate makes the buccal tissue robust and less prone to local toxicity or irritation from drugs, their dosage forms, and formulation excipients. The absence of Langerhans cells in the oral mucosal tissues imparts tolerance to potential allergens.[13] Therefore, prolonged administration of drugs (chronic use) to the oral cavity is less prone to induction of local tissue sensitization and allergic reactions. When compared to other mucosal delivery routes, buccal drug delivery offers a higher degree of control and reproducibility; it also allows the opportunity to remove the dosage form (i.e., chewing gums, lozenges, patches, etc.) to terminate drug absorption, if necessary.

D. Intraoral Drug Absorption

The mucosal tissue of the oral cavity provides a readily accessible site for drug delivery; however, it has a relatively small surface area in humans of approximately 100 cm^2. Although oral drug delivery begins with the placement of the dosage form in the mouth, which is the primary delivery portal to gain access to the large absorptive surface area (i.e., 101 m^2) of the GIT, the buccal tissues represent only a small fraction of the total absorptive surface area of the GIT (i.e., 0.01 percent or 0.01 m^2). However, due to the lower enzymatic activity, less hostile environment, better stability of drugs including proteins and peptides, and the high vascular supply of the oral cavity, it is an attractive route for local absorption of selected drugs and an ideal environment for the initial dissolution of the dosage form.[14–16] The estimated surface areas for drug absorption of the intraoral and GI routes of delivery, the local pH, relative enzymatic activity, and drug absorptive capability are summarized in Table 1.1.

The challenges of local delivery to the oral cavity are considerable and include overcoming the mucus-coating barrier and the high constant production and turnover of saliva. The high clearance of drugs from the oral cavity into the GIT by saliva is a major limitation, thus, strategies to improve retention of the drug in the oral cavity to take advantage of its high absorptive potential are a major technical challenge. In many cases delivery system designs can improve local retention, and thus enhance the delivery efficiency of the drug. Most of the research in the area of drug delivery to the oral cavity has focused on targeting delivery to the highly absorptive nonkeratinized tissues of the buccal and sublingual mucosa, as these regions are the primary absorptive tissues in the mouth. Strategies employed to maximize drug retention in the oral cavity have included use of bioadhesive dosage forms (i.e., solutions, tablets, patches), gums, lozenges, and dry powder, or liquid by needle injection. Most formulations include permeation enhancers to overcome the epithelial barrier properties of the mucosal tissue. In general, once the drug is dissolved in saliva it rapidly distributes throughout the mouth, and varying degrees of absorption may be expected by the entire surface area of the mucosal lining of the oral cavity.

The concept of mucoadhesion was introduced into the controlled drug delivery arena beginning in the early 1980s. The amount of drug absorbed across

Table 1.1 Characteristics of the Oral Cavity and Gastrointestinal Tract

Absorptive Site	Estimated Surface Area	Percent Total Surface Area	Local pH	Relative Enzymatic Activity	Relative Drug Absorption Capacity
Oral cavity (buccal)	100 cm^2 (0.01 m^2)	0.01	5.8–7.6	Moderate	High
Esophagus	—	—	6.0–7.0	Low	Low
Stomach	0.1–0.2 m^2	0.20	1.0–3.0	High	High
Small intestine Duodenum Jejunum Ileum	100 m^2 (taking intestinal microvilli area into account = 4500 m^2)	98.76	3.0–4.0	High	High
Large intestine Transverse colon Ascending colon Descending colon Cecum Sigmoid colon	0.5–1.0 m^2	0.99	4.0–6.0	Moderate	Low
Rectum	200–400 cm^2 (0.04 m^2)	0.04	5.0–6.0	Low	Low
Total surface area	101.25 m^2				

Source: Adapted from Reference 4.

a membrane, such as in the oral cavity, is directly proportional to the surface area of exposure, the applied drug concentration, the permeability coefficient of the drug, and the residence time. Although the total surface area of the buccal mucosa is about 100 cm^2; practically, the maximum size of a buccal delivery system should not be larger than 10 to 15 cm^2, with 1 to 3 cm^2 being most desirable. Given the normally short residence time for a solution of drug in the mouth, bioavailability of most drugs from the mouth is low. Fractional bioavailability via the buccal tissues is typically less than 10 percent with most drugs, if measurable absorption occurs at this site at all.[17] The ability of a mucoadhesive polymer to interact with biological membranes or surfaces and be retained on the membrane for an extended period of time makes it a good addition as a targeted mucosal drug delivery technology. Different formulations of oral mucosal drug delivery systems have been investigated and some have been commercialized. Conventional formulations of mucosal drug delivery systems are primarily tablets and patches. More recently, phase change polymers, which transform from a liquid to solid at body temperature, have broadened the design of oral mucosal drug delivery dosage forms. A preliminary discussion using phase change polymers as a new dosage form for oral mucosal drug delivery, strategies for developing and characterizing mucoadhesive dosage forms, as well as a review of certain conventional dosage forms such as patches and tablets are also presented in Chapter 2.

III. OVERVIEW: INTRAORAL DELIVERY SYSTEMS

A. Routes of Administration

There are a number of terms that have been defined by the Food and Drug Administration (FDA) to describe routes of administration for all approved dosage forms and drug products. The routes of administration have been summarized along with a brief definition, a short abbreviated name, and code taken from the *CDER Data Standards Manual.*[18] This FDA standard lists 95 routes of drug administration, including the definitions and numeric codes, which are used to compare approved drug products with therapeutic equivalence evaluations listed in the *Orange Book.*[19] The names, but not the definitions or codes have been harmonized with the E2B route of drug administration terms taken from the International Conference on Harmonization (ICH). The route of administration code consists of three digits and when possible the ICH E2B terms should take precedence and can be found in the ICH-M2 Codes.[20] These codes are also used as part of the National Drug Code Directory, which is a universal product identifier for human drugs.[21] Of the 95 routes of drug administration, 15 of them (16 percent) are related to pregastric intraoral delivery and are summarized in Table 1.2.

Examples of products corresponding to the various pregastric routes of drug delivery are presented in the various chapters of this book with a primary emphasis on the submucosal, sublingual, periodontal, dental, buccal, and intragingival routes of drug administration. In the future, as new and novel dosage forms are

Table 1.2 Pregastric Routes of Drug Administration

Number	Delivery Route Name	Definition	Short Name	CDER Code	ICH M2 Numeric Code
1	Buccal	Administration directly toward the cheek, generally from within the mouth	Buccal	030	002
2	Dental	Administration to a tooth or teeth	Dental	038	004
3	Endotracheal	Administration directly into the trachea	E-trache	401	007
4	Intracoronal, dental	Administration of a drug within a portion of a tooth covered by enamel and separated from the roots by a slightly constricted region known as the neck	I-Coronal	117	—
5	Intraesophageal	Administration within the esophagus	I-eso	072	—
6	Intragingival	Administration within the gingival	I-gingiv	307	—
7	Laryngeal	Administration directly upon the larynx	Laryn	364	
8	Oral	Administration to or by way of the mouth	Oral	001	048
9	Oropharyngeal	Administration directly to the mouth and pharynx	Oro	410	049
10	Periodontal	Administration around a tooth	P-odont	040	—
11	Sublingual	Administration beneath the mucous membrane	SL	024	060
12	Submucosal	Administration beneath the mucous membrane	S-mucos	053	—
13	Transmucosal	Administration across the mucosa	T-mucos	122	—
14	Transtracheal	Administration through the wall of the trachea	T-trache	355	—
15	Intratracheal	Administration into the trachea	—	—	039

developed and commercialized, there may be a need to expand the list of current official routes of delivery names. These might include the palatal route for those dosage forms that adhere to the upper hard palate and direct drug delivery into the palatal mucosa or locally into the oral cavity. Quick-dissolving films and wafers use the lingual route of administration, because they are applied to the tongue and rapidly dissolve to freshen the breath and disinfect the mouth. Examples of these consumer products are the breath fresheners (i.e., Listerine® Pocket-Packs oral care strips, and Myntz).[2]

B. Classification of Intraoral Dosage Forms

Dosage forms targeted for delivery to the intraoral cavity can be classified in terms of their dissolution or disintegration kinetics as either quick-dissolving (QD), slow-dissolving (SD), or nondissolving (ND), which release the drug over a period of seconds up to 1 minute, 1 to 10 minutes, and >10 minutes to hours, respectively. These IODs may be intended to present the drug for local release in the mouth (i.e., drug dissolved in saliva) or to the GIT (i.e., drug released as microparticles but not dissolved in saliva) for subsequent systemic absorption. The various IODs may further be targeted for release and local topical absorption by the tissues in the oral cavity for treatment of local diseases or for systemic absorption for treatment of diseases remote from the site of application. In addition, IODs may contain drugs such as nicotine (i.e., gums, lozenges, sublingual tablets, and inhalators for smoking cessation therapy) or nitroglycerin (i.e., sublingual tablets for treatment of angina pectoris) absorbed predominately by the tissues of the oral cavity (i.e., buccal or sublingual). Alternatively, and in most instances, only a small fraction of the drug released from an IOD is absorbed by the tissues of the oral cavity, a larger fraction may be absorbed along the pregastric segment of the GIT, and the majority is absorbed in the stomach and distal segments of the GIT. Furthermore, these dosage forms can modulate the pharmacokinetics of the drug and provide for rapid, slow, or delayed onset of action and short, intermediate, and prolonged duration of action depending upon the dosage form design. The IODs may be further categorized as noninvasive (i.e., QD formulations), semi-invasive (i.e., patches, periodontal filaments, microparticles, needleless injectors), or invasive (i.e., injection via needles, as used in dental anesthesia, or other drug delivery devices) depending upon their interaction with mucosal tissues and the degree to which they physically breach the membrane barrier to absorption.

The characteristic and biopharmaceutical properties of each of the IODs classified according to their rate of dissolution/disintegration as QD, SD, and ND dosage forms are discussed in detail in the following sections of this chapter.

1. Quick-Dissolving Delivery Systems

QD delivery systems undergo disintegration or dissolution in the saliva, generally within a few seconds to a minute, releasing the drug and inactive ingredients into the oral cavity. Other synonyms and definitions of QD delivery systems are provided in Chapter 14. The major fraction of the drug will eventually be swallowed with the saliva and transported along the GIT where the drug is subsequently absorbed. The technical advantages of these dosage forms include: ease of swallowing, administration without water anywhere and anytime, quick onset of action with some drugs, supervised administration, buccal or sublingual absorption, and local therapy of the oral mucosa. Therapeutic benefits of the mouth-dissolving dosage forms for patients may include: enhanced efficacy, improved convenience, and improved compliance. Pharmaceutical companies may benefit from these dosage forms due to product differentiation, life-cycle management, reduction of development costs, and outsourcing. The following therapeutic categories have been reported to have market opportunities for QD delivery systems: nonopioid analgesics, opioid analgesics, migraine headache, cough and cold, allergy, gastrointestinal, cardiovascular, central nervous system, urology, and other categories.

2. Slow-Dissolving Delivery Systems

SD intraoral delivery systems are generally dissolved in the oral cavity within 1 to 10 minutes and the following products are included in this category: chewable tablets, sublingual tablets, "lollipops," mucoadhesive tablets, and buccal tablets. Although there are many commercial products on the markets in the first three categories, only a few mucoadhesive or buccal tablets have recently been introduced (i.e., Striant®). However, there have been numerous reports of mucoadhesive tablets investigated in the scientific and patent literature, and some products are in the R&D pipeline. Numerous U.S. and foreign patents as well as patent applications have been published on buccal drug delivery with the first appearing in the late 1960s involving various mucoadhesive polymers. These polymers include but are not limited to polyacrylic acid, hydroxypropylmethyl cellulose, polyvinylpyrrolidone, hydroxypropyl cellulose, vinylpyrrolidone and its copolymers, polyethylene oxide, and sodium carboxymethyl cellulose.

3. Nondissolving Delivery Systems

ND intraoral dosage forms do not dissolve entirely when placed in the mouth and can provide for controlled drug delivery from 10 minutes to several hours, and up to a day or longer under the best circumstances. Examples of ND intraoral delivery systems include the following dosage forms: chewing gums, buccal and gingival patches, periodontal fibers, and drug delivery devices. Medicated chewing gum technology provides a new competitive advantage for product life-cycle management for mature and novel pharmaceuticals. This delivery system technology adds a new benefit in efficacy, convenience, and quality of life. Chewing gum formulations offer optimal patient-controlled dose titration, an appealing mouth feel and texture, controlled release of active substances, and can be manufactured under cGMPs.[22] The advantages of drug delivery in a gum formulation include fast onset of action, doses in the mg range, few side effects, drug release for up to 60 minutes, and reduced first-pass liver metabolism compared to oral dosage forms. A medicated chewing gum delivery system may be ideal for drugs having limited oral absorption in conventional tablets (i.e., nicotine) and indications where a quick onset of action is desirable, such as treatment of migraine, pain, allergy, angina, motion sickness, and smoking cessation may benefit.

Technologies similar to transdermal patches have been developed and designed to adhere to either the gingival or buccal mucosa. Mucoadhesive patches (Cydot®) have been developed and gingival local anesthetic patches have been commercialized (Dentipatch®) that can provide for drug delivery over a period of from 30 minutes to 24 hours depending upon the patch design and therapeutic indication. Other controlled drug delivery systems have been developed for the treatment of periodontal disease in the form of thin filaments and microparticles that are compacted into the periodontal pocket to provide for prolonged delivery of antibiotics such as tetracycline (Actisite® for up to 10 days).

IV. CURRENT STATUS: INTRAORAL DRUG DELIVERY

Table 1.3 presents a summary of the various selected intraoral drug delivery systems in the R&D pipeline and those that to the best of the authors' knowledge have been recently launched. Most IOD products approved prior to 1998 are not listed in this table, but are discussed in the other chapters in this book. The predominant IODs recently introduced into the marketplace have included quick-dissolve tablets (QDT) primarily for the prescription market (i.e., rizatriptan benzoate, olanzapine, mirtazapine, and ondansetron), and now more recently for the over-the-counter market (i.e., acetaminophen/caffeine and loratadine). Several other novel IODs have been introduced in recent years including slow-dissolving sublingual tablets (SDST), controlled-release intraoral delivery device (CRIOD), quick-dissolving wafer (QDW), rapid-melt tablets (RMT), Atrigel® delivery system (ADS), microparticle technology (MPT), resorbable gel (RG), medicated chewing gum (MCG), immediate-release spray (IRS), and lozenges (LO). Other novel IODs in the R&D pipeline include oral transmucosal tablet (OTT), liquid metered dose spray (LMDS), and mucoadhesive bioerodable preformed disc (MAD). These IODs are differentiated from each other and are based on proprietary formulation technologies. It should be noted, however, that the QD dosage forms rapidly dissolve or disintegrate in the saliva without water within seconds to a minute after application and can be generally described as mouth-dissolving dosage forms and include the following types listed above: ODT, QDW, and RMT.[23] The simplest definition of an ODT is: a single unit dose that disintegrates in the oral cavity and leaves an easy-to-swallow residue.[24,25] The Food and Drug Administration Center for Drug Evaluation and Research (CDER) defines an ODT as "a solid dosage form containing medicinal substances, which disintegrates rapidly, usually within a matter of seconds, when placed upon the tongue."[24] The definitions of the various IODs are reviewed in detail in Chapter 14.

Table 1.3 Summary of Intraoral Drug Delivery Systems: Recent Commercial Products and R&D Pipeline

Trade Name	Drug	Indication	Company (Marketer)	Dosage Form[a]	Recent Products Filed/Marketed	Doses (Mg)	Intraoral Technology	Approval/ Launch	Ref.
Maxalt-MTL®	Rizatriptan benzoate	Migraine	Merck	ODT	Europe U.S.	5, 10 5, 10	Freeze-dried orally disintegrating tablets	February 1998 June 1999	26
Zyprexa® Zydis®	Olanzapine	Schizophrenia	Eli Lilly	ODT	U.S.	5, 10	Freeze-dried orally disintegrating tablets	April 2000	27, 28
Uprima®	Apomorphine	Male erectile dysfunction	TAP	SDST	Europe approved U.S. NDA filed	2, 4 2, 3	Compressed sublingual tablet	May 2001 Pending	29
Actiq®	Fentanyl	Cancer pain	Anestha UK Cephalon	CRIOD	Europe U.S.		Lollipop	October 2000 December 2002	30
Risperdal® Quicklet®	Risperidone	Schizophrenia	Janssen-Cilag GmbH		Germany	0.5, 1, 2	Orodispersible tablet, dissolves on the tongue within seconds	October 2002	31
Risperdal® M-Tablet®	Risperidone	Schizophrenia	Janssen	ODT	U.S.	0.5, 1, 2	Orally disintegrating tablet, dissolves on the tongue within seconds	April 2003	
Excedrin® QuickTabs®	Acetaminophen Caffeine	Headache	Bristol-Myers Squibb Co.	QDT	OTC, U.S.	500 65	Quick-dissolving tablet Quick-dissolving tablet	October 2002	32
Tempra® Quicklets/ Tempra® FirsTabs	Acetaminophen	Headache	Bristol-Myers Squibb Co.	QDT	OTC, U.S.	500		—	—
Remeron™ SolTab®	Mirtazapine	Depression	Cima Labs Organon	QDT	Europe U.S.	15, 30, 45	Quick-dissolving tablet	July 2002 January 2001	33, 34
Zelapar	Selegiline HCl	Parkinson's Disease	Amarin Corp./ Elan Corp.	QDW	Filed U.S.		Quick-dissolving wafer	In development	35
Tramadol Flashtab®	Tramadol	Analgesic	Ethypharm/ Viatris	QDT	Europe filed	50	Quick-dissolving tablet	September 2001	36

Table 1.3 Summary of Intraoral Drug Delivery Systems: Recent Commercial Products and R&D Pipeline (continued)

Trade Name	Drug	Indication	Company (Marketer)	Dosage Form[a]	Recent Products Filed/Marketed	Doses (Mg)	Intraoral Technology	Approval/ Launch	Ref.
Fentanyl OraVescent™	Fentanyl	Narcotic analgesic	Cima Labs	OTT	U.S.	——	Effervescent oral transmucosal tablet	In development	37, 38
Claritin® RediTabs®	Loratadine	Antihistamine	Schering	ODT	OTC, U.S.	10	Orally disintegrating tablet	December 1996	39
Clarinex® RediTabs®	Deslortadine	Antihistamine	Schering	ODT	OTC, U.S.	5	Orally disintegrating tablet	June 2002	—
Buccal Delivery	Heparin	Anticoagulant, Deep vein thrombosis	Generex	LMDS	U.S.	—	Buccal delivery	IND pending	40
Buccal Delivery	Insulin	Diabetes	Generex Eli Lilly	LMDS	U.S.	—	Buccal delivery	IND pending	34
Zomig® Rapid-Melt Zomig®-ZMT	Zolmitriptan	Migraine	AstraZeneca	RMT	Europe Japan U.S.	2.5 2.5 2.5, 5.0	Rapid-melt tablet Rapid-melt tablet, orange flavored	November 1999 Filed July 2001 September 2001	41, 42, 43, 44
Buccal Delivery	Ondansetron	Chemo-therapeutic agent induced emesis	Glaxo Wellcome	ODT	U.K.	8.0	Orally disintegrating tablet	In development	45, 46
BEMA™ Fentanyl	Fentanyl	Cancer pain	Atrix Laboratories	MAD	U.S.	—	Mucoadhesive disc	In development	47
Atridox®	Doxycycline	Periodontal disease	Atrix Laboratories	ADS	Europe	2.8%	Periodontal gel	June 2001	48
Arestin®	Minocycline	Periodontal disease	Orapharma	MPT	U.S.	—	Microparticle technology	February 2001	49
Emdogain® Gel	Hydrophobic enamel	Periodontal disease	Biora AB	RG	U.S.	0.3 ml	Resorbable gel	March 2001	50

Table 1.3 Summary of Intraoral Drug Delivery Systems: Recent Commercial Products and R&D Pipeline (continued)

Trade Name	Drug	Indication	Company (Marketer)	Dosage Form[a]	Recent Products Filed/Marketed	Doses (Mg)	Intraoral Technology	Approval/ Launch	Ref.
Nicorette® Orange	Nicotine	Smoking cessation	SmithKline Beecham	MCG	U.S.	2, 4	Medicated chewing gum	October 2000	51
Alavert®	Loratadine	Antihistamine	Wyeth Consumer Healthcare	QDT	OTC, U.S.	10	Orally disintegrating tablet	December 2002	52
Fazaclo®	Clozapine	Schizophrenia	Alamo Pharma-ceutical	ODT	U.S.	25, 100	Orally disintegrating tablet	March 2004	—
Nitrolingual Pump Spray	Nitroglycerin	Angina	First Horizon Pharma-ceutical Corp.	IRS	U.S.	60-dose spray	Oral spray	November 2002	53
Commit™	Nicotine polacrilex	Smoking cessation	GlaxoSmith-Kline	LO	OTC, U.S.	2, 4	Lozenge	October 2002	54
Triaminic® Softchews (several formulations)	Pseudoephedrine HCl (other actives)	Pediatric Cough & cold	Novartis	QDT	OTC, U.S.	15	Quick-dissolving tablet	—	—
Benadryl® Fastmelt®	Pseudoephedrine HCl Diphenhydramine citrate	Allergy & sinus	Pfizer	QDT	OTC, U.S.	30 19	Quick dissolving tablet	—	—

[a] Dosage forms include: quick-dissolving tablets (QDT), quick-dissolving wafer (QDW), immediate release sprays (IRS), quick-dissolving film (QDF), oral transmucosal tablet (OTT), lozenge (LO), sublingual tablets (ST), slow-dissolving sublingual tablet (SDST); controlled release intraoral delivery device (CRIOD), rapid-melt tablet (RMT), mucoadhesive bioerodable preformed disc (MAD), Atrigel delivery system (ADS), liquid metered dose spray (LMDS), microparticle technology (MPT), resorbable gel (RG), medicated chewing gum (MCG). This table does not include chewable tablets intended for drug delivery to the stomach. See Reference 106 for details on OTC products.

A. Quick-Dissolving Delivery Systems

There are over 23 oral fast-dissolve drug products on the market in the United States, Europe has 32 products, and Japan has 31 launched products. There are more than 25 companies active in the development of fast-dissolve dosage forms; Cardinal Health (Zydis®) and CIMA (OraSolv®/DuraSolv®) are the market leaders with an 80 percent global market share. Table 1.4 summarizes the key innovator companies in the oral fast-dissolve tablet arena, and contrasts the different technologies and their disintegration, hardness, packaging, and drug loading characteristics.[55] All of these technologies have various advantages and limitations. For example, all can be taken without water and dissolve in the mouth in less than one minute. Due to the fragility and friability of most IODs they must be packaged in single-unit blister packs; only a few of these QD tablets are sufficiently hard and durable to allow them to be packaged in multidose bottles (i.e., DuraSolv®, AdvaTab®, and WOWTAB®).

The majority of the QD market consists of oral fast-dissolving tablets for OTC and prescription use, the market leader being OTC loratadine (Claritin® Reditabs®). Oral thin films are on the market mainly for consumer care applications (e.g., breath freshening), although films for therapeutic applications are now being commercialized. A number of new tablet and thin-film products are in the development pipeline, providing improved treatments for pain, allergies, insomnia, and anxiety. The fast-dissolve market is mainly comprised of the different technologies highlighted below including lyophilized wafers, orally disintegrating tablets, and thin-film or strip dosage forms.

1. Lyophilized Wafers

Lyophilized wafer technologies were developed by Cardinal Health (Zydis® technology) and Biotron (Kryotab™ technology) and are based on processes involving lyophilization or freeze-drying that result in fragile porous waferlike tablets. Lyophilization is generally a more expensive process that requires specialized processing equipment to mold the unit dose in a blister pouch resulting in a very fragile unit dose that cannot be packaged in bottles. These products and technologies are reviewed in detail in Chapters 11, 13, and 14.

2. Orally Disintegrating Tablets

These technologies were developed by Biovail (FlashDose® technology), Cima Labs (OraSolv®, DuraSolv®, and OraVescent® technology), Élan Pharmaceutical (fast melt formulation technology), Ethypharm (Flashtab® technology), Eurand (Ziplets® technology), KV Pharmaceutical (OraQuick™ technology), Shire Laboratories (Rapitrol™ technology), and Yamanouchi (WOWTAB® technology) and are based on either in situ molding of the dosage form in a unit dose blister pouch or by conventional high-speed tableting operations. These dosage forms are less fragile than those processed by lyophilization and some can be packed in multidose bottles. These technologies and the orally disintegrating products on the market are reviewed in detail in Chapters 11 through 14.

Table 1.4 Major Oral Fast-Dissolve Tablet Manufacturers and Technology Characteristics

Technology (Innovator)	Disintegration (sec)	Tablet Hardness/ Robustness	Packaging	Drug Loading Claim (mg)	Marketed Products
FlashDose® (Biovail)	5–15	Soft, friable	Blisters	Up to 600	1
Zydis® (Cardinal)	3–5	Very fragile	Blisters	Up to 400	15
OraSolv® (CIMA)	<30	Soft and fragile	Blisters (PakSolv)	Up to 750	3
DuraSolv® (CIMA)	<30	Robust, hard	Bottles or Blisters	Up to 500	3
FlashTab® (Ethypharm)	30–60	Relatively durable	Blisters	Up to 650	3
AdvaTab® (Eurand)	15–30	Robust, hard	Bottles or Blisters	Up to 700	2
OraQuick® (KV)	<20	Relatively durable	Bottles or Blisters	Up to 500	1
Lyoc® (Cephalon)	<10	Soft, friable	Blisters	Up to 1000	6
SATAB® (Sato)	<10	Relatively durable	Blisters	Up to 600	7
WOWTAB® (Yamanouchi)	<30	Relatively durable	Bottles or Blisters	Up to 500	14

3. Thin Films and Strips

The concept of using thin films or strips for delivery of active ingredients into the mouth is not new; early studies examined delivery of lidocaine, a local anesthetic, for dental applications from polymer films in the 1970s.[56,57] More recently, however, several thin-film or strip intraoral dosage form technologies have been developed as a means to quickly release an active agent upon placing the strip on the tongue. Thin-film and strip intraoral dosage forms have been developed by several companies including LTS Lohmann Therapie-Systeme AG, Zengen Inc., and Lavipharm Laboratories (Quick-Dis™ and Slow-Dis™ technology) and are based on a unique solution-coating process where the formulation is dispensed and metered to a controlled thickness onto a moving web and dried in precision temperature-controlled multizone ovens, die-cut, and packaged. The films are generally thin, flexible strips (2 × 3 cm) and can be packaged in multidose containers or individually pouched.

These films generally dissolve rapidly (within seconds), to release the active agents, but can be tailored to release the drug more slowly as well, depending upon their thickness, and selection of the polymer matrix. A film or strip can be defined as a dosage form that employs a water-dissolving polymer (generally a hydrocolloid, which may be a bioadhesive polymer), which allows the dosage form to quickly hydrate, adhere, and dissolve when placed on the tongue or in the oral cavity (i.e., buccal, palatal, gingival, lingual, or sublingual, etc.) to provide rapid local or systemic drug delivery. Drug release may be either quick (within seconds) or slower (within minutes) by varying the rate of dissolution of the films. These films are monolithic matrices and release the active ingredients multidirectionally when placed in the oral cavity.

The breath strip also known as the mouth-freshening strip market category was created by Pfizer's Warner-Lambert consumer healthcare division, which launched Listerine® PocketPaks™ breath freshener in October 2001.[2] This product consists of 24 small thin films or strips that dissolve when placed on the tongue to release an intense "minty" flavor for breath freshening. Since the development of Listerine® PocketPaks™, a number of other innovative thin-film products have recently been introduced into the consumer market for breath freshening such as FreshBurst™, Myntz™ Instastripz, Gel-A-Mint Sugar Free™, MagikStrips™, Altoids™ Strips (peppermint and cinnamon), and a number of other private label brands. These QD strips are available in a number of flavors and are packaged in a small disposable plastic or metal packlet with a hinged closure containing from 20 to 32 strips per pack. The nonadhering strips are stacked upon each other and a single strip can be easily dispensed by placing a finger on the strip and moving it forward with a sliding motion, thus dispensing the strip out of the packlet. The strips are intended to be placed on the tongue and dissolve within seconds to release their ingredients.

Chloraseptic® Relief Strips™ were the first oral thin-film product to incorporate a drug and were introduced in the United States in September 2003 by

Prestige Brands International for relief of sore throat.[58] Zengen, Inc. developed this new delivery technology, which is a medicated oral strip structured as a proprietary bilayer system. The process for manufacture of these strips eliminates the use of irritating solvents and reduces damage of the drug due to heat and moisture, maintaining the integrity and efficacy of the active ingredients. Certain auxiliary compounds used with the bilayer system can assist with the rate of dissolution, water solubility, mucous adhesion, and coating on the exterior film layer. For instance, depending on the auxiliary compounds the strips can be manufactured to dissolve quickly or over an extended period of time, regulating the release of active material.[59] Chloraseptic® Relief Strips™ are fast-acting oral strips that contain 3 mg of benzocaine and 3 mg of menthol (oral anesthetics and analgesics) and are provided in tamper-evident blister packs. Chloraseptic® Relief Strips™ contain the following inactive ingredients: acesulfame potassium, hydrocolloids (carboxymethylcellulose and modified pectin), flavoring agent (cherry flavor), plasticizer (glycerin), sweetener (sucralose), coloring agent (Red 40), lecithin, magnesium silicate, and water. The strips are available as an OTC product indicated for use in children age five or older and adults for the temporary relief of minor irritation, pain, sore throat, and sore mouth. Dosing involves taking one strip from the dispenser and placing on the back of the tongue to dissolve in the mouth; a second strip is used immediately after the first one dissolves and may be repeated every two hours as needed.

A number of companies have been involved as innovators in the development and manufacturing of patented thin-film strip technologies including LTS Lohmann Therapie-Systeme AG, and Lavipharm Laboratories, which developed the Quick-Dis™ and Slow-Dis™ technologies.[60]

Patents assigned to Lohmann Therapie-Systeme describe compositions containing therapeutic agents for delivery to the oral cavity in the form of a film, and are extensions of earlier technology patents on rapidly disintegrating sheetlike preparations, and quick-dissolving films for intraoral delivery.[61,62] The film composition comprises at least one water-soluble polymer, at least one polyalcohol, and at least one pharmaceutical ingredient, wherein the composition has mucoadhesive properties and is intended to quickly dissolve and disintegrate upon administration in the oral cavity.[63] Another patent describes a method for release of nicotine from a mucoadhesive film applied in the oral cavity where nicotine is released and delivered to the buccal mucosa.[64] Another patent describes a novel flexible quick-dissolving film containing nicotine that dissolves in the mouth without water.[65, 66] Unique containers have also been patented for dispensing this thin-film dosage form.[67] This represents a new class of dosage form for intraoral delivery that is designed to dissolve quickly (within seconds to minutes) when applied to the tongue or sublingually. The dosage form is prepared by a either a wet film coating or hot melt coating manufacturing process, where the formulation (i.e., solution for wet film coating or molten semisolid mass for hot melt coating) is cast onto a moving web, and dried or cooled, resulting in a thin flexible film, then die-cut (into any size or shape) and packaged into unit-dose blister packs or multidose containers.

Quick-Dis™ and Slow-Dis™ are examples of a unique intraoral delivery system (IODS) technology based on a unique solution coating process where the formulation is metered onto a moving web and dried in temperature-controlled multizone ovens, resulting in dried films that are die-cut and packaged. The resulting films are generally thin, flexible, and can be packaged in multidose containers or individually pouched. Quick-Dis™ allows for the precise control of drug release that dissolves quickly (i.e., within seconds) in the mouth and provides a unique method for oral delivery of active agents and is claimed to be a substantial improvement over existing technologies, such as fast-dissolving tablets. These QD films typically contain water-soluble hydrocolloids (i.e., hydroxypropyl methylcellulose, Pullulan, pectin, carboxymethylcellulose, etc.), an effective dose of an active agent, and other additives such as flavoring agents, plasticizers, and preservatives.[68] Quick-Dis™ formulations have been evaluated for the delivery of sildenafil, hydromorphone, estradiol, nicotine, and oxybutynin. The disintegration and dissolution characteristics of thin-film dosage forms are a function of their thickness (i.e., generally between 2 to 20 mils) and blend of hydrocolloids.[69] The benefits of these IODS include no water required for dosing, rapid dissolution, mucosal absorption (faster onset depending upon the physiochemical properties of the drug), taste masking (improves compliance), comfortable for sublingual or lingual use, and flexible thin film available in several shapes and sizes.

Other drugs have been evaluated in intraoral film dosage forms comprising a hydroxypropylcellulose and other hydrocolloid polymers as film bases such as lidocaine and lidocaine HCl for local anesthesia, and dissolution methods have been developed to characterize drug release profiles as well as techniques developed to evaluate the permeation of the drug through the oral mucosa of the hamster cheek pouch from these dosage forms.[70] Intraoral films and strips offer several advantages as a drug delivery system, including local delivery, rapid buccal absorption, no water required, rapid onset, dose titration, and product line extension, and future products are expected soon.

4. Quick-Dissolve Products

Many QD products have been developed and those that are categorized as ODT products in the U.S. market are summarized in Table 1.5. These are based on the technologies mentioned above and include products based on the DuraSolv®, OraSolv®, Flashtab®, Zydis®, and other proprietary technologies. Most of these drugs are product line extensions of already approved conventional tablet dosage forms, and primarily offer patient convenience. However, those products where compliance can be a problem in control of depression and schizophrenia in the institutional setting can benefit from the QD dosage form because they rapidly dissolve in the mouth and cannot be spit out as conventional tablets can, thus ensuring better control of the illness (i.e., Zyprexa® Zydis®, and Remeron® SolTabs®). Other dosage forms for control of migraine headache such as

Table 1.5 Summary of ODT Products in the United States

Technologies (Innovator)	Type of Dosage Form	Marketer	Drug	Indication	Products in the U.S. Market
DuraSolv®	Compressed	Wyeth	Loratadine	Allergy	Alavert®
OraSolv®	tablet	Schwarz Pharma	Hyoscyamine Sulfate	Irritable bowel	NuLev®
(CIMA)		Organon	Mirtazapine	Depression	Remeron® SolTabs®
		Alamo	Clozapine	Antipsychotic	Fazaclo®
		AstraZeneca	Zolmitriptan	Migraine	Zomig®-ZMT
Zydis®	Freeze-dried wafers	Schering	Loratadine	Allergy	Claritin® Reditabs®
(Cardinal Health)		Schering	Desloratadine	Allergy	Clarinex® Reditabs®
		Merck	Rizatriptan Benzoate	Migraine	Maxalt-MLT®
		GlaxoSmithKline	Ondansetron	Nausea and vomiting	Zofran ODT®
		Eli Lilly	Olanzapine	Schizophrenia	Zyprexa® Zydis®
		Janssen	Risperidone	Schizophrenia	Risperdal® M-Tab®
		Roche	Clonazepam	Seizures	Klonopin® Wafers
Other	Tablet	TAP Pharm. Inc.	Lansoprazole	Duodenal ulcers	Prevacid® SoluTab®

ondansetron are primarily prescribed as injectable products due to their poor oral bioavailability, thus the QDT dosage forms offer a less invasive alternative and provide additional convenience to the patient (i.e., Maxalt-MLT). On the other hand, zolmitriptan is available in a conventional oral tablet (Zomig® Tablets) for the treatment of migrane and a QDT tablet (Zomig® Rapidmelt®) provides comparable bioavailability to the conventional tablet with a slightly longer T_{max}, and offer little therapeutic advantage other than patient convenience. The rationale for developing QD dosage forms for the other indications are primarily for product line extension, extension of the patent, brand life, and patient convenience.

The following section provides a brief description of ODT and QD products that were recently approved and introduced into the market, as well as selected products filed with the FDA for approval.

Eli Lilly introduced Zyprexa®(olanzapine), the first of the newer generation of antipsychotics (atypical antipsychotics), for long-term therapy in schizophrenia in 1996 in a conventional tablet formulation. Zyprexa® has been prescribed to over 10 million patients in 84 countries worldwide since its market entry in 1996. The new orally disintegrating tablets (Zyprexa® Zydis®) based on the Zydis® freeze-dried tablet technology approved in the United States in 2000 now offer a more patient-convenient form of the drug with improved patient compliance.[27,28]

Merck introduced Maxalt-MLT® (rizatriptan benzoate) in the Netherlands in February of 1998 and in the United States in June 1998. The product used Zydis® technology and was the first migraine medicine approved using the Zydis® technology in Europe and the United States. Maxalt-MLT® is placed on the tongue and uses the saliva in the mouth to dissolve the tablet in seconds, enabling patients to take the mint-flavored tablet without water. The dosage form was designed for rapid absorption and onset of action and was one of the first quick-dissolving products on the market in the United States.[26]

Janssen-Cilag GmbH, a subsidiary of Johnson & Johnson, introduced Risperdal Quicklet® (risperidone) in October 2002 for the treatment of schizophrenia. The product is an oral formulation that dissolves within seconds on the tongue without water, making it almost impossible to spit out and convenient for patients who cannot swallow ordinary tablets.[31]

Bristol-Myers Squibb launched Excedrin® QuickTabs® in October 2002. This is the first and only adult over-the-counter headache medication that melts in the mouth and can be taken without water.[32]

A new drug application (NDA) for Zelapar® is still under review by the FDA.[71,72] Zelepar® is a novel and proprietary formulation of selegiline, an MAO-B inhibitor, being developed as an adjunct treatment for the symptoms of Parkinson's disease as an oral tablet using the patented Zydis® fast-dissolving technology of RP Scherer Corporation (now Cardinal Health). With only one eighth of the conventional tablet dose, the Zelapar® tablet dissolves in seconds and is significantly absorbed in the tissues of the mouth, without swallowing or the need for liquids, and may offer a convenience for patients who have trouble swallowing and may be an important new treatment option.[73]

Ethylpharm and Viatris have developed an innovative oral formulation of Tramadol for the relief of moderate to severe pain using Ethypharm's Flashtab® technologies. Tramadol Flashtabs® are mint-flavored tablets that melt rapidly in the mouth without water and combine several benefits of patient acceptability, accuracy of dosing, safety, ease of swallowing, and pleasant taste.[36]

Organon Inc. received U.S. marketing clearance for Remeron® SolTab®, an orally disintegrating antidepressant incorporating Cima's OraSolv® drug-delivery system with mirtazapine, the active agent in Remeron®. Remeron® SolTab® was the first orally disintegrating antidepressant approved by the FDA and was introduced during the first quarter 2001 through a joint promotional agreement with Solvay Pharmaceuticals Inc. [33,34]

Cima Labs signed a license and supply agreement with Wyeth Consumer Healthcare for a DuraSolv® loratadine OTC version of Claritin® RediTabs®. The drug is formulated into a new, orally disintegrating dosage form that dissolves quickly in the mouth without water or the need for chewing. [39]

Generex Biotechnology has completed a proof-of-concept study on buccal delivery of heparin using their oral spray platform technology, which provided comparable results to subcutaneous injection. Low molecular heparin is used both for the treatment of deep vein thrombosis and for the prevention of blood clots due to postsurgical complications.[40] The same technology is being used to develop a formulation for buccal delivery of insulin for the treatment of diabetes, and Generex entered into an agreement with Eli Lilly in September 2000 and with Elan in January 2002 to develop a buccal formulation for delivery of morphine for management of pain.[34] Cima Labs has developed a buccal delivery formulation of fentanyl using the OraVescent™ drug delivery system. The OraVescent™ technology is based on the effervescent quick-dissolving tablet formulation containing 200 μg of fentanyl citrate and was awarded a patent on the technology on May 29, 2002. The OraVescent™ fentanyl tablets are placed between the upper gum and cheek above the premolar. This formulation demonstrated a comparable pharmacokinetic profile compared to Actiq®, a commercial fentanyl oral transmucosal delivery system.[37,38]

AstraZeneca introduced Zomig® rapid-melt (RM) formulation, which was approved in Japan in 3Q of 2002 for the treatment of migraine.[41,42] The 2.5 mg strength of Zomig®-ZMT (zolmitriptan) was approved by the FDA in February 2001 and the 5.0 mg strength was approved on September 17, 2001[43] and followed the approval of Zomig®-ZMT 2.5 mg orally disintegrating tablets that dissolve on the tongue in seconds without the need for water. Zomig®-ZMT incorporates Cima Labs, Inc.'s DuraSolv® fast-dissolving drug delivery system.[44,74]

GlaxoWellcome has developed an ondansetron orally disintegrating tablet, which disperses rapidly when placed on the tongue and swallowed with water. Ondansetron 8 and 16 mg ODTs were shown to be equally effective to the conventional tablet in the treatment of radiotherapy and cyclophosphamide-induced nausea and emesis, and provide an effective alternative to the conventional ondansetron tablet.[45,46]

Wyeth Consumer Healthcare received FDA approval for loratadine, a nonsedating antihistamine (Alavert®) orally disintegrating tablet for OTC use in the treatment of allergy relief including sneezing, runny nose, itchy watery eyes, and itching of the nose and throat.[75] Alavert® was developed by Cima Labs based on its DuraSolv®, fast-dissolve formulation technology.[52]

Elan Corp. is developing fast-melt tablets that dissolve in the mouth, where the active ingredient is immediately absorbed. Elan's formulations include rapidly disintegrating effervescent formulations and a nanocrystal colloidal dispersion for poorly soluble drugs.[30]

5. Intraoral Liquid Delivery Technology

A number of liquid unit dose and metered spray quick-onset IODs have been developed and are highlighted in this section. Lingual (oral) aerosol and pump spray formulations developed by Flemington Pharmaceutical Corp. release the drug (i.e., nitroglycerin) in the form of a fine mist into the mouth for immediate absorption into the bloodstream via the mucosal membranes. The delivery system offers certain advantages including improving the safety profile of certain drugs by lowering the required dosage. In addition, the pump spray provides for a uniform unit dosage, improving reliability of local dose delivery, and allows medication to be taken without water. Drug absorption through the mucosal membranes of the mouth is generally rapid and minimizes the first-pass metabolism effect.[76,77]

A similar technology based on liquid metered-dose spray technology has been developed by Generex Biotechnology Corp. The spray formulation allows the drug to be absorbed by the mucosal tissues of the mouth and be rapidly absorbed into the bloodstream. The technology is based on a pharmaceutical agent, a unique liquid aerosol formulation, and the compact RapidMist™ device, which administers the formulation as a metered-dose spray. The technology is designed for protein-based drugs, which are encapsulated in mixed micelles made up of conventional enhancers. The enhancers selected are surface-active agents and serve a critical role to enhance the absorption of the drugs through the tissues of the oral mucosa.[78]

B. Slow-Dissolving Delivery Systems

Numerous slow-dissolving intraoral products have been developed based on several drug delivery technologies such as lozenges, sublingual tablets, mucoadhesive tablets, chewable tablets, and so on. Table 1.6 summarizes the recently introduced commercial slow-dissolving chewable tablets and their corresponding technologies.

1. Buccal Tablets

Striant®, developed by Columbia Labs, is a testosterone extended-release buccal tablet that delivers testosterone systemically for hormone replacement in hypogonadal

Table 1.6 Intraoral Drug Delivery Systems: Chewable Tablets

Proprietary Name	Drug	Indication	Company	Dosage Form[a]	Product Approval	Doses (Mg)	Technology	Ref.
Carbamazepine	Carbamazepine	Anticonvulsant	Caraco	CT	July 2001	100	Chewable Tablet	87
Tegretol®	Carbamazepine	Anticonvulsant	Novartis	CT	Prior to Jan. 1982	100	Chewable Tablet	87
Carbamazepine	Carbamazepine	Anticonvulsant	Taro Pharm INDs	CT	October 2000 July 2002	100, 200	Chewable tablet	87
Epitol®	Carbamazepine	Anticonvulsant	TEVA	CT	July 1992	100	Chewable tablet	87
Carbamazepine	Carbamazepine	Anticonvulsant	Warner Chilcott	CT	February 1988	100	Chewable tablet	87
Singulair®	Montelukast Sodium	Asthma	Merck	CT	March 2000 February 1998	4 5	Chewable tablet	87
Phazyme®	Simethicone	Anti-gas stomach relief	Block Drug	QDCT	2001	—	Quick Dissolve mint-flavored chewable tablets	88
Digestive Aid	Calcium	Acid-neutralizing stomach relief	Performance Labs	QDCT	2001	—	Quick-melt berry-flavored chewable tablet	89

[a] Dosage forms include: chewable tablets intended for drug delivery to the stomach (CT), and quick-dissolve chewable tablets (QDCT).

men.[79] Asftach® is a buccal tablet containing triamcinolone acetonide for treatment of apththous ulcers, and contains a bioadhesive layer and a dissolvable lactose nonadhesive backing layer.[80]

Other buccal adhesive delivery systems have been evaluated recently or are under various stages of development. For example, nifedipine/propranolol hydrochloride is a double-layer buccal tablet for systemic delivery and prolonged drug release to the buccal mucosa.[81] A bilayer buccal adhesive nicotine tablet has been evaluated that provides a drug release pattern combining fast-release and prolonged-release profiles and offers an improvement in smoking cessation therapy.[82] Other versions of this design of buccal adhesive tablets that control drug release, such as testosterone, based on varying the rate of hydration of the matrix have also been evaluated.[83] In addition, buccal adhesive tablets have been evaluated for the delivery of chlorhexidine diacetate for treatment of periodontal disease,[84] and for local delivery of antibiotics such as secnidazole, polymixin B, and tobramycin for treatment of oral infections.[85,86]

Buccal tablets offer several advantages as a drug delivery system including local delivery, rapid buccal absorption, rapid onset, prolonged release, "dialed in" drug release profiles for up to several hours, and product line extensions, and future products are expected soon.

2. Slow-Dissolve Products

Buccal delivery occurs in the mouth but the primary delivery route for many novel oral dosage forms is not by swallowing. Rather, the drugs are absorbed through the mouth's mucosal membranes and absorbed systemically. Buccal delivery is more convenient than injection, and does not require water; rather, the patient's own saliva dissolves the dosage form and solubilizes the drug to facilitate absorption. Many SD products that dissolve in the mouth within a few minutes have been developed (i.e., sublingual tablets, lozenges, and microparticles) and introduced in the European and U.S. market in recent years and several of these are highlighted in Table 1.3.

Uprima® (apomorphine hydrochloride), marketed by Tap Pharmaceutical Products, was granted marketing clearance in the European Union in May 28, 2001 for the treatment of erectile dysfunction.[29] Uprima® is available as a sublingual tablet in dosage strengths of 2, 3, and 4 mg. It is rapidly absorbed through the buccal mucosa to provide for a relatively quick onset of action. Uprima® has demonstrated a faster onset of action (as fast as 10 minutes) compared with Viagra® (1 hour); consecutive doses of Uprima® can be taken with less time in between (7 hours) than Viagra® (24 hours); and Uprima® has demonstrated an ability to safely and effectively treat men with varying severity of erectile dysfunction.

Cephalon Inc. reacquired all rights to Actiq®, a cancer pain product, in 12 European countries, from Elan's subsidiary, Elan Pharma International Ltd.[30] Cephalon markets Actiq® in the United States, and will market the drug in the United Kingdom, Germany, and France. Actiq® is a candy-based, "lollipop-like" product that slowly dissolves in the mouth releasing fentanyl for pain management.

The advantage of this product is the patient's ability to self-administer and titrate the dose to provide relief. For the six-month period ending June 30, 2002, Cephalon reported sales of $47.8 million for Actiq®.

Atrix Laboratories is developing BEMA™-fentanyl for treatment of cancer pain. This is an oral transmucosal technology based on a small semisoft disc that adheres to the mucosa, such as the side of the mouth, and delivers the drug as the film bioerodes. There is no need to remove the film because it dissolves with the moisture from the mucosa and saliva. [47]

Orapharma Inc. received approval in the United States on February 16, 2001 for Arestin® (minocycline) for the adjunctive treatment of periodontitis following scaling and root planing. The treatment uses the company's proprietary patented microsphere technology to deliver the dry-powder antibiotic minocycline beneath the gum, directly into the infected periodontal pocket. Arestin® does not require mixing or refrigeration and is easy to administer in a specially designed unit dose syringe. [49]

GlaxoSmithKline received approval on October 31, 2002 for the Commit™ lozenge, the first and only nicotine lozenge approved for over-the-counter sale.[54] The Commit™ lozenge contains 2 and 4 mg of nicotine polacrilex and helps control nicotine craving by delivering nicotine rapidly through the oral mucosal tissue. For initial therapy nine lozenges are used per day for the first six weeks of therapy. Those who smoke their first cigarette within 30 minutes of waking are directed to use the 4 mg lozenge strength and those who smoke after 30 minutes of waking are directed to use the 2 mg dose strength to improve the chances of quitting.

C. Nondissolving Delivery Systems

A wide variety of nondissolving intraoral drug delivery systems have been developed including chewing gums, mucoadhesive patches, controlled-release inserts, fibers, and medical devices. These delivery systems may be ideal for drugs having limited oral absorption from conventional tablets and indications such as treatment of migraine, pain, allergy, angina, motion sickness, smoking cessation, and local delivery to the oral mucosa may benefit. Selected recently introduced products are summarized below.

1. Medicated Chewing Gums

Chewing gums have been used for several decades as consumer products and represent over a $2.2 billion market in the United States alone.[90] The concept of using chewing gum as a matrix for drug delivery is not new since Aspergum® was introduced in 1928 as the first medicated gum containing acetylsalicylic acid, and later in the 1970s gums containing nicotine were introduced (Nicorette®), containing nicotine polacrylate for delivery of nicotine as an aid for smoking cessation.

SmithKline Beecham Consumer Healthcare launched Nicorette® Orange in the United States in October 2000. Nicorette® Orange is a sugar-free gum and is available in two strengths, 2 and 4 mg, and was approved for OTC sales on September 25, 2000. Nicorette® Orange helps to relieve cravings and nicotine withdrawal symptoms by providing a temporary alternative source of nicotine, easing the transition to become smoke free. Medicated and fluoride-containing gums are available in a number of countries (e.g., Fluogum® and Fluorett®) and a gum containing chlorhexidine (Vitaflo CHX®) has been available for treatment of gingivitis in Denmark and Sweden since 2001.[90] Other medicated chewing gums have been developed including a caffeine-containing gun (Stay Alert®) as a nonsleep aid, an antacid gum containing calcium carbonate (e.g., Surpass®, developed by Wrigley, and now discontinued), and an antiemetic gum (Travel Gum®), available in some European countries.[90,91]

2. Buccal Patches

Buccal patches have been evaluated over the past several decades as a viable means of drug delivery to the tissues of the oral mucosa (i.e., buccal, palatal, and gingival tissues). Buccal adhesive patches are generally modified-release dosage forms that have the potential to provide for controlled drug delivery from 1 to 24 hours depending upon their biopharmaceutical and dissolution characteristics (i.e., slow dissolving vs. nondissolving).[92] A buccal patch refers broadly to a dosage form that employs a bioadhesive, which allows the dosage form to adhere to the oral mucosal tissue (i.e., buccal, palatal, and gingival, etc.) for varying periods of time (i.e., hours to days) to provide prolonged local or systemic drug delivery and these may be either slow-dissolving or nondissolving tablets and patches. The FDA refers to these dosage routes and forms as buccal extended-release films and tablets, respectively, but are defined more generically here as buccal patches. Buccal patch dosage forms are solid matrices that are generally nondissolvable or slowly dissolvable and can be designed to deliver the drug unidirectionally (i.e., directly into the buccal tissue), bidirectionally (i.e., directly into the buccal tissue and into the saliva in the oral cavity), and multidirectionally (i.e., drug diffusion from all surfaces of the device).[93] In order to achieve these variations in targeted drug release three basic types of buccal patch designs have been developed, which include: (1) a monolithic matrix (Type I, for multidirectional drug release); (2) a multilayer matrix having an impermeable or semipermeable backing layer (Type II, for unidirectional drug release into the buccal mucosa, with a small fraction of lateral drug release); and (3) a multilayer matrix that has an impermeable layer over the back and sides of the device (Type III, for unidirectional drug release only into the buccal tissue).[92]

The concept of using buccal patches as a matrix for drug delivery is not new. For example, DentiPatch® has been developed by Noven, which is a lidocaine extended-release buccal patch that adheres to the gingival tissue to provide for local analgesia, and was approved in the United States in May 1996.[94]

ND dosage forms offer several advantages as a drug delivery system including local delivery, rapid buccal absorption, rapid onset, prolonged drug release, dose termination, and product line extension, and future products are expected soon.

V. INTRAORAL DRUG DELIVERY EXCIPIENTS

Fast-dissolving tablets and chewing gum drug delivery technologies are the most exciting recent developments in the category of intraoral dosage forms. A number of novel functional pharmaceutical excipients have been developed over the years upon which new IODs and technologies are based. Some of the more recent excipients and their functional characteristics are briefly summarized below.

Pharmaburst™ is a quick-dissolve excipient system developed by SPI Pharma for quick-dissolve dosage forms. It is a highly flexible, rapidly disintegrating excipient that imparts a smooth creamy mouth feel, and is manufactured under cGMPs.[95]

Calcium silicate (RxEXCIPIENTS™ FM1000) has been developed by Huber Engineered Materials as an effective dispersant and co-disintegrant for fast-melt direct compression tablets. This excipient when formulated at levels of between 0.5 to 25 percent with super disintegrants such as Crospovidone NF, Sodium Starch Glycolate NF, and Croscarmellose sodium results in tablets that dissolve within 5 to 60 seconds and have practical hardness and low friability.[96] Unlike other fast-dissolve excipients calcium silicate is nonhygroscopic and can be formulated and manufactured into fast-melt tablets by conventional rotary tableting machines, can dissolve in the mouth in less than a minute, and promote rapid release of the drug.

New innovations in chewing gum bases are being spurred by development of novel functional excipients such as compressed powder gum as a viable delivery system for drugs.[97] These innovations are being driven by potential new applications and markets that may develop as chewing gum technology advances and more convenient manufacturing options become available. For example, chewing gums that can be formulated in powder form with active ingredients without heating, and manufactured by direct compression into chewing gum tablets, offer new drug delivery options to the pharmaceutical formulator. Chewing gums can offer many advantages to pharmaceutical developers having these standard methods of manufacture in place. This technology can open up opportunities to deliver drugs that are heat-sensitive and that could not otherwise be manufactured as a stable product in conventional chewing gum bases, inasmuch as heating and melting are a required part of the traditional manufacturing process. Chewing gums have a number of advantages in terms of adding flavors and self-titration of drug release by the patient, which is dependent upon the rate and extent of chewing.

Pharmagum™ S is a directly compressible chewing gum delivery system.[95] Pharmagum™ can be directly compressed on a traditional tableting machine resulting in hard Chicklet®-type gum with a relatively high loading of otherwise poorly compactible ingredients. Pharmagum™ is a mixture of polyol(s) and/or sugar with a chewing gum base and complies with the Food Chemicals Codex specifications and contains ingredients that are "Generally Recognized as Safe" (GRAS). Pharmagum™ is a free-flowing powder with physical characteristics that allow easy handling, and compatibility with high-speed tableting machines.

Several companies are involved in developing new chemical permeation enhancers as novel functional excipients for improving drug delivery to skin and mucous membranes.[98] Some of these novel excipients useful in transdermal and dermal delivery may also find utility to improve oral mucosal absorption of poorly soluble and low permeable drugs as well, by modifying their diffusion, partitioning, or solubility in the oral mucosal tissue.[99]

In the future the use of new and improved permeation enhancers and other novel excipients may prove useful for drug delivery from various types of intraoral dosage forms to improve local delivery, buccal absorption, onset of action, facilitate product line extension, and provide patent protection, and future products using these approaches are expected soon.

VI. INTRAORAL DOSAGE FORM BARRIER PACKAGING

A number of innovative package designs for IODs have recently been introduced into the market. The novel multidose plastic dispensers for such products as FreshBurst™, Listerine® Pocketpacks™, and Myntz™ can accommodate up to 24 single quick-dissolving strips packaged in a small (0.5 × 3 × 5 cm) reusable, convenient disposable plastic packlet with a hinged closure cover. The nonadhering strips are stacked upon each other and a single strip can be easily dispensed. This is a packaging innovation that is an economical alternative to a 24-unit blister pack.

Blister barrier packaging is another alternative to provide protection for unit doses for fragile buccal wafers and tablets. A wide variety of barrier films is now available that can provide the required blister package rigidity and barrier properties to ensure the stability of unit doses. For example, quick-dissolving products such as Organon's Remeron® Sol-Tabs® (Cima Labs) and Merck's Maxalt-MTL® (Zydis® system) use multilayer foil-based barrier material to protect the dosage form, with the blister actually forming the tablet during the manufacturing process.[100] Both Cima Labs and Merck have filed patents where the packaging is mentioned as part of the process, and shows the importance of packaging materials not only to ensure physical protection and stability of the dosage form, but also to provide protection from an intellectual property position.[101,102] Thus, the various fast-dissolve systems can be very fragile, and regular push-through blister packaging would destroy the tablet upon removing it from the blister, so the packaging requires peelable closure films.

There is a wide variety of packaging materials and polymers available for selection with a wide range of physical properties, gas and moisture vapor transmission rates, and cold and heat seal properties. Some of those used for commercial thermoformable packaging materials include rigid polymers such as PVC, PETG, APET, CPET, PP, or COC. These thermoplastic polymers are usually combined by coextrusions, lamination, or coating to form multilaminate barrier films.[102]

VII. FUTURE TRENDS: INTRAORAL DRUG DELIVERY

A. Patent Domain

Hundreds of patents have been issued over the past several decades, which provide proprietary protection for the products described in this chapter. Much of the patent literature relates to mucoadhesive buccal and fast-dissolving intraoral dosage forms. However, many other patents describe novel slow- and controlled-release dosage forms targeted for drug delivery to the oral cavity as well.

There are over 23 therapeutic indications being investigated and claimed in the patent literature on the buccal drug delivery system since 1960 and at least 50 patents issued on buccal drug delivery. The five most frequently cited therapeutic classes of drugs claimed in these patents include organic nitrates, nonsteroidal anti-inflammatory drugs (NSAIDs), local anesthetics, bronchodilators, and antibacterial/antibiotics. Over 47 therapeutic agents delivered by various intraoral dosage forms were claimed in the patent literature since 1960. The five most frequently cited therapeutic agents include nitroglycerin, theophylline, isosorbide dinitrate, nifedipine, and lidocaine. The most prevalent indications included the following categories: angina, inflammation, gingivitis, arrhythmia, and asthma.

B. Products in Development

In addition to the dozens of products introduced over the past decade targeted for dissolution or disintegration in the oral cavity, there are a multitude of new dosage forms and products under development that will be introduced over the next several years. These fall into primarily three classes including the quick, slow, and nondissolving (i.e., controlled-release technologies) dosage forms.

Numerous products utilizing the slow-dissolving tablets, wafers (QD systems), and sublingual tablets are already on the market, and convenience of use without the need to drink or swallow has contributed to their success. Chewable tablets will have continued success particularly for the pediatric and geriatric segments of the OTC market. Currently, chewable tablet multivitamin supplements are the dominant product category in the OTC market.

Although QD tablets are the most widely used of the quick-dissolve IODs, other new, more elegant quick-dissolving flexible thin-film dosage forms (Quick-Dis™ and Slow-Dis™) utilizing cellulosic excipients as the drug carrier matrix are under development.[65] The Quick-Dis™ technology is based on wet film web-coating technologies adapted from transdermal adhesive film manufacturing processes.[60] These new thin-film drug delivery systems are ideal for delivery of highly potent drugs requiring a rapid onset of action (i.e., treatment of erectile dysfunction, ulcers, smoking cessation, etc.) by way of buccal absorption or absorption throughout the entire GIT. Thus, for most drugs the onset of action is expected to be faster compared to a conventional tablet, which is available for absorption from the stomach to the colon.

SD dosage forms will continue to be explored in special market niches for those drugs that are poorly absorbed from the GIT and where absorption from the mucosal tissue of the oral cavity is greater or where local delivery is an advantage to treat diseases of the mouth. These formulations consist of slow-dissolving tablets, which either adhere to the mucosal tissue of the oral cavity, or various nonmucoadhesive formulations such as lozenges and sublingual tablets or lollipoplike devices such as Oralet® for the slow release of fentanyl as a preanesthetic medicament.

Nondissolving formulations are a special class of controlled-release dosage forms that consist of chewing gums, patches, and devices. Nicorette® is an example of a chewing gum for delivery of nicotine for smoking cessation therapy. The drug is released by the chewing action and mastication of the gum, and release can be titrated by the patient over a period of an hour or so. Other medicated gums are under development for the delivery of sildenafil, diphenhydramine, nicotine in second-generation products, and others for various indications where a rapid onset or local action is required.[103–105] Chewing gums offer several advantages as a drug delivery system including local delivery, rapid buccal absorption, rapid onset, dose titration, dose termination, and product line extension, and future products are expected soon. Cydot® is an example of a patch technology where the patch adheres to the buccal mucosa for a period of up to 24 hours to slowly release melatonin for normalizing circadian rhythms. Other drugs may also benefit from intraoral delivery from patches where controlled or long-term delivery (i.e., once a day dosing) is desired. Other patches have recently been introduced for teeth whitening and other mucosal patches are under development for buccal delivery of selected drugs.

Finally, several innovative self-actuated drug delivery devices designed for administration of drugs to the tissue of the intraoral cavity and the buccal mucosa are being developed including aerosol sprays, liquid pump sprays, or activated mists (i.e., RapidMist™ device), and needleless injectors (i.e., PowderJect® device). These devices are being developed in multidose formats for a variety of dugs requiring control of diabetes (i.e., insulin), pain (e.g., fentanyl and morphine), anticoagulants (i.e., heparin), and flu vaccines (i.e., influenza) delivered noninvasively to the oral cavity for buccal absorption.

VII. SUMMARY

Presently, intraoral dosage forms are commercialized worldwide and represent a multibillion-dollar business with products that span a multitude of delivery systems and indications for local and systemic therapy. The majority of the commercial products are based on drugs that have been approved by other routes of administration (i.e., injectable, conventional oral tablets and capsules, etc.) and their delivery has been improved for either local therapeutic indications (i.e., lidocaine) or to improve systemic oral bioavailability (i.e., nitroglycerin, nicotine, fentanyl, etc.) whereas others simply provide convenience (i.e., quick-dissolve tablets), taken without water, anywhere, anytime, and provide product line extensions for conventional drugs for multiple indications.

In this chapter we have introduced the various classifications of IODs targeted to the oral cavity. Sublingual tablets have been available for over half a century. Most of the other QD and SD technologies have only more recently been developed and introduced as commercial products. The continuing pressures for improved compliance and convenience of taking medications by the consumer are opening new market opportunities for improved and convenient to administer products. The unmet market needs have resulted in an explosion of new quick-dissolving and chewable formulations, which do not require drinking water or swallowing tablets or capsules, and many of these have been introduced as OTC products for improved patient convenience of dosing and product line extensions.[106] The need for more rapid onset of action and improved absorption is an additional impetus for the continual development and improvement of QD drug delivery systems. Although SD intraoral delivery systems are not met with the same patient convenience factors as QD products, because they are required to remain in the mouth longer and may interfere with speech, there are select market niches that will continue to be taken advantage of nonetheless. The category of nondissolvable dosage forms is the recent newcomer in terms of pharmaceutical technology, however, chewing gum has been well accepted for decades in numerous consumer products, and only more recently has been used as a matrix for drug delivery. The use of gum as a matrix for controlled drug delivery to the oral cavity is in its infancy and there may be many other products available in this type of dosage form in the future as the technology advances.

Many challenges remain for the noninvasive intraoral delivery of protein and peptide pharmaceuticals. Their low bioavailability by the intraoral route using conventional delivery systems and technologies will continue to call into question the commercial feasibility of this route of delivery for these classes of compounds.[11] The limitations of retention and stability of large molecular weight and polar compounds in the oral cavity, as well as poor permeability and low bioavailability, continue to remain a challenge. However, in the future, one may envision methods for the delivery of peptide and protein therapeutics that may overcome these limitations. Perhaps the use of semi-invasive delivery technologies based on transdermal microneedle-array patch delivery systems that adhere with bioadhesives or mechanical insert devices containing drugs, devices that adhere to the teeth, or other mechanisms of chemically or mechanically enhanced absorption and prolonged retention of the delivery system, may provide the means to solve these drug delivery challenges in the future.

ACKNOWLEDGMENTS

Deborah Bingham, Technology Catalysts, is kindly acknowledged for providing market research data on the intraoral drug delivery systems. The artwork provided by Subhra Ghosh and Jang-Ik Lee is also gratefully acknowledged.

REFERENCES

1. Rathbone, M.J. Oral Mucosal Drug Delivery. Marcel Dekker: New York, 1996.
2. Bruney, C. Market report, evolution of breath strips, Soap and Cosmetics, 10–15, August/September, 2003.
3. Sastry, S.V., Ostrander, K., Chu, J.S., Agarwal, V., and Ghosh, T. K. Annual AAPS Symposium: Oral Fast Dispersing Delivery Systems—Recent Advances, Salt Lake City, UT, October 29, 2003.
4. DeFelippis, M.R. Overcoming the challenges of noninvasive protein and peptide delivery, Am. Pharm. Review, 6(4):21–30, 2003.
5. Sobrero, A. Compts Rendus Hebdomadaires des Sciences. De l'Acadamie des Sciences, 24:247–248, 1847.
6. Murrell, W. Nitroglycerin as a remedy for angina pectoros, Lancet, 151:225–227, 1879.
7. de Vries, M.E., Bodde, H.E., Verhoef, J.C., and Junginger, H.E. Developments in buccal drug delivery, Crit. Rev. Ther. Drug Carrier Syst., 8(3), 271–303, 1991.
8. Shojaei, A. Buccal mucosa as a route for systemic drug delivery: A review, J. Pharm. Pharmaceut. Sci., 1:15–30, 1998.
9. Squier, C.A. The permeability of oral mucosa, Crit. Rev. Oral Biol. Med., 2(1):13–32, 1991.
10. Veillard, M.M. Buccal and gastrointestinal drug delivery systems. In Bioadhesion — Possibilities and Future Trends, Gurny, R. and Junginger H.E., eds. Wissenschaftliche Verlagsgesellschaft mbH: Stuttgart, 1990.
11. Zhou, X.H. Overcoming enzymatic and absorption barriers to non-parenterally administered protein and peptide drug, J. Control Rel. 29:239–252, 1994.
12. Gutniak, M.K., Larsson, H., Sanders, S.W., Juneskans, O., Holst, J.J., and Ahren, B. GLP-1 Tablet in type 2 diabetes in fasting and post-prandial conditions, Diabetes Care, 20:1874–1879, 1997.
13. Bodde, E.H., de Vries, M.E., and Junginger, E.H. Mucoadhesive polymers for the buccal delivery of peptides, structure-adhesiveness relationships, J. Control Release, 13:225–231, 1990.
14. Veuillez, F., Kalia, Y.N., Jacques, Y., Deshusses, J., and Buri, P. Factors and strategies for improving buccal absorption of peptides, Eur. J. Pharmaceut. Biopharm., 51:93–109, 2001.
15. Senel, S., Kremer, M., Nagy, K., and Squier, C. Delivery of bioactive peptides and proteins across oral (buccal) mucosa, Curr. Pharm. Biotech., 2:175–186, 2001.
16. Walker, G.F., Langoth, N., and Bernkop-Schnurch, A. Peptidase activity on the surface of porcine mucosa, Int. J. Pharmaceutics, 233:141–147, 2002.
17. Harris, D. and Robinson, J.R. Drug delivery via the mucous membranes of the oral cavity, J. Pharm. Sci., 81:1–10, 1992.
18. CDER Data Standards Manual, Route of Administration, C-DRG-00301, version 002, http://www.fda.gov/cder/dsm/DRG/drg00301.htm (April 14, 1992).
19. Electronic Orange Book, Approved Drug Products with Therapeutic Equivalence Evaluations. 2002. http://www.fda.gov/cder/ob/default.htm, Food and Drug Administration, November 27.
20. E2B Route of Administration List (Attachment 2) with ICH-M2 Codes, Ver 1.4, Based on ICH E2b Step 4, 1997. http://www.fda.gov/cder/m2/icsr/routsas4.pdf, December.

21. The National Drug Code Directory, Internet NDC Directory, 2002. http://www.fda.gov/cder/ndc; http://www.fda.gov/cder/ndc/tblroute.txt, September 30.
22. Medical chewing gum, Drug Delivery Technology, 2(5):8, July/August, 2002.
23. Parakh, S.R. and Gothoskar, A.V. A review of mouth dissolving tablet technologies, Pharm. Tech., 92–100, November, 2003.
24. Brown, D. Oral delivery, orally disintegrating tablets—taste over speed, Drug Delivery Tech. 3(6):58–61, September, 2003.
25. Klancke, J. Dissolution testing of orally disintegrating tablets, Dissolution Technol., 6–8, May, 2003.
26. Research and development expenditure in 1998, R&D Directions, 6(1), January, 2000.
27. Zyprexia Zydis, package insert, Eli Lilly, www.lilly.com, 2003.
28. Product performance, Med. Ad. News, 20(2), February, 2001.
29. Drug approvals, Med. Ad. News, 20(8), August, 2001.
30. Elan enhancers cash position, Med. Ad. News, 21(12):11, December, 2002.
31. Fast-dissolving Risperdal in Europe, Med. Ad. News, 21(12):6, December, 2002.
32. Melts-in-the-mouth Excedrin, Med. Ad. News, 21(12):6, December, 2002.
33. First European markets, Remeron SolTab, Med. Ad. News, 21(10):6, October, 2002.
34. Remeron SolTab, R&D Directions, 8(3):59, March, 2002.
35. Amarin announces acceptance for filing of new drug application for Zelepar by Elan, Drug Delivery Tech., 2(5):17, July/August, 2002.
36. Ethylpharm, Viatrix develop oral version of tramadol for Europe, Drug Delivery Tech., 2(6):16, September, 2002.
37. DePalma, A. Oral delivery: Adding value to old and new drugs, Drug Discovery Devel., 5(7):93–94, July/August, 2002.
38. Pather, S.I., Siebert, J.M., Hontz, J., and Khankari, R.K. Enhanced buccal delivery of fentanyl using the OraVescent drug delivery system, Drug Delivery Technol., 1:54–57, October, 2001.
39. Cima Labs announces license & supply agreement with Wyeth, Drug Delivery Technol., 2(6):22, September, 2002.
40. Generex Biotechnology completes proof-of-concept study for buccal heparin, Drug Delivery Technol., 2(4):23, June, 2002.
41. Drug approvals, Med. Ad. News, 19(1):85, January, 2000.
42. Zomig rapid-melt formulation approved in Japan, SCRIPT 2738:27, April 17, 2002.
43. New medicines of 2001 by dose form, Med. Ad. News, 21(2):43, February, 2002.
44. For migraine, R&D Directions, 8(1):75, January, 2002.
45. Walter, P., Curtis, P., Davidson, N., Erikstein, B., L' Esperance, B., Miller, I., Rapoport, B., and Ruff, P. Comparison of an orally disintegrating ondansetron tablet with the conventional ondansetron tablet for cyclophosphamide-induced emesis in cancer patients: A multicenter, double-masked study, Clin. Therapeutics, 21(3):492–502, 1999.
46. LeBourgeois, J.P., McKenna, C.J., Coster, B., Feyer, P., Franzen, L., Goedhals, L., Marzecki, Z., Souhami, L., Stewart, A., Tonnessen, F., Haigh, C., Mitchell, T., Wilkinson, J.R., and Graham, E. Efficacy of an ondansetron orally disintegrating tablet: A novel oral formulation of this 5-HT(3) receptor antagonist in the treatment of fractionated radiotherapy-induced nausea and emesis. Emesis Study Group for the ondansetron orally disintegrating tablet in radiotherapy, Clin. Onco., 11(5):340–347, 1999.

47. Drug in development, R&D Directions, 8(2):53, February, 2002.
48. Atrix's dental product gains new markets, Current Drug Discovery, p. 18, June, 2001.
49. For periodontitis, R&D Directions, 7(4):64, April, 2001.
50. Marketing approval, Appl. Clin. Trials, 10(3):19, March, 2001.
51. Product launch, Med. Ad. News, 19(12):4, December, 2000.
52. Waiting in the gate, Med. Ad. News, 21(8):12, August, 2002.
53. For angina, Med. Ad. News 22(1):84, January, 2003.
54. Press release, U.S. Food and Drug Administration approves effective new tool to help smokers quite, GlaxoSmithKline, October 31, 2002.
55. Written communications from Technology Catalyst Inc. International, Falls Church, VA, 2003.
56. Roller, N.W. Lidocaine topical film strip for oral mucosal biopsies, J. Oral Med., 30:55–58, 1975.
57. Danjo, K., Higuchi, F., and Otsuka, A. Release of lidocaine from polymer film dosage forms, Chem. Pharm. Bull. (Tokyo), 43:1759–1763, 1995.
58. Patient information, Chloraseptic relief stripes, cherry, Prestige Brands, International, 2003, http://www.chloraseptic.com.
59. Zengen, Inc. press release. Zengen's new technology first to deliver drug active ingredients in an oral strip, advancement to revolutionize oral drug delivery. August 22, 2003, http://www.eurekalert.org/pub_releases/2003-08/z-znt082203.php.
60. Borsadia, S.B., O'Halloran, D., and Osborne, L. Oral film technology, Drug Delivery Technol., 3(3):63–66, May, 2003.
61. Horstmann, M., Laux, W., and Hungerbach, S. Rapidly disintegrating sheet-like presentations of multiple dosage units, US Patent 5,629,003, May 13, 1997.
62. Zerbe, H.G., Guo, J.-H., and Serino, A. Water soluble film for oral administration with instant wettability, US Patent, 5,948,430, September 7, 1999.
63. Zerbe, H.G., Guo, J.-H., and Serino, A. Water soluble film for oral administration with instant wettability, US Patent, 6,177,096, January 23, 2001.
64. Zerbe, H.G., Guo, J.-H., and Serino, A. Water soluble film for oral administration with instant wettability, US Patent 6,592,887, July 15, 2003.
65. Chen, L.-L., Pfister, W.R., Renn, D.W., Buranachokpaisan, T., Osborn, J., Tran, H.S., and Tao, L. Compositions and methods for mucosal delivery, US Patent 6,552,024B1, April 22, 2003.
66. Chen, L.H., Buranachokpaisan, L.T. and Pfister, W. Effect of substitution and viscosity of hydroxypropyl methylcellulose on the properties of a quick-dissolving intraoral delivery system, AAPS Eastern Regional Meeting, abstract 6:41, 1999.
67. Cremer, K., Ludwig, K., Anhauser, D., Schumann, K., Steinborn, P., and Bungarten, U. Primary packaging unit for film-like or oblate-like administered shapes. US Patent 6,655,112, December 2, 2003.
68. Chen, L.H., Liang, A., Zheng, X., and Wu, H-L. Quick dissolving oral mucosal drug delivery device with moisture barrier coating, US Patent application, 20030118653A1, June 26, 2003.
69. Chen, L.H., Pfister, W.R., Renn, D.W., Buranachokpaisan, T., Osborne, J., Tan, H.S., and Li, T. Compositions and methods for mucosal delivery, US Patent Application 200300683786, April 10, 2003.

70. Okamoto, H., Taguchi, H., Iida, K., and Danjo, K. Development of polymer film dosage forms of lidocaine for buccal administration. I. Penetration rate and release rate, J Control. Rel., 77:253–260, 2001.
71. Amarin Corporation press release, Amarin Corporation announces acceptance for filing of a new drug application for Zelapar™ by the FDA, May 23, 2002.
72. Amarin Corporation press release, Amarin Corporation announces Zelapar™ receives approvable letter from FDA, February 18, 2003.
73. Clarke, A., Brewer, F., Johnson, E.S., Mailard, N., Hartig, F., Taylor, S., and Corn, T.H. A new formulation of selegiline: improved bioavailability and selectivity for MAO-B inhibition, J. Neural Transm., 110:1241-1255, 2003.
74. Zomig ZMT, Med. Ad. News, 20(6):70, June, 2001.
75. Press release, Multimedia available: FDA approves over-the-counter loratadine drug Alavert, Business Wire, Friday, December 20, 2002.
76. Lingual (Oral) spray, Drug Delivery Technol., 2(5):79, July/August, 2002.
77. The buccal route for systemic delivery, World Pharm. Devel., 2:139, 2001.
78. RapidMist™ Device, www.generex.com., 2003.
79. Striant (testosterone buccal system) mucoadhesive CIII, www.rphlink.com/striant.html, October, 2003.
80. Nagai, T. and Machida, Y. Buccal delivery systems using hydrogels, Adv. Drug Del. Rev. 11(1–2):179–191, 1993.
81. Remunan-Lopez, C., Portero, A., Vila-Jato, J.L., and Alouso, M.J. Design and evaluation of chitosan/ethylcellulose mucoadhesive bilayer devices for buccal drug delivery, J. Control. Rel., 55(2–3):143–152, 1998.
82. Park, C.R. and Munday, D.L. Development and evaluation of a biphasic buccal adhesive tablet for nicotine replacement therapy, Int. J. Pharm., 237(1–2):215–226, 2002.
83. Bologna, W.J., Levine, H.L., Cartier, P., and de Ziegler, D. Bioadhesive progressive hydration tablets and methods of making and using the same, US Patent 6,248,358, June 19, 2001.
84. Juliano, C., Cossu, E.G., and Giuchedi, P. In vitro/in vivo studies of sodium alginate mucoadhesive buccal tablets for the local delivery of an antimicrobial agent, Cont. Rel. Soc. 30th Annual Meeting Proced. #751, 2003.
85. Arora, S., Kohi, K., and Ali, A. Microbiological evaluation of mucoadhesive buccal tablets of secnidazole against anerobic strains for the treatment of periodontal disease, Cont. Rel. Soc. 30th Annual Meeting Proced. #376, 2003.
86. Carr, D., Buchanan, E., Robertson, S., and Halliday, J.A. Controlled release buccal insert for sustaining drug concentrations in saliva, Cont. Rel. Soc. 30th Annual Meeting Proced. #541, 2003.
87. Electronic Orange Book: Approved drug products with therapeutic equivalence evaluations, current through November 2002, http://www.fda.gov/cder/ob/default.htm, November 27, 2002.
88. Phazyme anti-gas/simethicone quick dissolve chewable tablets, Product Alert, 18(10): May 28, 2001.
89. Acid free acid-neutralizing digestive aid chewable tablets that dissolve quickly and produce no chalky or gritty texture, Product Alert, 18(10): May 28, 2001.
90. Conway, B. Chewing gum as a drug delivery system, The Drug Delivery Companies Report, Autumn/Winter: 33–35, 2003.

91. Kamimori, G.H., Karyekar, C.S., Ottersetter, R., Cox, D.S., Balkin, T.J., Belenky, G.L., and Eddington, N.D., Pharmacokinetics of diphenhydramine in healthy volunteers with a dimenhydrinate 15 mg chewing gum formulation, Int. J. Pharm., 234:159–167, 2002.
92. Hao, J., and Heng, W.S. Buccal delivery systems, Drug Dev. Ind. Pharm., 29(8):821–832, 2003.
93. Anders, R. and Merkle, H.P. Evaluation of laminated mucoadhesive patches for buccal drug delivery, Int. J. Pharm., 49:231–240, 1989.
94. Electronic Orange Book: Approved drug products with therapeutic equivalence evaluations, current through November 2003. U.S. Department of Health and Human Services, Food and Drug Administration, Center for Drug Evaluation and Research, Office of Pharmaceutical Science, Office of Generic Drugs, Updated: December 22, 2003, http://www.fda.gov/cder/ob/default.htm.
95. SPI Pharma Group, PharmagumTMS. Directly compressible chewing gum delivery system. Technical bulletin, 2003.
96. Augsburger, L., Madieh, S., Mehra, D.K., Chittamuru, R., Leigh, R., and Wellman, J. The role of calcium silicate in developing directly compressible "fast melt" tablets. Poster No. W4220, American Assoc. Pharm. Sci. Annual Meeting, October 27–30, 2003.
97. Tablet & capsule trends, Nutriceutical World, 7(1):29–33, January, 2004.
98. Hadgraft, J. Perspectives on dermal and transdermal delivery, Drug Delivery Syst. Sci., 1(4):107–112, 2001.
99. Hadgraft, J. Modulation of the barrier function of the skin. Skin Pharmacol., Appl. Skin Physiol., 14 (Suppl 1): 72–81, 2001.
100. Van Weeren, R. and Gibboni, D.J. Barrier packaging as an integral part of drug delivery, Drug Delivery Technol., 2(4):48–53, June, 2002.
101. Kearny, P. et al. US Patent No. 5,631,023, Method for making freeze dried drug dosage forms, RP Scherer Corp, May, 1997.
102. Katzner et al. US Patent No. 6,155,423, Blister package and packaged tablet, CIMA Labs, December, 2000.
103. Munder, C.B., Li W., and Ream, R.L., Sildenafil citrate chewing gum formulations and methods of using the same, US patent 6,592,850, 2003.
104. Valoti, M., Frosini, M., Dragoni, S., Fusi, F., and Sgaragli, G., The rate of absorption and relative bioavailability of caffeine administered in chewing gum versus capsules to normal healthy volunteers, Methods Fund Exp. Clin. Pharm., 25:377–381, 2003.
105. Morjaria, Y., Irwin, W.J., Conway, B.R., Barnett, P.X., and Chan, R.S., Chewing gum as a drug delivery system, Proceed. Intern. Symp. Control. Rel. Bioactive Mat., 30:651, 2003.
106. Sifton, D.W., Physicians Desk Reference for Nonprescription Drugs and Dietary Supplements, PDR 24 Ed, Thompson Medical Economics, Montvale, NJ.

2

Preclinical Assessment of Oral Mucosal Drug Delivery Systems

Bing Li and Joseph R. Robinson

CONTENTS

I. INTRODUCTION

The first evidence of drug absorption via the oral mucosa was noted over 100 years ago (1). Subsequently, in 1879, sublingual administration of nitroglycerin was reported to successfully alleviate the symptoms of classic angina pectoris (2). Since then, oral mucosal drug delivery has drawn more and more attention because of its potential advantages over other routes of delivery. Following delivery through the oral sublingual mucosa, drugs are rapidly absorbed into the reticulated vein and are transported through the facial veins, the internal jugular vein, and then the brachiocephalic vein, and are finally drained into the general circulation (3). Via this pathway, hepatic first-pass metabolism is bypassed and rapid systemic delivery with improved bioavailability is achieved. The oral mucosa provides accessibility to allow precise dosage form localization. It provides the opportunity to directly modify tissue permeability, inhibit protease activity, or decrease immunogenic responses. Naturally, there are also a number of disadvantages to drug delivery by the oral mucosal route. Relatively small surface area, as compared to the skin, and significant loss of the applied dose due to swallowing and salivary flow are two of the major important loss routes.

The major barrier to oral drug delivery is low tissue permeability due to the presence of the oral epithelia lining on the outermost layer of the oral mucosa. Most drugs do not readily penetrate this epithelial barrier at a sufficient rate to generate therapeutic levels. Various in vitro and in vivo models have been developed to assess oral mucosal permeability. The first part of the present chapter discusses the permeability barrier of the oral mucosa and reviews the in vitro and in vivo methods of assessing oral mucosal drug permeability.

The mouth represents the initial opening to the gastrointestinal tract and thus substances placed in this area are expected to be swallowed. Moreover, the mouth is continuously flushed with salivary fluid, presumably to help the swallowing of food and to clean the oral cavity. On this basis it is not surprising that maintaining drugs in the oral cavity through the use of an appropriate drug delivery system is a significant challenge. Historically, lozenges and troches, which were sucked on by the patient, were the traditional way of delivering drugs to the oral cavity. More recently, bioadhesive tablets and gels have been used to give longer contact time with the absorbing tissue and thus minimize drug loss through salivary flow.

The concept of mucoadhesion was introduced into the controlled drug delivery area beginning in the early 1980s. The amount of drug absorbed across a membrane, such as in the oral cavity, is directly proportional to the surface area of exposure, the applied drug concentration, the permeability coefficient for the drug, and the residence time. Given the normally short residence time for a solution of the drug in the mouth, bioavailability of most drugs from the mouth is low. The ability of a mucoadhesive polymer to interact with biological membranes or surfaces and be retained on the membrane for an extended period of time makes it a good addition to a mucosal drug delivery system. A variety of methods has been reported to study mucoadhesion. The second part of this chapter reviews the in vitro and in vivo methods that have been utilized to assess mucoadhesion.

Different formulations of oral mucosal drug delivery systems have been investigated and some have been commercialized. Conventional formulations of mucosal drug delivery systems are primarily tablets and patches. More recently, phase change polymers, which change from liquid to solid at body temperature, have broadened the design of oral mucosal drug delivery dosage forms. A preliminary discussion of using phase change polymers as a new dosage form for oral mucosal drug delivery, as well as a review of certain conventional dosage forms such as patches and tablets, is presented in the third part of this chapter.

Safety is always a major concern for any drug delivery system. In the last part of this chapter, some major safety issues that need to be considered when oral mucosal drug delivery is used are discussed.

II. ANATOMICAL AND PHYSIOLOGICAL DETAILS OF THE ORAL MUCOSA

A. Tissue Types and Characteristics

The oral cavity contains four anatomically distinct sites with unique tissue types, which include the palatal, gingival, sublingual, and buccal mucosae. The gingival tissue surrounds and holds the teeth in pair. The palatal tissue refers to the roof of the mouth. The sublingual tissue is composed of the floor of the mouth and ventral tongue, and buccal mucosa refers to the cheek and inside of the upper and lower lip as shown in Figure 2.1.

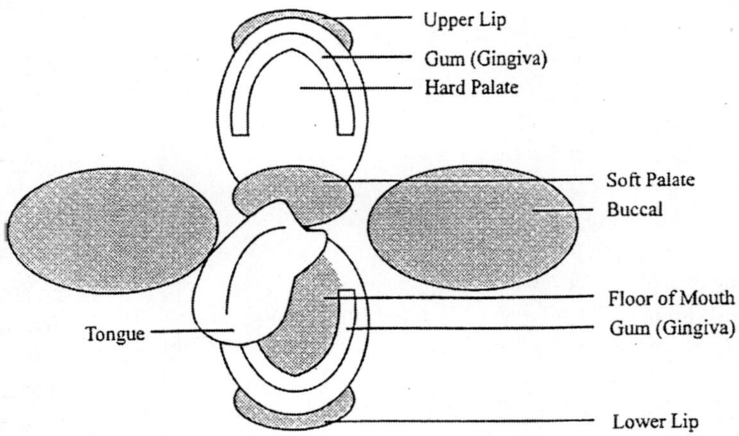

Figure 2.1 Depiction of anatomical locations within the oral cavity (from Reference 44).

The regional variability in biophysical and biochemical characteristics of the oral mucosae lining reflects the difference in barrier properties. Generally speaking, there are two types of mucosal epithelium that can be differentiated based on the presence and molecular weight of keratins in the epithelial cell surface: keratinized and nonkeratinized epithelium. Keratins are tonofilaments that are composed of at least seven component proteins, with molecular weights ranging from 40 to 70 kDa. Cells of nonkeratinized mucosa contain mostly low molecular weight keratins, whereas in keratinized mucosa, the higher molecular weight keratins predominate (4). Usually, keratinized epithelium is less flexible, has a more flattened shape, and is devoid of organelles. The epithelial of the gingiva and hard palate are keratinized and therefore have a cornified surface which very much resembles that of the upper epidermis in skin. This type of mucosa occupies approximately 24 percent of the total oral mucosa surface (5,6). The buccal mucosa and sublingual mucosa, on the other hand, have a surface lining composed of a nonkeratinized, stratified squamous epithelium, and occupy approximately 60 percent of the oral mucosa surface (7). The epithelial thickness of these different regions varies from one to another, rendering different permeabilities and barrier functions (Table 2.1). Therefore, they may require a different drug delivery system design.

Table 2.1 Characteristics of Some Human Epithelia

Location[a]	Thickness (μm)	Keratinization
Buccal	500–600	No
Sublingual	100–200	No
Gingival	200	Yes
Palatal	250	Yes

[a] Modified from Reference 10.

Currently, the sublingual and buccal areas are the most commonly used routes for oral mucosal drug delivery. The sublingual route is relatively permeable, giving rapid absorption and acceptable bioavailabilities for some drugs, and is convenient, accessible, and generally well accepted. It is generally accepted that the sublingual region cannot easily be used for sustained-release dosage forms because of the difficulty of maintaining a dosage form in this area for an extended time and lack of acceptance by the patient. The buccal area also has potential for drug delivery. However, it is less permeable than the sublingual area, and generally is not able to provide rapid absorption and relatively good bioavailability compared with sublingual administration. Both of these routes have been investigated clinically for the delivery of a variety of drugs as shown in Table 2.2.

B. Generalized Structure of Oral Mucosa and Its Barrier Function

The oral mucosa consists of an outermost layer of stratified squamous epithelium, below which lies a basement membrane, and underneath is a lamina propria followed by the submucosa, as shown in Figure 2.2.

Table 2.2 Clinical Applications of Systemic Delivery via the Oral Mucosae

Mucosa	Drug	Brand Name/ Manufacturer[a]	Comment
Sublingual	Nitroglycerin	Nitrostat, Parke-Davis, etc.	Delivery route of choice, relatively fast blood level
	Buprenorphine	Temgesic, Reckitt & Colman	Alternative to injectable form, relatively fast blood level
	Nifedipine	Adalat, Bayer; Procardia, Pfizer	Gives more rapid onset of action than enteral form
Buccal	Methyltestosterone	Metandren, Ciba; Oreton	Avoids first-pass hepatic metabolism
	Prochlorperazine	Schering Buccastem,[a] Reckitt & Colman	Alternative to enteral tablet

[a] Modified from Reference 17.

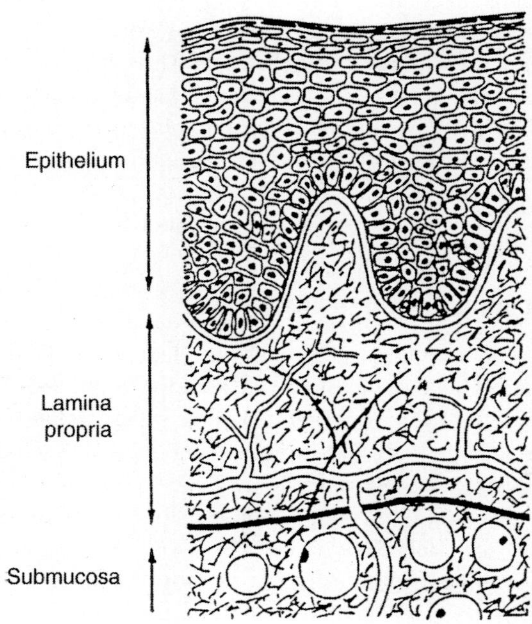

Figure 2.2 Depiction of buccal mucosa tissue (from Reference 17).

1. Epithelium

The epithelium of the oral mucosa serves as a protective covering for the tissues beneath and a barrier to the entry of foreign materials. These functions are reflected in the organization of the epithelium in which individual epithelial cells are closely opposed and stratified so there are a number of layers that show a sequence of differentiation. The uppermost layers (top third) form a surface that is resistant to physical insult and to penetration by foreign substances (7).

Like the epithelium of the gastrointestinal tract and the skin, the epithelial lining of the mucosa is composed of a constantly renewing cell population which is produced by cell division in the basal region. Cells increase in size and become flattened as they progressively mature and migrate from the basal layer towards the epithelial surface, showing increasing levels of protein tonofilaments and declining levels of some cytoplasmic organelles. An approximate expression of this dynamic event is to calculate the turnover time, which is defined as the time taken for the total number of cells in the tissue to be shed and replaced by an equal number of cells through division (7). In general, the oral epithelium turns over faster than skin epidermis but more slowly than the gastrointestinal epithelium. Estimates for the turnover time of human buccal epithelium vary between 4 and 14 days (8). Due to the long turnover time, a properly designed buccal adhesive delivery system may remain in place for many hours to days.

The barrier function of the epithelium depends primarily on cohesion between individual epithelial cells and adhesion of the whole epithelium to the underlying tissue. Cell connections represent the paracellular route for drug absorption and hence a major pathway for water soluble drugs. Three major types of junctions have been found in oral epithelium: desmosome, gap junction, and tight junction (8). Tight junctions are occluding junctions that play a critical role in maintaining the concentration differences of small hydrophilic molecules across epithelial cell sheets by sealing the plasma membranes of adjacent cells together to create a continuous, impermeable, or semipermeable barrier for diffusion across the cell sheets.

2. Basement Membrane and Connective Tissue

Basement membrane refers to an irregular continuous interface between the epithelium and connective tissue. This basal complex anchors the epithelium to the connective tissue and supplements the barrier function of the superficial layers of the epithelium to prevent some large molecules from passing the oral mucosa. The bulk of connective tissue consists of a collagen fiber network, the organization of which determines mechanical stability, resistance to deformation, and extendibility of the tissue. Most likely, the connective tissue, along with the basement membrane, is not considered to influence the diffusion of most compounds of pharmacological interest although these two regions may limit the movement of some macromolecules and complexes. Vascular drainage from the oral mucosa is principally via the lingual, facial, and retromandibular veins, which open into the internal jugular vein (9). This is important from a drug delivery point of view for many drugs that require avoidance of hepatic first-pass clearance.

3. Salivary Flow

There are three major salivary glands and numerous minor glands that continuously secrete saliva into the mouth. Major components of saliva are mucins, inorganic salts, proteins, lipids, and water. It has moderate levels of various esterases, carbohydrases, and phosphatases, but little to no protease. Usually, the pH level of saliva is slightly acidic (i.e., pH 5.8 to 7.1), and goes up to 7.6 when it is stimulated (10). The average rate of saliva secretion is 0.2 to 0.4 ml/min when resting and becomes 2 ml/min when it is stimulated. There is about a 70 μm thick mucus coating throughout the mouth which is constantly cleared by swallowing. Daily output of saliva in humans is 750 to 1000 ml. Due to this continuous salivary flow, drug solutions are easily cleared from the mouth and thus provide a challenge for designing oral mucosal drug delivery systems.

III. EVALUATION OF MUCOSAL PERMEATION

A. In Vitro Mucosal Permeation Techniques

The most commonly used in vitro method to study oral mucosal permeability is the use of a permeability chamber (11–16). Two types of permeability cells have been used: side-by-side horizontal (i.e., Ussing-type) and vertical (i.e., Franz-type). Diffusion cells are very useful to measure the transmembrane flux of a substance across a mucosa and to study the effects of absorption enhancers on the membrane. These cells are also easily adapted to measure transepithelial electrical resistance (TEER) and thus monitor integrity of the membrane. They are also useful for physiological and toxicological studies. In both cell types, a well-defined area of mucosa from an animal is clamped between the donor and receiver compartments of each cell. A known concentration of penetrant can be introduced into one cell (donor) and its concentration measured as a function of time in the other cell (receiver). The temperature of this system can be controlled by circulating water from a thermostat constant temperature water bath and the fluid in the chambers can be agitated by either stirring bar or air bubbles to prevent formation of a stagnant diffusion layer.

The advantage of an in vitro permeability study is that the experiment is performed in a controlled environment under well-defined conditions. The primary drawback of such a study is that the tissue is removed from its natural environment and placed in a highly artificial environment. It is difficult to verify that the permeability characteristics of the isolated tissue are identical to those of the living animal. Another specific concern in using excised sublingual mucosa is that numerous ducts from the submandibular and sublingual salivary glands open to the mucosal surface of this region (17). Therefore, to find a sufficiently large area of sublingual mucosa, which is not perforated by these ducts, is difficult.

Using cell cultures is an alternative to assess oral tissue permeability (18). Tavakolli-Saberi and Audus (18) studied enzyme activity and permeation properties of cultured buccal epithelium, which is an in vitro model derived from the hamster cheek pouch. They demonstrated that this cultured epithelial membrane has similar permeability properties to freshly excised tissue. There are a number of advantages in using a cell culture system. Such a system, although structurally simple, is functionally complex and its integrity is essential for normal permeability studies. In addition, the use of cell cultures offers the advantages of homogeneity of cell type, ease of handling, ability to study transport under controlled conditions, and manipulation of parameters influencing transport. It obviously simplifies the experimental procedure of using excised animal mucosa and provides an alternative to the time-consuming and more expensive animal studies. Moreover, a cell culture system is very useful in investigating drug trafficking at a cellular or subcellular level as well as specific drug transport mechanisms. In addition, it may provide the opportunity to use human rather than animal tissues. However, cell culture systems need to be used with caution. A number of factors, such as whether the cells are a primary culture or a stable cell line, whether they are normal or transformed, the differentiation potential of the cells, their passage number, cell heterogeneity, cell viability, and their phenotypic stability, and so on, need to be considered in order to compare the morphological, biochemical, and transport properties of an in vitro cell culture model with the equivalent in vivo model (19).

B. In Vivo Methods

The easiest and simplest intact mucosal permeability studies utilized biological response of the living organism locally or systemically to assess mucosal permeability (20). Nitroglycerin, an organic nitrate for treatment of angina pectoris, was the most commonly used drug applied to the oral mucosa and its systemic pharmacological effects were monitored to gauge absorption (21). Similar studies have been reported for a local effect in which penetration of an anesthetic was monitored by measuring changes in sensation of the mucosa to pain and pressure (22). Results collected from human subjects make this method direct and avoid the need for human correlations that are necessary when using other animals as models.

One of the simplest measurements of penetration through oral mucosal tissue is the so-called "buccal absorption test," sometimes called the "swirl and spit test." It was first introduced in 1967 by Beckett and Triffs (23) and has been widely utilized to determine drug absorption via the oral mucosa (24–29). In this method, a buffered drug solution, of known volume and concentration, was placed in the subject's mouth and was swirled in the mouth for a defined length of time. The solution was then expelled and assayed. The amount of drug absorbed was determined by the difference between the amounts of drug introduced and recovered. Beckett and Hossie (30) have discussed the outcome of such studies relative to pH effects, drug structure–absorption relationships, and kinetic interpretation. It was mentioned in their paper that individuals who produce abnormally large volumes of saliva are less suitable subjects for this buccal absorption test than those with normal salivary production.

Drug excipient interactions were also studied by this general test (31,32). This method has been modified in several ways. First, Dearden and Tomlinson (33) took salivary secretion into account and improved the kinetic data analyses by assuming a linear relationship for salivary secretion and time. Second, Schurmann and Turner (34) included a nonabsorbable marker (phenol red) within the drug solution to monitor for accidental swallowing of the solution. The most recent modification by Tucker (35) was more sophisticated. In this method, multiple samples were withdrawn from the mouth to generate kinetic data. Phenol red was used as a marker enabling salivary secretion to be compensated for and also to monitor for accidental swallowing. The biggest advantage of this method is that it allows drug disappearance from the mouth and appearance in the plasma to be determined simultaneously. These experiments are relatively straightforward, easy to perform, and permit collection of quantitative data directly from a human subject.

However, there are also limitations. The accuracy of this experiment is limited by sensitivity of the equipment that is used to measure drug concentration. The percent absorption should be substantial in order to get reliable data. In addition, estimation of absorption from pharmacological data requires a detailed knowledge of the drug's pharmacodynamics. Also, the buccal absorption test cannot provide information as to the relative permeability of different regions of the oral cavity because all the oral mucosa are in contact with the drug solution.

A number of in vivo perfusion studies have been reported as a modification of the buccal absorption test (36–40). Generally speaking, perfusion experiments are done by attaching the perfusion chambers to the oral mucosa of an anesthetized dog. Drug solutions are circulated through the device and collected at different time intervals. Blood samples are withdrawn periodically during the experiment to generate pharmocokinetic data. This method is commonly used in pharmacokinetic studies. Moreover, because the chamber can be applied to various sites in the oral cavity (e.g., buccal membrane, floor of mouth, dorsum of the tongue, or ventral tongue), this experiment can be performed at several sites on the oral mucosa membrane at the same time. However, local drug metabolism is the major problem that needs to be considered when this experiment is performed. To avoid this problem, an IV infusion experiment is usually performed along with the perfusion experiment. The pharmacological effects of the two experiments can be compared to obtain the flux and permeability coefficient of the drug across the oral mucosa into the general circulation (41). This approach is only useful for drugs whose biological effects can be measured and their biological effects must be proportional to drug concentration. A good example is the drop in blood glucose by administration of insulin. Therefore, a suitable choice of drugs to conduct this experiment is those drugs that are bioactive at low plasma levels and are able to demonstrate a satisfactory pharmacological effect at low plasma drug levels.

C. Animal Models for Permeability Measurement

A variety of lab animals has been used for oral mucosal permeability studies. Among these, the most commonly used animal models are dogs, rabbits, and pigs. A general criterion for selecting an in vivo animal model is the resemblance of the animal mucosa to the oral mucosa of human beings in both ultrastructure and enzyme activity, which represent the physical and metabolic barriers of the oral mucosa. Based on histological examinations, rodent species, such as the rat, guinea pig, and hamster, would constitute poor models because of the extensive keratinization of their buccal mucosa (42). Rabbits have been suggested as a good animal model for studying nasal, rectal, and vaginal mucosal permeability according to histological studies (3). Generally, the absorption behavior in large animals can be assumed to be fairly similar to that of humans (39). Using large animals takes advantage of conducting several repeat experiments using the same animal so that the data obtained will be more accurate (i.e., the intrasubject variability is usually less than the intersubject variability). It is also useful to study differences in absorption behavior between various histologically different oral mucosa. In this sense, the dog appears to be the best choice. This has been supported by Barsuhn's buccal absorption studies of flurbiprofen in humans and the dog (37). Their results reveal that the permeability of human and dog buccal mucosa are similar.

For in vitro studies, not only resemblance of the animal mucosa and humans needs to be considered, but also whether there is an in vivo and in vitro correlation. Many comparative studies between animal models have been examined (15,43). It is suggested that dogs, pigs, and rhesus monkeys are reasonably good animal models for studying oral mucosal drug absorption, whereas rats and hamsters should be used with care because of their highly keratinized tissue characteristics and low permeability (44).

IV. EVALUATION OF MUCOADHESION

Bioadhesive controlled-release systems have been thoroughly studied in the drug delivery field. Robinson (45) defined bioadhesion as the attachment of synthetic or natural polymers to a biological substrate and if the substrate is mucus or a mucus membrane, it is also called mucoadhesion. The potential advantages of bioadhesive-based delivery systems include prolonged drug delivery, localization of drug therapy, targeting to specific diseased tissues, and improving the viability of nonparental nonoral routes of drug administration, which presumably will lead to greater drug bioavailability (45). Polymers such as polyethylene glycol (46), cellulose derivatives (47), polyacrylic acid (48,49), and thermally modified starch (50) have been described as promising candidates for bioadhesive drug delivery systems. Second-generation mucoadhesives, including lectin and lectinlike molecules that bind to cell surface glycoconjugates, have recently been investigated for their binding to the oral mucosa (51,52). To assess the mucoadhesive potential of different polymers, several in vitro techniques have been reported. For a thorough review of the mucoadhesion assessment methods, the reader is referred to Duchene et al. (53), Park and Park (54), and Ahuja et al. (55). The following are typical examples of the most important points and some recent developments in this area.

A. In Vitro Assessment of Mucoadhesion

Various parameters have been used to evaluate mucoadhesion properties. They include, but are not limited to, adhesion strength, adhesion number, and duration of adhesion. Among these, the mechanical measurement of adhesion strength is the most direct method. Tensile, shear, and peel stress are the most commonly used indicators to assess the strength of adhesiveness. Adhesion number is usually used as a parameter to evaluate the adhesive properties of small particles. The ultimate parameter to assess the bioadhesive property of a particular bioadhesive candidate may be the duration of adhesion, which is very commonly used in in vivo mucoadhesion assessment.

1. Methods Based on Measurement of Adhesion Strength

The adhesion strength method measures the force required to break the adhesive bond between a model membrane and the test bioadhesive. The general procedure of this method can be described by the following examples. Ponchel et al. (48,49) used a tensile apparatus to determine the bond strengths of poly(acrylic acid)-hydroxypropyl methylcellulose tablets to bovine sublingual tissue. In the bioadhesion test, animal tissue surface and the tablet surface were brought together with an initial force and separated under constant rate of extension. The detachment force or adhesion strength and the work of adhesion were determined when the tablet and tissue were pulled apart. The work of adhesion is defined as the area under the curve of the detachment force versus extension. Bouckaert et al. (56,57) utilized this method to determine the bioadhesive characteristics of tablets consisting of thermally modified starch and poly(acrylic acid). To assess the significance of detachment force and work of adhesion for the in vitro prediction of their formulations, they also performed an in vitro/in vivo correlation study. They concluded that this in vitro bioadhesion test provided information only on the initial adhesion, and not on the residence time of the tablet in the oral cavity. It is unclear which parameter, detachment force, or work of adhesion has to be considered when evaluating buccal bioadhesion. They suggested that care has to be taken when classifying polymers according to their in vitro bioadhesion properties, because differences seen in vitro do not necessarily induce significant differences in vivo. A summary of examples of tensile strength measurement is given in Table 2.3.

2. Methods Based on Measurement of Shear Strength

Shear stress measures the force that causes the bioadhesive to slide with respect to the mucus layer in a direction parallel to their plane of contact (55). Smart et al. (58,59) developed this technique for measuring mucoadhesion. Their initial test system was developed wherein a suspension of ion exchange resin particles flowed over the inner mucosal surface of a section of guinea pig intestine, and the adhering weight was determined. Because this method resulted in poor data reproducibility, it was modified. In the improved method, mucoadhesion was investigated by measuring the maximum force recorded when a polymer-coated glass plate was slowly pulled from an "homogenized" guinea pig intestinal mucus gel after several minutes contact. The in vitro results of this method were compared with the related in vivo results previously obtained by Chen and Cyr (46). The rank order correlation between the two results suggested that materials that adhere to an isolated mucus gel would also adhere to an oral mucosal membrane. Table 2.4 gives a summary of the methods involving shear strength measurement. Schematic pictures of the different apparatus used to measure the adhesion strength and shear strength of bioadhesives can be seen in the Duchene et al. paper (53).

3. Bioadhesion Test Using Nonbiological Substances

Chen and Cyr (46) had performed in vitro bioadhesion testing using nonbiological substrates. A sample intraoral bandage was attached to a wet dialyzing cellophane membrane representing the gingiva. The force for separation was measured as a quantitative expression of wet adhesive strength. The results suggested some correlation between duration and adhesive shear strength. However, they were criticized for using a synthetic substrate and whether their results were a true measure of bioadhesion was questioned (60).

4. Fluorescent Probe Method

Park and Robinson (61) developed a fluorescent probe method to study polymer interaction with a conjunctival epithelial cell membrane. Briefly, the experiment consisted of adding a fluorescent liposoluble probe, pyrene, which localizes in the lipid bilayer of the cell membrane, to a suspension of cultured human conjunctival epithelial cells. The cells were then mixed with various bioadhesive polymers. The addition of a polymer, which binds to the cell membrane, resulted in compression of the lipid bilayer, causing a change in fluorescence proportional to polymer binding. Measurement of pyrene fluorescence provides a quantitative way to compare the interaction of polymers with the cell membrane.

Table 2.3 Bioadhesion Measurement of Adhesion Strength

Bioadhesive[a]	Instrument	Substrate	Ref.
CP934, HPC	Spring balance	Mouse peritoneal membrane	81
Gelatin capsule	Modified prescription balance	Porcine esophagus	82, 83
Cross-linked PAA, PMA, PHEMA	Modified surface tensiometer	Rabbit stomach tissue	84–86
PAA, HPMC	Tensile apparatus	Bovine sublingual mucosa	48, 49, 87
CP 934, CP EX 55, HPMC, HPC	Modified pan balance	Fresh intestine from male Wistar rats	88
CP, HPC	Modified spring balance	Mouse peritoneal membrane	89
Modified starch, PAA, PEG, NaCMC	Modified tensile apparatus	Porcine attached gingiva	50
CP 934, PHEMA, Eu RL 100	Modified Du Nouy tensiometer	Porcine intestinal mucosa	90
CP, Hyaluronic acid	Electronic digital microbalance	Porcine gastric mucin gel	91
PAA, HPC	Tensile tester	Porcine buccal mucosa	92
Chitosan, polycarbophil, CMC, pectin, xanthan gum	Modified tensiometer	Pig intestinal mucosa	93
Alginate, CMC, chitosan	Precise microbalance (dynamic contact analyzer)	Intestinal tissue from Sprague–Dawley rats	94
Copolymers of dextran, polyacrylamide, PAA	Tensile apparatus	Cellulose paper disk impregnated with porcine mucine gel	95
Modified starch PAA	Tensile tester	Porcine gingiva	57
NaCMC, HPC	Modified tensile tester	Rabbit stomach/intestinal tissue	96, 97
CP 934, HPMC, chitosan, acacia	Tensile tester	PVP/cellulose acetate hydrogel	98, 99
Na alginate, PEG	Tensile apparatus	Guinea pig ileum mucosa	100
Chitosan, Na alginate	Spring tension gauge	Rat peritoneum membrane	101
Copolymer of ε-caprolactone and ethylene oxide	Tensile tester	Rat duodenum mucosal tissue	102
CP 934	Dynamic contact analyzer	Porcine gastric mucin	103

[a] PMA = Polymethacrylic acid; PHEMA = polyhydroxyethyl methacrylic acid; Eu = Eudragit; CP = carbopol; HPC = hydroxy propylcellulose; HPMC = Hydroxy propylmethylcellulose; CMC = carboxy methylcellulose; PEG = Polyethylene glycol; PAA = polyacrylic acid. Modified from Reference 55.

Table 2.4 Bioadhesion Measurement of Shear Strength

Bioadhesive[a]	Instrument	Substrate	Ref.
CP 934, NaCMC, HPMC, gelatin, PVP, acacia, PEG, pectin, tragacanth, Na alginate	Wilhelmy plate method (microforce balance)	Mucus from guinea pig intestine	59
CP 934 ointment	Shearing stickiness test apparatus	Glass plates	104
Polyethylene gel, NaCMC, hydrolyzed gelatin	Instron model 1114	Polymethacrylate	47
Ca polycarbophil, NaCMC, HPMC, Eudragit	Modified Wilhelmy plate surface tension apparatus	Homogenized mucus from pig intestine	105

[a] CP = carbopol; CMC = carboxy methylcellulose; HPMC = Hydroxy propylmethylcellulose; PVP = polyvinyl pyrrolidone. Modified from Reference 55.

5. In Situ Method

Ranga Rao and Buri (62) developed a simple, quantitative, and realistic in situ method to test the bioadhesive potential of polymers. In this technique, glass spheres or drug crystals were first coated with the polymer to be tested. Later, known amounts of these coated particles were placed on rat jejunum or stomach and kept in a humid environment. The tissue was then washed with phosphate buffer or dilute HCl at a constant rate. The percent of particles retained on the tissue was considered as an index of bioadhesion.

6. Organ Culture Technique

Needleman and Smales (60) utilized an organ culture technique to examine the duration of adhesion on carefully maintained mucosal tissue. In their study, an organ culture was used to maintain hamster cheek pouch mucosa, submerged on stainless steel grids in a growth medium. Small quantities of adhesive gels were syringed onto the mucosal surface. The duration of adhesion was assessed by retention to the gel. In this way, the effect of mucin can also be explored by simply placing mucin on the tissue before application of the adhesive gel.

7. Ligand Receptor Binding Method

For natural bioadhesive polymers, because some possess unique characteristics as bioadhesives (e.g., lectinlike molecules can bind to cell surface glycoconjugates), histological methods that determine lectin receptors presenting on the oral mucosal surfaces can be utilized to assess lectin mucoadhesive property. Nantwi et al. (52) studied the binding of lectin to mucosal surfaces by using a solution containing a range of lectins exposed to the surface of unprocessed oral mucosal cells (isolated human buccal cells and whole rat epithelial surfaces) and concluded that a range of receptors are present on the oral mucosa to which a lectin-containing drug delivery system may be targeted.

8. Flow Channel Methods

Flow channel methods utilize a thin channel filled with bovine submaxillary mucin. A particle of a bioadhesive polymer was placed on the mucin gel, and its static and dynamic behavior monitored (63).

9. Falling Liquid Film Method

Small intestinal segments from the rat were placed at an inclination on a tygon tube flute. The adhesion of particles to this surface was monitored by passing a particle suspension over this surface (64).

10. Colloidal Gold Staining Method

This technique utilizes red colloidal gold particles that have been stabilized by an adsorbed mucin molecule (mucin–gold conjugate). Bioadhesive hydrogels develop a red color on the surface upon interaction with the mucin–gold conjugate. The interaction was measured either by intensity of the red color on the hydrogel surface or by a decrease in the concentration of conjugates from the absorbance changes (65).

11. Thumb Test

A simple way that can be used to identify mucoadhesives is the difficulty of pulling the thumb from the adhesive and as a function of pressure and contact time (66).

12. Viscometric Method

Viscosity of a porcine gastric mucin dispersion was measured in different polymer solutions. The mucin–polymer bioadhesive bond strength was quantified and the force of adhesion calculated (67).

B. In Vivo Assessment of Mucoadhesion

In vivo techniques for measuring the mucoadhesive strength to the oral mucosa are relatively few. They mainly focus on the duration time of adhesion of the bioadhesives using humans as subjects. In this case, the criteria of choosing a particular subject, the mucosal sites chosen, the bioadhesive formulation, and a variety of other factors that may affect the in vivo results need to be considered when the experiment is designed.

Anders and Merkle (68) have studied the duration of buccal mucosal adhesion of adhesive patches in vivo. A 26-year-old male was selected as a subject and a self-adhesive patch was attached to the subject's right or left buccal mucosa and a blank patch (as a nonadhesive control) on the other side. The size of the patch used was chosen to cover the maximum buccal area available. The duration of mucosal adhesion was the time required until the adhesive patch completely lost its adhesive contact with the mucosa. This test requires well-trained and motivated subjects and during the test the subject is not allowed to drink or eat.

Bouckaert and Remon (57) investigated the bioadhesive properties of a buccal bioadhesive miconazole slow-release tablet. In this study, a blind crossover study with seven healthy subjects was performed. The volunteers were instructed to test one tablet per day, with an interval of two days, and to insert the tablet in the morning after breakfast. The tablet was placed on the attached gingiva in the region of the right upper canine and fixed for one minute with a slight manual pressure on the lip. The adhesion time of the tablet was recorded and was defined as the time after which the bioadhesive tablet was no longer visible under the lip. This study suggested that dry tablet forms can adhere for long periods and may be particularly useful for systemic drug delivery via the buccal mucosa.

Needleman et al. (69,70) studied the bioadhesion of different formulations of gel delivery vehicles on the periodontal and oral mucosa. Twelve subjects were selected to participate in their study. Two locations were chosen to examine oral bioadhesion: buccal maxillary anterior gingivae and alveolar mucosa buccal in the mandibular molar region. These sites were selected to provide some contrast between a relatively dry and protected, nonmobile mucosa in the upper jaw and a mobile mucosa in the lower jaw where salivary pooling might be expected to occur. The same sites were used on the left and right side of the mouth to give a total of four experimental sites per subject. The mucosa of each of the four sites was gently dried with gauze. The gel was then syringed onto a film which was cut from a tissue culture coverslip plastic and covered one side of the film. The film with adherent gel was applied to the dried site and seated into position with gentle finger pressure for about two seconds. The subjects were then observed for retention of the film. No food or drink was permitted in the first three hours. The in vivo results indicate that there exist considerable inter- and intrasubject variation. By comparing the in vivo with the in vitro results, as well as the in vivo results with other studies, they concluded that the application site of the bioadhesive, the size of the area where the bioadhesive is applied, as well as the taste of the formulation all play important roles in determining mucoadhesion time.

V. ORAL MUCOSAL DRUG DELIVERY SYSTEMS

Several bioadhesive dosage forms for oral mucosal drug delivery have been developed and some are already commercially available. They are mostly in the form of patches and tablets, and some are in the form of ointments, films, and powders. The newest approach in this area has been the use of multifunctional polymers (e.g., polymers possessing both bioadhesive and phase change characteristics) and/or polymers that are bioadhesive but also capable of being penetration enhancers, or peptide stabilizers. The following section focuses on bioadhesive drug delivery systems such as patches and tablets, as well as the potential application of phase change polymers.

A. Bioadhesive Tablets

Bioadhesive tablets for oral mucosal drug delivery are usually based on an erodible tablet (71). An example of this uses crosslinked hydroxypropyl cellulose (Synchron^{R}) as both a bioadhesive and a drug release regulator (72). The drug (nitroglycerin) is mixed with the bioadhesive and then compressed into tablets. Due to adhesion of the gradually eroding polymer to the buccal mucosa, this dosage form may remain in place for three hours and the drug release profile showed zero-order release kinetics. They claim that the patient may talk, drink, and even eat with the tablet in place.

A two-layered tablet with hydroxypropyl cellulose and carbopol 934 as the bioadhesive layer and a lactose nonadhesive as the backing layer was designed by Nagai et al. (73). The role of the upper (lactose) layer is to prevent drug diffusing away from the absorption site and to allow easy placement of the tablet in the mouth. The lower layer, which constitutes an erodible bioadhesive, together with triamcinolone acetonide as an active ingredient provides sustained release of the drug for the treatment of local aphtous stomatitis.

Nagai et al. (74) and Ishida et al. (75) also proposed a delivery system to deliver local anesthesia for toothache. This three-layer tablet consists of a core containing the drug and bioadhesives (hydroxypropyl cellulose and Carbopol 934 as before). This core is then coated on the upper layer with the bioadhesive mixture as a second layer, and finally a third layer, consisting of a mixture of hydroxypropyl cellulose and Carbopol, is applied as a cap layer. This form is reported to give a long-acting local anesthetic action.

B. Bioadhesive Patches

Although bioadhesive patches pose a relatively new technology to pharmacy, they have developed very quickly with the recent development of bioadhesive technology. Generally speaking, four different types of adhesive patches have been studied for oral mucosal drug delivery, as shown in Figure 2.3.

In these cases, the adhesive polymer serves either as a drug carrier itself, or an adhesive layer link between a drug-loaded layer and the mucosa, or a shield to cover a drug-containing disc. The design of these patches provides either unidirectional or bidirectional release of the drug. The size of such systems typically varies from 1 to 16 cm^2, depending on the specific purpose of the application. Usually, 1 to 3 cm^2 patches are commonly used because of convenience and comfort. However, the administration site is also a factor. Large-size patches can be administered at the central position of the buccal mucosa, (i.e., center of the cheek), whereas the sublingual and gingival sites require a rather small-sized patch.

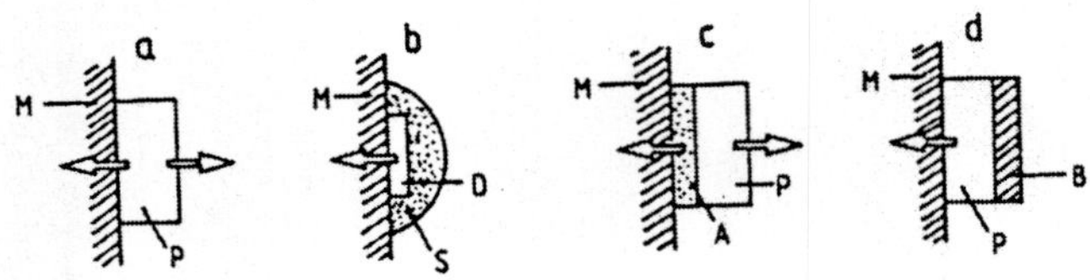

Figure 2.3 Variation in buccal patches: (a) bidirectional release from adhesive patch by dissolution or diffusion; (b) unidirectional release from patch embedded in an adhesive shield: (c) bidirectional release from a laminated patch; (d) unidirectional release from a laminated patch. M: mucosa; P: polymer with peptide; D: drug depot; S: adhesive shield; A: adhesive layer; B: impermeable backing layer (from Reference 80).

A variety of polymers can be used for oral mucosal patches. This includes water-soluble and insoluble polymers of both ionic and nonionic types (71). With soluble polymer systems, drug release is accompanied by dissolution of the polymer; therefore the overall drug release rate and duration are determined by both polymer dissolution and drug diffusion, whereas in a nonsoluble hydrogel system, drug release follows fickian or nonfickian diffusion kinetics, depending on design.

The duration of mucosal adhesion of different bioadhesive patches varies from minutes to days depending on the type of polymer used, its amount per patch, and additional factors such as the drying technique used to prepare the patches. Anders and Merkle (68) evaluated adhesive patches consisting of two-ply laminates with an impermeable backing layer and a hydrocolloid polymer layer containing the drug. The polymers they investigated were hydroxyethylcellulose, hydroxypropylcellulose, poly(vinylpyrrolidone) and poly(vinylalcohol). Their in vivo results indicated the adhesion time for all the above polymers on human buccal tissue is within the range of several minutes. Veillard et al. (36) developed a patch consisting of a mucoadhesive basement membrane, a rate-limiting center membrane, and an impermeable facing membrane. The mucoadhesive polycarbophil permits tight attachment to the buccal mucosal and allows the patch to remain in place for approximately 17 hours in dogs and humans regardless of eating and drinking. Lee and Chien (76) recently reported on a bilayer mucoadhesive polymer system (Carbopol 934 and PVP). It consists of a fast-release layer and a sustained-release layer to achieve sustained release of a peptide drug. This system is claimed to adhere to gingival/alveolar mucosa for over 24 hours.

A safety issue with patches is their potential to cause choking should they become displaced in the mouth. Suitable precautionary measures may include a fast-dissolving component of the patch so that if it does come off it is in a disintegrated form.

C. Phase Change Polymers

Phase change polymers are a relatively recent candidate for drug delivery design. This type of polymer undergoes phase change as a result of changes in the environment. Environmental triggers include chemical stimuli (e.g., pH, specific ions, and concentration) and physical stimuli (e.g., temperature, electric/magnetic field). Among responsive polymeric drug delivery systems, the temperature-sensitive one may be the most appealing provided that the physical change occurs quickly before the product is swallowed. If a small temperature change (e.g., from room to body temperature) can cause polymers to undergo a phase change from a liquid to a semisolid, the advantage it may bring is greater residence time and hence better bioavailability. In situ gel formers also have the advantage over solid systems in that being instilled as a liquid drop can result in better patient compliance due to ease of administration. Especially for the patient with a local oral disease (e.g., aphthous stomatitis), a slight pressure on the aphthae, caused by administering a tablet or patch, may result in a sharp pain to the patient. Therefore, such formulations are usually less acceptable than a drop of solution. In addition, this phase change polymer drug delivery system can provide controlled drug release by modifying the polymer structure or altering the polymer composition to achieve maximum therapeutic effect.

It is worth mentioning that for oral mucosal drug delivery, the phase change property may not be enough for the devices to reside in place for extended times. To meet the requirement of prolonging the residence time long enough to achieve a therapeutic effect, the system should also have a mucoadhesive property. Indeed, a variety of phase change polymer drug delivery systems has been investigated. The majority of this research has focused on studying phase change properties with different stimuli; in rare cases studies have been done to further explore their bioadhesive property. Chen and Hoffman (77) reported a graft copolymer composed of side chains of a temperature-sensitive polymer (poly N-isopropyl acrylamide) grafted onto a pH-sensitive backbone polymer (polyacrylic acid). They report that one of the components, polyacrylic acid, possesses bioadhesive properties when high molecular weight fractions are used. Therefore, random copolymers of acrylic acid and N-isopropyl acrylamide will not work because of the loss of temperature sensitivity when the acrylic acid content is increased to the point where bioadhesive properties are obtained. However, no bioadhesive data of the graft copolymer was published.

One limitation of a phase change polymer system may be the relatively slow response to external stimuli (78). A recent study on comb-type grafted hydrogels reported a rapid deswelling of the polymer in response to temperature changes, with polymer collapse occurring in about 20 minutes (79). The state of the mucoadhesive phase change polymer drug delivery system field is still in its early development. Considerable technical barriers must be overcome before this technology can be taken forward to the clinic.

VI. PRODUCT SAFETY

The usual systemic safety issues for components of a buccal drug delivery system must be considered. However, above this standard concern is the issue of local irritation. Many buccal delivery systems contain penetration enhancers which when left in intimate contact with oral cavity tissue for an extended period lead to irritation of the tissue and discomfort to the patient. At times the mere contact of the drug or polymer with the tissue for extended times can lead to significant irritation. On this basis, local irritation and patient acceptance should be addressed very early in the development process.

VII. SUMMARY

It is evident that due to the success and advantage of drug delivery through the oral mucosal tissue for some drugs, there is now renewed interest and active product development activity for the next generation oral mucosal drug delivery systems. This chapter has reviewed some of relevant issues in the field of oral mucosal delivery. Progress in molecular biology is giving us more and more detailed information about the structure and function of the oral mucosal tissues. Developments in polymer science and chemical engineering are introducing new bioadhesives to meet the requirements of drug delivery applications. With further studies of phase change polymers and the extensive research being conducted in the area of conventional oral mucosal dosage form development, we can confidently predict that there will be many novel oral mucosal drug delivery systems in the near future.

REFERENCES

1. Sobrero A. Comptes Rendus Hebdomadaires des Sciences. de l'Acadamie des Sciences 1847; 24:247–248.
2. Murrell W. Nitro-glycerin as a remedy for angina pectoris. Lancet 1879; 151:225–227.
3. Chien YW. Biopharmaceutics basis for transmucosal delivery. S.T.P. Pharma Sciences 1995; 5:257–275.
4. Tseng SCG, Jarvinen MJ, Nelson WG, Huang J-W, Woodcock-Mitchell J, and Sun T-T. Correlation of specific keratins with different types of epithelial differentiation: Monoclonal antibody study. Cell 1982; 30:361–372.
5. Collins LM and Dawes C. The surface area of the adult human mouth and thickness of the salivary film covering the teeth and oral mucosa. J Dent Res 1987; 66:1300–1302.
6. Wertz PW and Squier CA. Cellular and molecular basis of barrier function in oral epithelium. Critical Reviews in Therapeutic Drug Carrier Systems 1991; 8:237–269.
7. Squier CA, Johnson NW, and Hopps RH. Human Oral Mucosa: Development, Structure and Function. Victoria, Australia: Blackwell Scientific, 1976.

8. Hill MW. Cell renewal in oral epithelial. In: Meyer J, Squier CA, Gerson SJ, eds. The Structure and Function of Oral Mucosa. New York: Pergamon, 1984:53–81.
9. Squier CA and Wertz PW. Structure and function of the oral mucosa and implications for drug delivery. In: Rathbone MJ, ed. Oral Mucosal Drug Delivery. New York: Marcel Dekker, 1996:1–26.
10. Veillard MM. Buccal and gastrointestinal drug delivery systems. In: Gurny R, Junginger HE, eds. Bioadhesion—Possibilities and Future Trends. Stuttgart: Wissenschaftliche Verlagsgesellschaft mbH, 1990.
11. Le Brun PPH, Fox PLA, De Vries ME, and Dodde HE. In-vitro penetration of some β-adrenoreceptor blocking drugs through porcine buccal mucosa. Int J Pharm 1989; 49:141–145.
12. Dowty ME, Kunth KE, Irons BK, and Robinson JR. Transport of thyrotropin releasing hormone in rabbit buccal mucosa in-vitro. Pharm Res 1992; 9:1113–1122.
13. Longer MA. Characterization of buccal epithelia relevant to peptide drug delivery. Ph.D. dissertation, University of Wisconsin, Madison, WI, 1988.
14. Siegel IA. Effect of chemical structure on nonelectrolyte penetration of oral mucosa. J Invest Dermatol 1981; 76:137–140.
15. Siegel IA, Izutsu KT, and Watson E. Mechanisms of non-electrolyte penetration across dog and rabbit oral mucosa in vitro. Arch Oral Bio 1981; 26:357–360.
16. Tolo K and Jonsen J. In-vitro penetration of tritiated dextrans through rabbit oral mucosa. Arch Oral Biol 1975; 20:419–422.
17. Harris D and Robinson JR. Drug delivery via the mucus membranes of the oral cavity. J Pharm Sci 1992; 81:1–10.
18. Tavakoli-Saberi MR and Audus KL. Cultured buccal epithelium: An in-vitro model derived from the hamster pouch for studying drug transport. Pharm Res 1989; 6:160–166.
19. Borchardt RT, Hidalgo IJ, Hillgren KM, and Hu M. Pharmaceutical applications of cell culture: An overview. In: Wilson G, Davis SS, Illum L, Zweibaum A, eds. Pharmaceutical Application of Cell and Tissue Culture to Drug Transport. New York: Plenum, 1991:1–14.
20. Squier CA and Johnson NW. Permeability of oral mucosa. Br Med Bull 1975; 31:169–175.
21. Gibaldi M and Kanig JL. Absorption of drugs through the oral mucosa. J Oral Ther Pharmacol 1965; 1:440–450.
22. Siegel IA, Hall SH, and Stambaugh R. Current Concepts of the Histology of Oral Mucosa, Proceedings of a Symposium, Chicago, IL, 1971.
23. Beckett AH and Triggs EJ. Buccal absorption of basic drugs and its application as an in-vivo model of passive drug transfer through lipid membranes. J Pharm Pharmacol 1967; 19 (suppl.):31S–41S.
24. Beckett AH and Moffat AC. The influence of alkyl substitution in acids on their performance in the buccal absorption test. J Pharm Pharmacol 1968; 20 (suppl.):239S–247S.
25. Beckett AH and Moffat AC. The influence of substitution in phenylacetic acids on their performance in the buccal absorption test. J Pharm Pharmacol 1969; 21 (suppl.):139S–143S.
26. Beckett AH and Moffat AC. Correlation of partition coefficient in n-heptane aqueous system with buccal absorption data for a series of amines and acids. J Pharm Pharmacol 1969; 21 (suppl.):144S–150S.

27. Beckett AH and Moffat AC. Kinetics of buccal absorption of some carboxylic acids and the correlation of the rate constant and n-heptane: aqueous phase partition coefficients. J Pharm Pharmacol 1970; 22:15–19.
28. Evered DF and Vadgama JV. Absorption of amino acids from the human buccal cavity. Biochem Soc Trans 1981; 9:132–133.
29. Evered DF and Mallett C. Thiamine absorption across human buccal mucosa in vivo. Life Sci 1983; 32:1355–1358.
30. Beckett AH and Hossie RD. Buccal absorption of drugs. Handbook Exper Pharma 1971; 28:25–46.
31. McElnay JC and Temple DJ. The use of buccal partitioning as a model to examine the effects of aluminum hydroxide gel on the absorption of propranolol. Br J Clin Pharmacol 1982; 13:399–403.
32. McElnay JC, Sidahmed AM, and D'Arcy PF. Examination of the chloroquine-kaolin drug absorption interaction using the buccal partitioning model. J Clin Hopt Pharm 1982; 7:269–273.
33. Dearden JC and Tomlinson E. Correction for effect of dilution on diffusion through the membrane. J Pharm Sci 1971; 60:1278–1279.
34. Schurmann W and Turner P. A membrane model of the human oral mucosa as derived from buccal absorption performance and physicochemical properties of the beta-blocking drugs actenolol and propranolol. J Pharm Pharmacol 1978; 30:137–147.
35. Tucker IG. A method to study the kinetics of oral mucosal drug absorption from solutions. J Pharm Pharmacol 1988; 40:679–683.
36. Veillard MM, Longer MA, Martens TW, and Robinson JR. Preliminary studies of oral mucosal delivery of peptide. J Control Rel 1987; 6:123–131.
37. Barsuhn CL, Olanoff LS, Gleason DD, Adkins EL, and Ho NF. Human buccal absorption of flurbiprofen. Clin Pharmacol Ther 1988; 44:225–231.
38. Zhang J, Ebert CD, McJames S, Gijsman H, and Stanley TH. Transbuccal permeability of isoproterenol in the dog model. Pharm Res 1989; 6:S135.
39. Yamahara H, Suzuki T, Mizobe M, Noda K, and Samejima M. In situ perfusion system for oral mucosal absorption in dogs. J Pharm Sci 1990; 79:963–967.
40. Zhang J, Streisand J, Niu S, McJames S, Freimann V, Hague B, Maland L, Natte R, and Stanley TH. Estimation of buccal fentanyl absorption bioavailability by measuring drug depletion from vehicle solution: validation of the methods in dogs. Pharm Res 1992; 9 (10 suppl.):S177.
41. Zhang J, Niu S, Hague BI, Hamilton CA, and Coleman DC. Studies on the bioavailibilities of buccally administered etomidate in dogs. Pharm Res 1993; 10 (10 suppl.):S202.
42. Ebert CD, John VA, Beall PT, and Rosenzweig KA. Transbuccal absorption of diclofenac sodium in a dog model. In: Lee PI, Good WR, eds. Controlled Release Technology: Pharmaceutical Applications. Washington, DC: American Chemical Society, 1987:310–321.
43. Eggerth RM, Rashidbaigi ZA, Mahjour M, Goodhart FW, and Fawzi MB. Evaluation of hamster cheek pouch as a model for buccal absorption. Proc Int Symp Control Rel Bioact Mater 1987; 14:180–181.
44. Zhang H. Drug transport pathway in rabbit oral mucosa, Ph.D. dissertation, University of Wisconsin, Madison, WI, 1994.

45. Robinson JR. Rationale of bioadhesion/mucoadhesion. In: Gurny R, Junginger HE, eds. Bioadhesion: Possibilities and Future Trends, Stuttgart: Wissenschaftliche Verlagsesellschaft, 1990:13–15.
46. Chen JL and Cyr GN. Compositions producing adhesion through hydration. In: Marly RS, ed. Adhesion in Biological Systems. New York: Academic, 1970:163–181.
47. Gurny R, Meyer JM, Peppas NA. Bioadhesive intraoral release systems: design, testing and analysis. Biomaterials 1984; 5:336–340.
48. Ponchel G, Touchard F, Duchene D, and Peppas NA. Bioadhesive analysis of controlled release systems. I. Fracture and interpenetration analysis in poly(acrylic acid)-containing systems. J Control Rel 1987; 5:129–141.
49. Ponchel G, Touchard F, Wouessidjewe D, Duchene D, and Peppas NA. Bioadhesive analysis of controlled-release systems. III. Bioadhesive and release behavior of metronidazole-containing poly(acrylic acid)-hydroxypropyl methylcellulose systems. Int J Pharm 1987; 38:65–70.
50. Bottenberg P, Cleymaet R, de Muynck C, Remon JP, Coomans D, Michotte Y, and Slop D. Development and testing of bioadhesive, fluoride-containing slow-release tablets for oral use. J Pharm Pharmacol 1991; 43:457–464.
51. Kompella UB and Lee VHL. Delivery systems for penetration enhancement of peptide and protein drugs: design considerations. Adv Drug Delivery Rev 1992; 8:115–162.
52. Nantwi PKK, Cook DJ, Rogers DJ, and Smart JD. Lectins for drug delivery within the oral cavity-investigation of lectin binding to oral mucosa. J Drug Target 1997; 5:45–55.
53. Duchene D, Touchard F, and Peppas NA. Pharmaceutical and medical aspects of bioadhesive systems for drug administration. Drug Dev Ind Pharm 1988; 14:283–318.
54. Park K and Park H. Test methods of bioadhesion. In: Lenaerts V, Gurny R, eds. Bioadhesive Drug Delivery Systems. Boca Raton, FL: CRC, 1990:43–64.
55. Ahuja A, Khar R, and Ali J. Mucoadhesive drug delivery systems. Drug Dev Ind Pharm 1997; 23:489–515.
56. Bouckaert S, Lefebvre RA, and Remon J. In vitro/in vivo correlation of the bioadhesive properties of a buccal bioadhesive miconazole slow-release tablet. Pharm Res 1993; 10:853–856.
57. Bouckaert S and Remon J. In-vitro bioadhesion of a buccal, miconazole slow-release tablet. J Pharm Pharmacol 1993; 45:504–507.
58. Smart JD and Kellaway IW. In vitro techniques for measuring mucoadhesion. J Pharm Pharmacol 1982; 34:70p.
59. Smart JD, Lellaway IW, and Worthington HEC. An in-vitro investigation of mucosa-adhesive materials for use in controlled drug delivery. J Pharm Pharmacol 1984; 36:295–299.
60. Needleman IG and Smales FC. In vitro assessment of bioadhesion for periodontal and buccal drug delivery. Biomaterials 1995; 16:617–624.
61. Park K and Robinson JR. Bioadhesive polymers as platforms for oral controlled drug delivery: method to study bioadhesion. Int J Pharm 1984; 19:107.
62. Ranga Rao KV and Buri P. A novel in situ method to test polymers and coated microparticles for bioadhesion. Int J Pharm 1989; 52:265–270.

63. Mikos AG and Peppas NA. Scaling concepts and molecular theories of adhesion of synthetic polymers to glycoproteinic networks. In: Lenaerts V, Gurny R, eds. Bioadhesive Drug Delivery Systems. Boca Raton, FL: CRC, 1990:25–42.
64. Teng CLC and Ho NFH. Mechanistic studies in the simultaneous flow and adsorption of polymer-coated latex particles on intestinal mucus I: Methods and physical model development. J Control Rel 1987; 6:133–149.
65. Park K. A new approach to study mucoadhesion: colloidal gold staining. Int J Pharm 1989; 53:209–217.
66. Kamath KR and Park K. Mucosal adhesive preparation. In: Swarbrick J, Boylan JC, eds. Encyclopedia of Pharmaceutical Technology. Vol. 10. New York: Marcel Dekker, 1994:133–163.
67. Hassan EE and Gallo JM. A simple rheological method for the in-vitro assessment of mucin-polymer bioadhesive bond strength. Pharm Res 1990; 7:491–495.
68. Anders R and Merkle HP. Evaluation of laminated muco-adhesive patches for buccal drug delivery. Int J Pharm 1989; 49:231–240.
69. Needleman IG, Martin GP, and Smales FC. An investigation of bioadhesion for periodontal and oral mucosal drug delivery. J Clin Periodontol 1997; 24:394-400.
70. Needleman IG, Martin GP, and Smales FC. Characterization of bioadhesives for periodontal and oral mucosal drug delivery. J Clin Periodontol 1998; 25:74–82.
71. Merkle HP, Anders R, and Wermerskirchen A. Mucoadhesive buccal patches for peptide delivery. In: Lenaerts V, Gurny R, eds. Bioadhesive Drug Delivery Systems. Boca Raton, FL: CRC, 1990:105–136.
72. Schor JM, Davis SS, Nigalaye A, and Bolton S. Susadrin transmucosal tablets. Drug Dev Ind Pharm 1983; 9:1359–1377.
73. Nagai T, Machida Y, Suzuki Y, and Ikura H. Method and preparation for administration to the mucosa of the oral or nasal cavity. U.S. Patent 4226848, 1980.
74. Nagai T and Machida Y. Mucoadhesive dosage forms. Pharm Int 1985; 6:196–200.
75. Ishida M, Nambu N, and Nagai T. Mucosal dosage form of lidocaine for toothache using hydroxypropyl cellulose and carbopol. Chem Pharm Bull 1982; 30:980–984.
76. Lee Y and Chien YW. Oral mucosa controlled delivery of LHRH by bilayer mucoadhesive polymer systems. J Control Rel 1995; 37:251–261.
77. Chen G-H and Hoffman AS. Graft copolymers that exhibit temperature-induced phase transitions over a wide range of pH. Nature 1995; 373:49–52.
78. D'Emanuele A. Responsive polymeric drug delivery systems: Meeting the patient's needs. Clin Pharmacokinet 1996; 31:241–245.
79. Yoshida R, Uchida K, Kaneko Y, Sakai K, Kikuchi A, Sakurai Y, and Okano T. Comb-type grafted hydrogels with rapid de-swelling response to temperature changes. Nature 1995; 374:240–242.
80. Merkle HP, Anders R, Sandow J, and Schurr W. Drug delivery of peptides: the buccal route. In: Davis SS, Illum L, Tomlinson E, eds. Delivery Systems for Peptide Drugs. New York: Plenum, 1986:159–176.
81. Ishida M, Machida Y, Nambu N, and Nagai T. New mucosal dosage form of insulin. Chem Pharm Bull 1981; 29:810–816.
82. Marvola M, Vahervou K, and Rajaneimi M. Development of a method for studying the tendency of drug products to adhere to the esophagus. Pharm Sci 1982; 71:975–977.

83. Marvola M, Rajaneimi M, Marttila E, Vahervuo K, and Sothmann A. Effect of dosage form and formulation factors on the adherence of drugs to the esophagus. J Pharm Sci 1983; 72:1034–1036.
84. Ch'ng HS, Park H, Kelly P, and Robinson JR. Bioadhesive polymers as platforms for oral controlled drug delivery II: Synthesis and evaluation of some swelling, water-insoluble bioadhesive polymers. J Pharm Sci 1985; 74:399–405.
85. Park H and Robinson JR. Physico-chemical properties of water insoluble polymers important to mucin/epithelial adhesion. J Control Rel 1985; 2:47–57.
86. Park H and Robinson JR. Mechanisms of mucoadhesion of poly(acrylic acid) hydrogels. Pharm Res 1987; 4:457–464.
87. Lejoyeux F, Ponchel G, Wouessidjewe D, Peppas NA, and Duchene D. Bioadhesive tablets influence of the testing medium composition on bioadhesion. Drug Dev Ind Pharm 1989; 15:2037–2048.
88. Smart JD. An in-vitro assessment of some muco-adhesive dosage forms. Int J Pharm 1991; 73:69–74.
89. Satoh K, Takayama K, Machida Y, Suzuki Y, Nakagaki M, and Nagai T. Factors affecting the bioadhesive property of tablets consisting of hydroxypropyl cellulose and carboxyvinyl polymer. Chem Pharm Bull 1989; 37:1366–1368.
90. Lehr CM, Bouwstra JA, Tukker JJ, and Junginger HE. Intestinal transit of bioadhesive microspheres in an in-situ loop in the rat-A comparative study with copolymers and blends based on poly(acrylic acid). J Control Rel 1990; 13:51–62.
91. Saettone MF, Chetoni P, Torracca MT, Burgalassi S, and Giannaccini B. Evaluation of muco-adhesive properties and in-vivo activity of ophthalmic vehicles based on hyaluronic acid. Int J Pharm 1989; 51:203–212.
92. Chen WG and Hwang GC-C. Adhesive and in-vitro release characteristics of propranolol bioadhesive disc system. Int J Pharm 992; 82:61–66.
93. Lehr CM, Bouwstra JA, Schacht EH, and Junginger HE. In vitro evaluation of mucoadhesive properties of chitosan and some other polymers. Int J Pharm 1992; 78:43–48.
94. Chickering D, Jacob J, Panol G, and Mathiowitz E. A tensile technique to evaluate the interaction of bioadhesive microspheres with intestinal mucosa. Proc Int Symp Control Rel Bioact Mater 1992; 19:88–89.
95. Thermes F, Grove J, Rozier A, Plazonnet B, Constancis A, Bunel C, and Vairon JP. Mucoadhesion of copolymers and mixtures containing polyacrylic acid. Pharm Res 1992; 9:1563–1567.
96. Jimenez-Castellanos MR, Zia H, and Rhodes CT. Assessment of an in-vitro method for measuring the bioadhesiveness of tablets. Int J Pharm 1993; 89:223–228.
97. Jimenez-Castellanos MR, Zia H, and Rhodes CT. Mucoadhesive drug delivery systems. Drug Dev Ind Pharm 1993; 19:143–194.
98. Guo JH. Investigating the surface properties and bioadhesion of buccal patches. J Pharm Pharmacol 1994; 46:647–650.
99. Guo JH. Bioadhesive polymer buccal patches for buprenorphine controlled delivery: Formulation, in-vitro adhesion and release properties. Drug Dev Ind Pharm 1994; 20:2809–2821.
100. Save T, Shah UM, Ghamande AR, and Venkitachalam P. Comparative study of buccal adhesive formulations and sublingual capsules of nifedipine. J Pharm Pharmacol 1994; 46:192–195.

101. Miyazaki S, Nakayama A, Oda M, Takada M, and Attwood D. Chitosan and sodium alginate based bioadhesive tablets for intraoral delivery. Biol Pharm Bull 1994; 17:745–747.
102. Martini L, Attwood D, Collect JH, and D'Emanucle A. The bioadhesive properties of a triblock copolymer of ε-caprolactone and ethylene oxide. Int J Pharm 1995; 113:223–229.
103. Rillosi M and Buckton G. Modeling mucoadhesion by use of surface energy terms obtained by the Lewis acid-Lewis base approach. Int J Pharm 1995; 117:75–84.
104. Ishida M, Nambu M, and Nagai T. Ointment-type oral mucosal dosage form of Carbopol containing prednisolone for treatment of aphtha. Chem Pharm Bull 1983; 31:1010–1014.
105. Sam AP, Van den Heuij JTM, and Tukker JJ. Mucoadhesion of both film-forming and non-film-forming polymeric materials as evaluated with the Wilhelmy plate method. Int J Pharm 1992; 79:97–105.

3

Modulation of Oral Transmucosal Permeability: Permeation Enhancers

Bhaskara R. Jasti, Venugopal Marasanapalle, and Xiaoling Li

CONTENTS

I. INTRODUCTION

The oral route is the most preferred route of drug delivery as it is convenient, inexpensive, and versatile. However, drug delivery by this route has certain disadvantages such as first-pass metabolism by the liver and gastrointestinal enzymatic degradation of the drug. Therefore, other transmucosal routes such as nasal, rectal, vaginal, ocular, and oral mucosae are being considered as alternatives to conventional oral dosage forms for drug delivery to avoid the above disadvantages associated with conventional oral delivery (i.e., tablets, capsules, syrups, etc.). Of these routes of delivery, the buccal oral mucosa has emerged as one of the target sites for administration of drugs in a wide variety of dosage forms, particularly for those drugs targeted for local delivery in the oral cavity and systemic absorption.

The buccal route of drug delivery provides the opportunity for drug absorption through the buccal epithelial lining of the oral cavity (mucosa of the cheek) for it to exhibit its action locally or systemically. The noninvasive nature of administration, ease and convenience of administration, precise localization, and increased permeability of the buccal mucosa compared to other transepithelial routes makes this a promising route of delivery. Also, the rich supply of blood vessels and lymphatics in the buccal mucosa results in rapid onset of drug action for those that have the requisite physicochemical profile (de Vries, et al., 1991; Shojaei, 1998; Squier, 1991).

Drugs absorbed from the buccal mucosa may directly enter the systemic circulation by way of the jugular vein, minimizing the first-pass liver metabolism and gastric acid- or enzyme-mediated degradation (salivary fluid has lower enzymatic activity than gut). The presence of food or variations in the gastric emptying rate has little or no influence on drug delivery by the buccal route. The continuous exposure of the oral mucosal tissues to a multitude of substances and its high cellular turnover rate makes the buccal tissue robust and less prone to local toxicity or irritation from drugs, their dosage forms, and formulation excipients. The absence of Langerhans cells in the oral mucosal tissues imparts tolerance to potential allergens (Bodde et al., 1990). Therefore, prolonged administration of drugs (chronic use) to the oral cavity is less prone to induction of local tissue sensitization and allergic reactions. When compared to other mucosal delivery routes, buccal drug delivery offers a higher degree of control and reproducibility; it also allows the removal of the systemic dosage form (i.e., chewing gums, lozenges, patches, etc.) to terminate drug absorption if necessary.

This chapter reviews the advantages and disadvantages of intraoral mucosal drug delivery to the mouth and the local anatomical sites within the oral cavity for targeted localized drug delivery and absorption. The kinetics of drug delivery to the oral cavity are discussed with a focus on mechanisms of enhancement and strategies to optimize drug absorption by the oral mucosal tissues. Finally, selected oral mucosal drug delivery systems are presented and technologies and strategies for optimizing drug delivery and local absorption from these dosage forms through the mucosal tissues of the oral cavity using permeation enhancers are explained.

II. ORAL MUCOSAL DRUG DELIVERY

The rationale and key advantages of drug delivery to the oral cavity for local and systemic absorption include:

1. Alternative to injection
2. Improved patient compliance, "patient-friendly"
3. Convenience
4. Targeted delivery to treat local diseases of the oral cavity
5. Potential route for delivery of proteins/peptides/vaccines
6. Opportunity for product line extension of quick-dissolving dosage forms
7. Dosing or administration anywhere at anytime without water
8. Increase drug bioavailability in some cases

There are, however, a number of technical challenges to overcome in order to effectively deliver drugs to the oral cavity for local oral mucosal absorption and systemic delivery. The absorption of drugs from the oral mucosal tissues has the following disadvantages.

1. Drug transport is by passive diffusion, drug absorption is low, which results in a low bioavailability (Shojaei, 1998).
2. Buccal mucosa, like the small intestine, offers a lipoidal barrier and this route is usually practical for small lipophilic molecules (Washington et al., 2001b).
3. Only a few hydrophilic drugs or compounds such as certain amino acids and monosaccharides have been reported to be transported via a carrier-mediated process (Harris and Robinson, 1992; Rathbone and Hadgraft, 1991; Veuillez et al., 2001).
4. Taste masking may be necessary for drugs that are bitter or irritable.
5. The total area for drug absorption is small (100 to 170 cm^2) when compared to the total area of gastrointestinal absorption (Washington et al., 2001a).
6. The dosage form must be kept in place for effective absorption because salivary flow may wash away the dissolved drug and the dosage form may be swallowed prior to drug dissolution.

III. ANATOMY OF THE ORAL CAVITY

The different anatomical regions of the oral cavity and mucosal tissues (excluding the odontal structures) are shown in Figure 3.1. The various target sites for drug delivery and absorption may include the upper and lower lip, gums (gingiva),

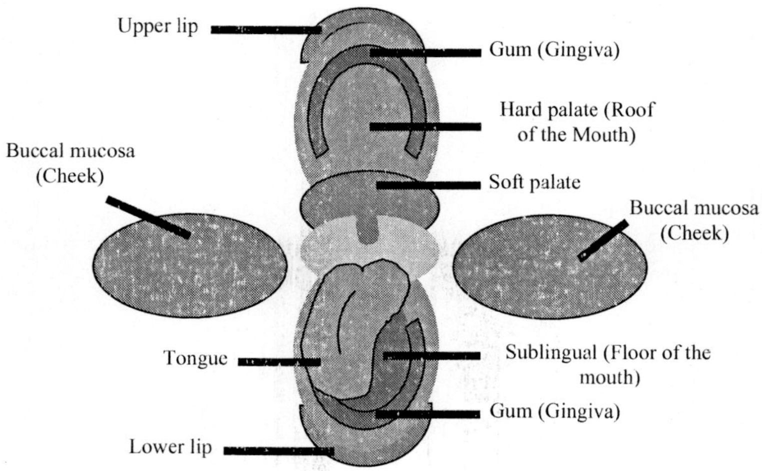

Figure 3.1 Different anatomical regions of the oral cavity (reproduced with permission from Hoogstraate et al., 1996a).

hard palate, soft palate, floor of the mouth (sublingual), tongue, and buccal mucosal tissue (cheek).

The oral mucosal tissues can be divided into two types, namely, *keratinized* epithelium of the masticatory regions consisting of the gums (gingiva), palatal mucosa, and the inner side of the lips and *nonkeratinized* regions consisting of the floor of mouth (sublingual) and the buccal mucosa (Chen et al., 1999). The differences between the two types of epithelia are: (1) the superficial layer of the nonkeratinized layer is rougher when compared to keratinized epithelium (Morgan, 2000), and (2) the elongated *rete processes*, which provide the attachment of epithelium to the underlying connective tissue, are deeper and narrower in keratinized epithelium as opposed to nonkeratinized epithelium. The buccal mucosa has two components: the *epithelium* and the underlying connective tissue, *lamina propria*, which is interfaced by the basal complex. A cross-sectional structure of the buccal mucosa is shown in Figure 3.2.

A. Buccal Epithelium

The buccal epithelium is composed of 40 to 50 layers of nonkeratinized stratified squamous cells. It is 500 to 800 μm in thickness with varying degrees of maturity. The uppermost superficial layer of cells is comprised of flattened compact differentiated cells of about 150 μm in thickness (Gandhi and Robinson, 1988; Harris and Robinson, 1990; Shojaei, 1998). The epithelial cohesion in the superficial layers is achieved by the lipid and glycolipid contents extruded from the cellular membrane coating granules (MCGs) in the intercellular space (Gandhi and Robinson, 1994). Deeper into the epithelium is the Malpighian layer, which

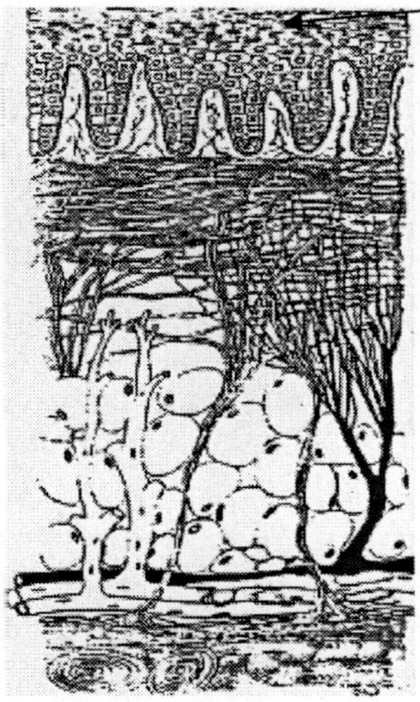

Figure 3.2 Cross-sectional structure of buccal tissue (reproduced from Veuillez et al., 2001 with permission).

consists of cells at various stages of differentiation, which are more tightly held together by desmosomes compared to the upper layer and there is less tortuosity. The epithelium terminates with the basal lamina, a proteinaceous fibrous matrix of 1 to 2 μm in thickness, which is a permeability barrier for drugs. The papillary contour of the basal region permits efficient vascularization to the cells.

B. Lamina Propria

The lamina propria consists of collagen fibrils, a supporting layer of connective tissue, blood vessels, and smooth muscle. The structure of the lamina propria is not dense and it is not a barrier to drug permeation (Veuillez et al., 2001). It has a hydrated matrix, which presumably facilitates permeation of hydrophilic substances and large molecules (Harris and Robinson, 1992).

C. Submucosa

The submucosa is a relatively dense connective tissue that contains a few accessory salivary glands, *mucus acinus* (Burkitt et al., 1993). Mucus acini are surrounded by myoepithelial cells that aid in the secretion of saliva.

D. Biochemical Structure

The oral mucosa contains various extracellular materials, which contribute both to elasticity and to the permeability barrier. The mucosa mostly contains polar lipids such as phospholipids, cholesterol sulfate, and glycosyl ceramides and imparts fluidity to the membrane (Veuillez et al., 2001). The nonkeratinized regions have a higher permeability to water and hydrophilic compounds compared to the keratinized regions.

IV. PHYSIOLOGY AND ENVIRONMENT OF THE ORAL CAVITY

The oral cavity has a protective role during mastication, where the keratinized areas of hard palate and gingival tissue resist the shear forces and abrasion of food materials. Similar to skin, it is also host to a multitude of microorganism species. The oral mucosa offers a significant barrier to the absorption of toxins produced by the microorganisms (Washington et al., 2001b).

The saliva is secreted primarily by parotid, submandibular (submaxillary), and sublingual glands. About one to two liters of saliva per day is secreted in the human mouth. A basal secretion at the rate of 0.5 ml/min always occurs and it can increase up to 7 ml/min when stimulated by the thought, smell, or taste of food. Saliva is viscous, colorless, opalescent, and hypotonic in nature. The pH of saliva varies between 6.2 and 7.4. The presence of bacteria and their metabolism of carbohydrates can reduce the pH of a microenvironment of 3 to 4. Saliva is composed of water, mucus, proteins, mineral salts, and the enzyme amylase. Mucus contains a glycoprotein called mucin. The major ions that are present in the saliva are sodium, potassium, chlorine, and bicarbonate. In addition to amylase, ptyalin is also present in the saliva. Saliva also contains thiocyanate, antibodies, and lysozyme that help to kill oral bacteria.

V. MODES OF TRANSPORT ACROSS BUCCAL MUCOSA

The physicochemical properties of a drug are important for its passive transport across the mucosa of the oral cavity. For drug absorption to take place through the buccal mucosa of the oral cavity, the dosage form must dissolve in saliva liberating the drug into a solution. Then the drug will partition into the mucus covering the buccal mucosa at which time it is available for permeation.

There are two pathways by which passive drug transport across the buccal mucosa can take place for it to reach the local adjacent structures and systemic circulation: transcellular and paracellular routes that enable the drug to reach systemic circulation. Drugs can travel through these two routes simultaneously, but one route is preferred over the other depending upon the physicochemical properties of the molecules (i.e. molecular weight, polarity, etc.).

A. Transcellular Pathway

Drug permeation through the epithelial cells involves transport across the apical cell membrane, the intracellular space, and the basolateral membrane as shown in Figure 3.3. Drug transport through the transcellular pathway, also known as the intracellular pathway, may be by passive transport (diffusion, pH partition)

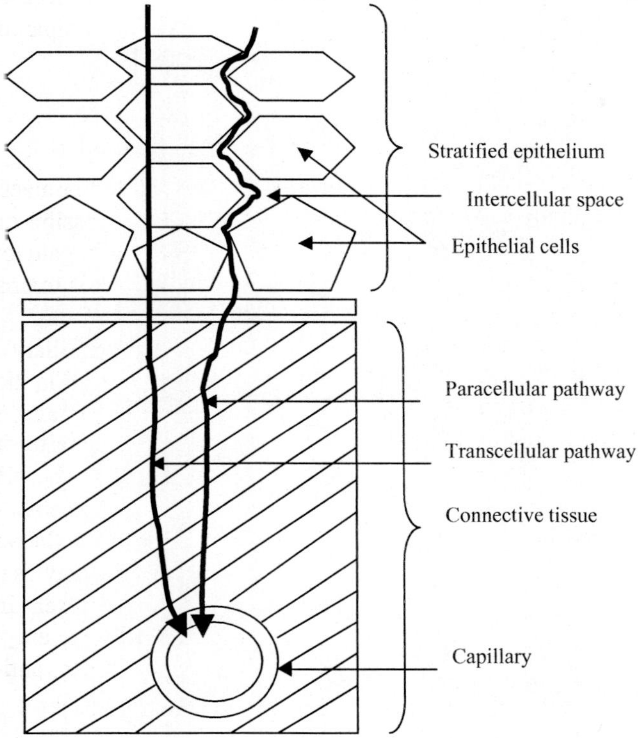

Figure 3.3 Transport pathways of molecules across buccal tissue (reproduced from Senel and Hincal, 2001).

of small molecules or by active transport (facilitated and carrier-mediated diffusion) of ionic and polar compounds and endocytosis and transcytosis of macromolecules. Drug transport through the transcellular pathway is a complex phenomenon that is dependent on various physicochemical parameters of the drug, including molecular weight, lipophilicity, hydrogen bond potential, charge, and conformation. Lipophilic compounds and small hydrophobic molecules predominantly undergo transcellular transport. Transcellular diffusion is inversely proportional to the amount of membrane coating granules present in the intracellular spaces (Gandhi and Robinson, 1994).

Because the cell membrane is lipophilic in nature, hydrophilic drugs will have difficulty permeating the cell membrane due to a low partition coefficient. Passive transport of hydrophilic compounds, including macromolecules such as polypeptides and proteins, can be enhanced by the interaction of the absorption-enhancing excipients with both the phospholipid bilayer and the integrated proteins. Some small water-soluble molecules such as amino acids, ions, and sugars can be transported through the aqueous pores in the cell membrane.

B. Paracellular Pathway

Drug permeation through the epithelial cells also involves transport through the lipids or in-between the epithelial cells as shown in Figure 3.3. The paracellular pathway (also known as the intercellular pathway) can be of two types: one is an essentially hydrophobic route, through the lipid bilayer, and the other is a hydrophilic route associated with the narrow aqueous regions adjacent to the polar head groups of the lipid bilayers (Veuillez et al., 2001). For compounds transported through the paracellular route, tortuosity and intercellular space are the main hindrances to permeability. A substance with equal solubility in aqueous and lipid media can permeate by both para- and transcellular pathways. However, because the intercellular spaces and cytoplasm are hydrophilic in character, lipophilic compounds would have low solubility in this environment and thus this route will be preferred by hydrophilic compounds. Paracellular transport is of interest especially in peptide and protein drug delivery because the intercellular space does not contain peptidases.

C. Active (Carrier-Mediated) Transport System

An active carrier-mediated transport system is reported to be present, especially for small molecules such as monosaccharides and amino acids. Small peptides such as di- and tripeptides may also permeate by this mechanism. The selectivity in absorption of one type of stereoisomer (from D- and L-forms) of amino acids and glucose by the buccal mucosa, similar to the small intestine, indicated the presence of active or carrier-mediated transport in buccal mucosa (Veuillez et al., 2001). However, characterization of active transport, transporter capacity, and specificity in each region of the oral cavity has not been reported (Veuillez et al., 2001).

VI. KINETICS OF TRANSPORT MECHANISMS

A. Paracellular Pathway

Hydrophilic drugs are transported across the buccal mucosa by the paracellular route and the steady-state flux (J_P) of the drug is modeled as:

$$J_P = \frac{D_P \varepsilon}{h_P} C_D$$

(Shojaei et al. 1998),
where:

ε is the area fraction of the paracellular route;
h_p is the length of the paracellular route;
D_p is the diffusion coefficient in the intercellular spaces; and
C_D is the concentration of drug in the donor chamber (Gandhi and Robinson, 1994).

B. Transcellular Pathway

Lipophilic compounds are generally transported through the transcellular pathway and the steady-state flux (J_T) of a drug across the buccal mucosal membrane can be determined as

$$J_T = \frac{(1-\varepsilon)D_T K_P}{h_T} C_D$$

(Shojaei et al. 1998),
where:

K_p is the partition coefficient between the lipophilic phase (cell membrane) and the aqueous hydrophilic donor phase;
h_T is the length of the transcellular route; and
D_T is the diffusion coefficient in the lipophilic phase (Zhang and Robinson, 1996).

Because the oral epithelium is stratified, drug permeation may involve a combination of these two routes. The route that predominates, however, is generally the one that provides the least hindrance to diffusion. Transcellular resistance is proportional to the volume of the membrane coating granules in the intracellular space whereas paracellular resistance is the result of the extrusion of the lipid contents in the intercellular space (Gandhi and Robinson, 1994). Regardless of the transport route, the physicochemical characteristics of the drug play an important role in determining the rate and extent of permeation and local or systemic therapeutic levels.

C. Intraoral Drug Absorption and Transport Pathways

It has been shown that nicotine in its neutral form crosses the buccal mucosa by the transcellular pathway, but switches over to a paracellular pathway upon protonation, at lower pH. Greater transport through keratinized as compared to nonkeratinized regions of the oral mucosa was explained on the basis of differences in lipids that are extruded into the intercellular space from the MCGs (Chen et al., 1999). Similarly, acyclovir was reported to be transported through the paracellular pathway across the buccal epithelium. Acyclovir exists in three different ionic states and all three states contribute to its transport through the paracellular pathway (Shojaei et al. 1998).

Nonselective passive transport through the wide-open porous pathway is usually undergone by multiple charged species. Molecules such as dextran, which are quite hydrophilic, permeate predominantly by paracellular pathway. Solubilization of the intercellular lipid matrix and, in higher concentrations, of the cell membrane lipids by sodium glycocholate can result in the transformation of the predominantly paracellular pathway to a transcellular pathway (Hoogstraate et al., 1996b).

VII. PERMEABILITY OF BUCCAL MUCOSA

The mucosal membranes of the oral cavity have a total area of about 100 cm^2 and show differences in structure, thickness, and blood flow, depending upon their anatomical location in the oral cavity. The oral mucosa is a complex tissue with a series of layers that differ in their permeabilities. Tight junctions are rare throughout the buccal epithelium. Mucus present on the epithelial layer has positively and negatively charged functional groups, which may act as barriers for peptides, but hydrated mucus is believed to facilitate the permeation of peptides through the buccal mucosa. Thus, it is derived that mucus is not as significant a barrier as the other underlying layers in buccal tissue. The permeability barrier to drugs is the intercellular region of the superficial layers of the epithelium due to the intercellular material derived from MCGs (Gandhi and Robinson, 1994). In contrast, the lamina propria is not very dense and it may not hinder the permeability of even large molecules, and a hydrated matrix may even facilitate movement of molecules (Harris and Robinson, 1992; Veuillez et al., 2001). Mucosal permeability also varies as a function of the anatomical site, particularly by those sites having a keratinized layer that hinders the permeability of hydrophilic molecules. The volume and organization of the intercellular region and the presence of lipids such as ceramides and acylceramides extruded by MCGs contribute to the permeability barrier (Chen et al., 1999). Therefore, it is important to consider these factors in the selection of the transmucosal route for drug delivery in the oral cavity.

VIII. METHODS OF TRANSPORT ENHANCEMENT

A drug molecule that travels through a particular route must overcome many barriers upon administration of the intraoral delivery system. Different strategies may be employed for the enhancement of drug transport through the tissues of the oral cavity. For the buccal and oral mucosal routes, these strategies include the use of: (1) permeation enhancers, (2) enzyme inhibitors, and (3) vehicles/cosolvents.

A. Chemical Methods

Modulation of drug permeation through the oral mucosa is usually achieved by using chemical penetration enhancers. These excipients should be safe, nontoxic, pharmacologically and chemically inert, nonirritant, and nonallergenic (Senel and Hincal, 2001). In addition, the tissue should be able to revert back to its normal integrity and barrier properties upon removal of the enhancers. Various classes of penetration enhancers are used for buccal penetration; they include bile salts, surfactants, fatty acids and derivatives, chelators, cyclodextrins, chitosan, and enzyme inhibitors. Among these, bile salts are the most commonly used enhancers (Senel and Hincal, 2001). Penetration enhancers are nonspecific and may allow the entry of other compounds not associated with the active pharmaceutical ingredient. The degree of enhancement depends on the composition of the delivery vehicle, pretreatment with enhancer, and so on. There are several mechanisms by which enhancers improve drug permeation, as follows:

1. Altering the mucus rheology by reducing the viscosity and/or elasticity of mucus layer.
2. Interacting with the lipid or protein membrane components of cell membrane and increasing the membrane fluidity and facilitating transcellular transport of substances.
3. Solubilizing the intercellular lipids and thus facilitating paracellular transport.
4. Inhibiting the endo- and exopeptidases that can degrade peptide drugs.
5. Increasing the thermodynamic activity or solubility of drugs, which can be achieved by changing the vehicle composition, and micellization, and also by ion-pair formation. Increasing the thermodynamic activity (e.g., achieving a supersaturated state) of drugs in the dosage form increases the flux of the drug across the buccal mucosal membrane.

Permeation enhancers can be classified based upon their structure, mechanism of action, and the type of drug whose permeability is enhanced (Senel and Hincal, 2001). There are many proposed mechanisms by which permeation enhancers may increase drug absorption. Some examples are described below for each class of enhancer, which are summarized in Table 3.1.

B. Bile Salts

Bile salts belong to a class of natural or semi-synthetic surfactants and include sodium glycodeoxycholate, sodium glycocholate, sodium taurodeoxycholate, and sodium taurocholate. They have been widely used as absorption enhancers of drugs through different mucosae by co-administering them along with or incorporating them into the drug formulations (Veuillez et al., 2001). Bile salts are believed to work through the extraction of membrane protein or lipids, membrane fluidization, and reverse micellization in the membrane, thus creating aqueous channels (Veuillez et al., 2001). Although there are slight variations in the mechanism of enhancement of diffusion of different drugs, generally bile salts diffuse into the buccal epithelium and accumulate to a certain level to alter the integrity of the cell membrane that opens up the intracellular domain, which in turn facilitates the transcellular pathway of the drug and shortens the diffusion pathway significantly (Hoogstraate et al., 1996b; Senel and Hincal, 2001). However, the changes associated with the application of bile salts on the buccal mucosa are reversible. The mechanism of permeation enhancement in buccal mucosal tissue is nevertheless complicated due to the fact that some bile salts also have inhibitory effects on the mucosal membrane peptidases. A dual action of alteration of membrane integrity and inhibition of peptidases could be responsible for enhanced diffusion of proteins and peptides (Veuillez et al., 2001; Yamamoto et al., 1992).

Table 3.1 Enhancers Used for Increasing the Oral Mucosal Permeability of Different Drugs

Class and Enhancer	Drug	Reference
Bile Salts		
Sodium taurocholate		Veuillez et al., 2001
Sodium taurodeoxycholate		
Sodium deoxycholate		
Sodium glycocholate and EDTA		
Sodium glycodeoxycholate	Acyclovir	Shojaei et al., 1998
	Buserelin	Hoogstraate et al., 1996b
	Dextran	Hoogstraate et al., 1996a
	Dideoxycytidine	Xia et al., 2002
	ISIS Oligonucleotide 3082	Jasti et al., 2000
	Acyclovir	Shojaei, et al., 1998
Sodium glycocholate	Buserelin	Hoogstraate, et al., 1996b
	ISIS Oligonucleotide 3082	Jasti et al., 2000
	Morphine HCl	Senel et al., 1998
Surfactants		
Sodium lauryl sulfate	Calcitonin	Nakada et al., 1988
	Insulin	Oh and Ritschel, 1990
Sucrose laurate	Lidocaine HCl	Ganem-Quintanar et al., 1998
Fatty Acids		
Sodium laurate	Insulin	Senel and Hincal, 2001
Sodium myristate	Calcitonin	Veuillez et al., 2001
Oleic acid	Lidocaine HCl	Ganem-Quintanar et al., 1998
Lauric acid and propylene glycol	Insulin	Aungst and Rogers, 1994
Vehicles and Adjuvants		
Ethanol	Peptides	Veuillez et al., 2001
Propylene glycol	Buspirone	Birudaraj, 2001
Chelators		
EDTA (ethylene diamine tetraacetic acid)		Veuillez et al., 2001
Salicylates		
Sodium citrate		
Polyacrylates		
Cyclodextrins		
α-, β-, γ-cyclodextrins		Veuillez et al., 2001
Methylated β-cyclodextrins		
Hydroxypropyl cyclodextrin	Buspirone	Birudaraj, 2001
Chitosan	Transforming growth factor-β	Senel et al., 2000
	Hydrocortisone	
Enzyme Inhibitors		
Aprotinin	Peptides	Veuillez et al., 2001
Bestatin		
Bile salts		

Sodium salts of bile acids were found to increase the transmucosal permeability of a number of molecules (Duchateau et al., 1986; Gordon et al., 1985; O'Hagan and Illum, 1990). These bile salts are chemical derivatives of cholesterol and are amphiphilic in nature. The proposed mechanisms of permeation enhancement by bile salts include: (a) altering the cell membrane integrity, (b) formation of micelles in aqueous solutions, which act as drug carriers along the concentration gradient, and (c) stabilizing the drugs to enzymatic metabolism, which is important with proteins and peptides (Duchateau et al., 1986; Gumbiner and Simons, 1986). The effect of bile salts on enhancing drug delivery in nasal and rectal mucosae was reported and the order of enhancing the permeability was deoxycholate > glycocholate > taurocholate, which correlated well with the lipophilicity of the bile salts (O'Hagan and Illum, 1990; Yamamoto et al., 1992). Enhancement of the buccal permeability of morphine hydrochloride, which depended on the concentration of glycocholate, was reported (Senel et al., 1998). Also, a synergistic effect of glycocholate and EDTA on mucosal permeability when used concurrently was reported (Yamamoto et al., 1992). The perturbing effect of bile salts on the integrity of membranes is reversible and the severity is less than other enhancers such as nonionic surfactants. In light of its potential advantages, use of sodium glycocholate (NaGC), a trihydroxy bile salt, in the buccal permeation enhancement of a mono-nucleotide, acyclovir, ISIS 3082 a 20-mer antisense oligonucleotide and a peptide buserelin is discussed in the following sections.

Sodium glycodeoxycholate (GDC) was found to increase the permeation of ionic compounds such as dideoxycytidine, acyclovir, oligonucleotides, and macromolecules like dextran (Hoogstraate et al., 1996b; Jasti et al., 2000; Shojaei et al., 1998; Xiang et al., 2002). It was reported to solubilize membrane lipids resulting in permeability enhancement.

1. Acyclovir

The in vitro permeation of acyclovir, a synthetic guanosine analogue with a molecular weight of 225, was reported by Shojaei et al. (1998). The steady-state flux of acyclovir across porcine buccal mucosa was shown to be dependent linearly upon the concentration of acyclovir in the donor chamber, indicating that its permeation was through a passive diffusion process over the investigated concentration range (Figure 3.4). The permeation of acyclovir was believed to occur through a paracellular pathway, although the permeation of acyclovir varied with pH because changes in pH did not result in a change in the partition coefficient. Also, the permeability coefficients of anion and cation were of the same magnitude as that of water, suggesting a wide-open, nonselective porous pathway for acyclovir transport.

Co-administration of acyclovir with sodium glycocholate increased the steady-state flux of acyclovir by 2 to 9 times across porcine buccal mucosa. The enhancement ratio increased steadily with increasing sodium glycocholate and reached a plateau at 20 mM concentration, which is just above the critical micellar

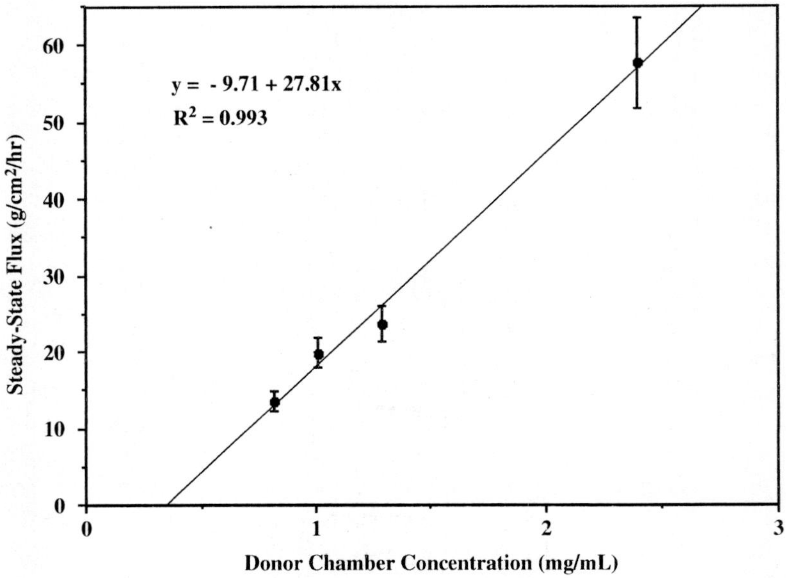

Figure 3.4 Effect of donor chamber concentration of acyclovir on steady-state flux through porcine buccal mucosa (reproduced from Shojaei et al., 1998).

concentration of sodium glycocholate as measured based on capillary height (Figure 3.5). Also, the permeation enhancement effect of sodium glycocholate through porcine buccal mucosa was shown to be independent of pH (Figure 3.6), suggesting that its three ionic species, namely, anion, cation, and zwitterion, are transported by the paracellular route.

2. ISIS Oligonucleotide

The buccal permeation of a model antisense oligonucleotide, ISIS 3082 (a 20 mer with a sequence of 5′-TGC ATC CCC CAG GCC ACC AT-3′), was reported (Jasti et al., 2000; Li et al., 1997). The steady-state flux of ISIS 3082 (shown in Table 3.2) was found to increase with increasing concentrations of ISIS 3082 in the donor chamber. The co-administration of 100 mM sodium glycocholate enhanced ISIS 3082 permeation 17.4 times. Because ISIS 3082 is a multiple charged molecule similar to acyclovir, the authors suggested that the enhancement effect of sodium glycocholate was due to increase in permeation via the paracellular route.

3. Buserelin

The influence of sodium glycocholate on the in vivo buccal permeation of buserelin, a luteinizing hormone releasing hormone (LHRH) agonist, D-Ser(Bu)-LHRH

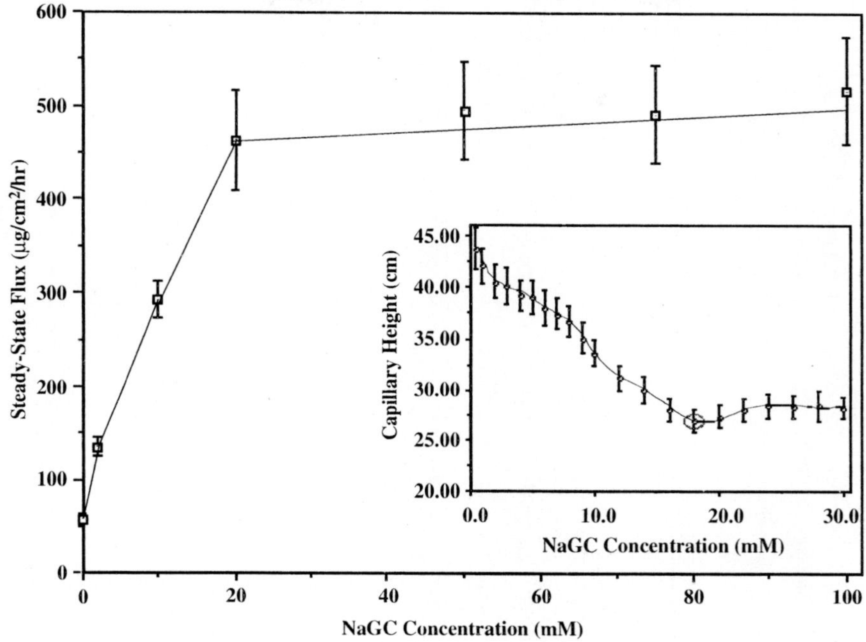

Figure 3.5 Critical micelle concentration of sodium glycocholate at 25°C in Krebs buffer (figure inset) and the effect of sodium glycocholate concentration on steady-state flux of acyclovir through porcine buccal mucosa (reproduced from Shojaei et al., 1998).

nonapeptide-ethylamide, was reported by Hoogstraate et al. (1996a). In this study, co-administration of sodium glycocholate increased the steady-state plasma concentration from 4.4 to 23.2 nM. The bioavailability increased from 1.0 to 5.3 percent. Due to lack of detailed plasma monitoring, no mechanism of action for sodium glycocholate enhancement effect was offered. The authors concluded that the permeation modulation of sodium glycocholate was transient, because no visible mucosal damage was observed.

4. Surfactants

Surfactants can be categorized into groups: anionic, cationic, and nonionic. Anionic surfactants include sodium lauryl sulfate and sodium laurate, cationic surfactants include cetylpyridinium chloride, and nonionic surfactants include poloxamer, Brij, Span, Myrj, and Tween. Surfactants are believed to enhance drug transport by perturbing the entire membrane architecture, affecting both the protein domain integrity as well as lipid structures (Veuillez et al., 2001). Sodium lauryl sulfate (SLS) has been shown to enhance the buccal absorption of human calcitonin and insulin (Nakada et al., 1988; Oh and Ritschel, 1990; Veuillez et

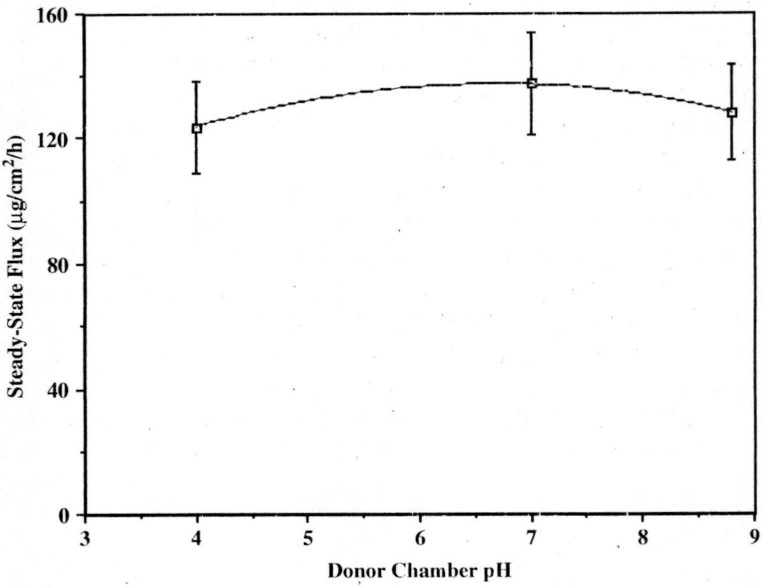

Figure 3.6 Effect of donor chamber pH on sodium glycocholate permeation enhancement of acyclovir across porcine buccal mucosa in presence of 2 mM NaGC (reproduced from Shojaei et al., 1998).

Table 3.2 Steady-State Flux and Porcine Buccal Permeability of ISIS 3082 (n = 3)

Donor Conc. (%)	Steady-State Flux (μg/cm^2-hr)	Permeability Coefficients ($\times 10^8$ cm/sec)
5	0.19 ± 0.16	0.105 ± 0.90
5[a]	4.13 ± 5.55	3.25 ± 2.42
3[a]	0.87 ± 0.29	1.83 ± 1.27
1[a]	0.51 ± 0.33	1.43 ± 0.93

[a] With 100 mM sodium glycocholate.
Source: Reproduced from Jasti et al., 2000.

al., 2001). The effect of a nonionic surfactant on the mucosal permeation of lidocaine hydrochloride was studied (Ganem-Quintanar et al., 1998). Sucrose laurate (L-1695) enhanced the permeability of lidocaine 14 to 22 times across palatal and buccal mucosae. The chain length was crucial in determining the enhancing effects of surfactants. Sucrose laurate was more effective in enhancing the permeability of lidocaine hydrochloride when compared to sucrose palmitate, sucrose stearate, and sucrose oleate.

5. Fatty Acids and Derivatives

The effect of fatty acids on enhancement of mucosal drug delivery is not well characterized. However, some studies have shown that the effect of fatty acids depends on the presence and the position of double bonds, isomer types (*cis* or *trans*), chain length, and degree of branching (Veuillez et al., 2001). Unsaturated fatty acids are usually more disruptive than the saturated counterparts having the same carbon number (Veuillez, et al., 2001). Fatty acids such as oleic acid, caprylic acid, sodium laurate, and myristate were studied as penetration enhancers (Senel and Hincal, 2001). Sodium laurate and myristate have been shown to enhance the buccal absorption of insulin and calcitonin (Veuillez et al., 2001). Oleic acid/hydroalcoholic solutions enhanced the permeation of lidocaine hydrochloride solution across buccal mucosa (Ganem-Quintanar et al., 1998). Oleic acid was shown to act as an enhancer having a proposed mechanism of insertion between the alkyl chains of membrane lipids, causing a disturbance of the lipid packing order in the deep lipid bilayer and polar head regions resulting in increased fluidity of the phospholipid domains (Ganem-Quintanar et al. 1998).

6. Vehicles and Adjuvants

The permeability of drugs can be enhanced by dissolving or dispersing the drug in a suitable solvent or vehicle to increase the dissolved concentration of the drug in contact with the mucosal tissue. Alternatively, the vehicle used to deliver the drug should be able to achieve a supersaturated state thereby increasing the thermodynamic activity of the drug and partitioning into the mucosal tissue. The vehicle may also increase the solubility of the drug in the epithelial mucosal barrier thereby increasing the partitioning of the drug from the vehicle to the mucosa. A combination of solvents may be used to achieve this goal (Veuillez et al., 2001). Some examples of solvents used to improve drug solubility and permeability are lauric acid in propylene glycol, ethanol, and dimethylsulfoxide (DMSO), as well as N-methylpyrrolidone (NMP) (Veuillez et al., 2001).

It was reported that use of 10 percent lauric acid in propylene glycol effectively enhanced the buccal permeability of insulin (Aungst and Rogers, 1994). Similarly, ethanol at different concentrations (15 to 30 percent) enhanced the transport of peptides (Veuillez et al., 2001). A co-solvent system of propylene glycol and isotonic McIlvaine buffer resulted in significant increase in buspirone flux through buccal mucosa when the concentration of propylene glycol was increased from 10 to 40 percent (Birudaraj, 2001).

7. Chelators

Chelators have been used as penetration enhancers including EDTA, salicylates, sodium citrate, and polyacrylates. They have been investigated for their potential use as enhancers in rectal, ophthalmic, and transdermal drug delivery. They are believed to interfere with the calcium efflux of the membrane (Senel and Hincal, 2001).

8. Cyclodextrins

Cyclodextrins are cyclic oligosaccharides consisting of (α-1,4)-linked α-D-glucopyranose units with a lipophilic central cavity and a hydrophilic outer surface. Examples of cyclodextrins include α-, β-, and γ-cyclodextrins and methylated β-cyclodextrins; they are believed to work as enhancers through the inclusion of membrane compounds (Senel and Hincal, 2001). Cyclodextrins are able to form inclusion complexes with many drugs by incorporating drug molecules or the lipophilic moiety of the molecule into the cavity (Thorsteinn and O'Fee, 2002). Hydroxypropyl cyclodextrin (HPCD) at a concentration of 10 mM increased the flux of buspirone 35 times (Birudaraj, 2001).

9. Chitosan

Chitosan was shown to increase buccal mucosal absorption of a large bioactive peptide-transforming growth factor-β (Senel et al., 2000). Chitosan, being a positively charged biodegradable polymer, can interact with the negatively charged mucosal surface and enhance drug uptake (Senel and Hincal, 2001). Its interaction with the proteoglycan matrix by ionic interactions may lead to a transient widening of the intercellular filaments (Portero et al., 2002). Chitosan was shown to increase the permeability of macromolecules, fluorescein isothiocyanate labeled dextrans (FD) in a TR146 buccal epithelium model (Portero et al., 2002). In addition to widening of the tight junctions, chitosan's interference with the extracellular lipid and glycolipid contents extruded by MCGs was thought to be responsible for its permeation enhancement effects (Portero et al., 2002). The in vitro flux of hydrocortisone (HC) and transforming growth factor-β (TGF-β) across buccal mucosa was increased sixfold by chitosan (Senel et al., 2000). More hydrophilic TGF-β showed higher permeability into the deeper tissue layers when compared to the hydrophobic HC. These results indicated a direct permeability enhancing effect of chitosan on the organized intercellular lipid lamellae.

10. Enzyme Inhibitors

Peptidase inhibitors can be used alone or in combination with permeation enhancers to stabilize peptide and protein drugs to overcome both enzymatic and physicochemical barriers to permeation. Protease inhibitors such as aprotinin, bestatin, and bile salts have been shown to stabilize peptides against buccal mucosal enzymes (Veuillez et al., 2001). The mechanism of increased stability of peptides by enzyme inhibitors is by altering the conformation of peptides and making them less susceptible to enzymatic degradation and thus imparting higher stability (Lee and Yamamoto, 1990; Veuillez et al., 2001).

IX. DELIVERY SYSTEM DESIGN

The absorption of drugs by the oral mucosal route can be improved by increasing the contact time of the drug delivery system or residence time of the drug in the oral cavity. Designing a delivery system that can release the drug over an extended period of time staying in close contact with the buccal mucosa is necessary for drugs that have to be administered over an extended period of time for local or systemic action. Bioadhesion involves the bonding of two materials (at least one of which is biological) by interfacial forces. If the biological membrane is covered by mucus, the term *mucoadhesion* is used. Bioadhesive systems have been shown to increase the buccal permeation for peptides, chiefly by increasing direct physical contact with the mucosa to provide for targeted delivery into and through the mucosal tissue. Water-soluble and insoluble hydrocolloids and hydrogels have been used as bioadhesive systems. Bioadhesives have been formulated into tablets, for example, Susadrin® (Pharmax Ltd.), which contains nitroglycerin, gels, and patches. Hydrogels, which release the drug by swelling and thereby allowing drug transport through the spaces in the polymer network, are being widely studied for their use in bioadhesive gels. Polyacrylic-based hydrogels have also been extensively studied. An example of a commercially available device is the OTS (oral transmucosal system, TheraTech), which has been used to deliver glucagon-like insulintropic peptide.

X. IMPACT OF ENHANCERS ON DOSAGE FORM

The enhancement of drug permeation by different chemical classes of enhancers has been reported (Ganem-Quintanar et al., 1998; Hoogstraate et al., 1996a; Jasti et al., 2000; Portero et al., 2002; Shojaei et al., 1998). However, enhancers alone may not be effective to facilitate the permeation of all drugs; the type of dosage form also plays an important role in determining the type of enhancer required to optimize delivery. Permeation enhancement has a limited role to play if the drug is delivered in a dosage form such as a fast-dissolving tablet or a chewable gum and the drug is intended for absorption by the gastrointestinal route. Fast-dissolving tablets are taken without water and dispersed or dissolved in saliva when placed on or under the tongue. These dosage forms are expected to dissolve within minutes when put in the oral cavity; some examples are Alavert® fast dissolving tablets (Wyeth Consumer Healthcare) and Excedrin® Quicktabs (Bristol-Myers Squibb Company). Drug absorption from fast-dissolving tablets takes place across the buccal mucosa and via the gastrointestinal tract. Enhancers may, however, play a major role in permeation of drugs from bioadhesive dosage forms for targeted buccal delivery where there is prolonged contact time with the tissue of the oral cavity.

For many drugs that are susceptible to first-pass metabolism and enzymatic degradation following oral delivery, the prospect of using buccal delivery depends on the ability to increase and enhance drug permeation across the buccal mucosa. Therefore, the role of enhancers is essential to make the buccal mucosa a potential site for administration of these drugs. Advances in permeability modulation and formulation with appropriate enhancers can provide for effective and feasible buccal drug delivery for many drugs, which otherwise must be injected or ingested with water such as in conventional dosage forms.

REFERENCES

Aungst, J.B., and Rogers, J.N. (1994). Comparison of the effects of various transmucosal absorption promoters on buccal insulin delivery. Int. J. Pharm., 53, 227–235.

Birudaraj, R. (2001). Transbuccal Drug Delivery: In Vitro Characterization of Transport Pathway of Buspirone and Bioadhesive Drug Delivery System. University of the Pacific, Stockton, CA.

Bodde, H.E., de Vries, M.E., and Junginger, E.H. (1990). Mucoadhesive polymers for the buccal delivery of peptides, structure-adhesiveness relationships. J. Control. Release, 13, 225–231.

Burkitt, G.H., Young, B., and Heath, J.W. (1993). Wheater's Functional Histology (3d ed.). Edinburgh: Churchill Livingstone.

Chen, H.L., Chetty, J.D., and Chien, W.Y. (1999). A mechanistic analysis to characterize oral mucosal permeation properties. Int. J. Pharm., 184, 63–72.

de Vries, M.E., Bodde, H.E., Verhoef, J.C., and Junginger, H.E. (1991). Developments in buccal drug delivery. Crit. Rev. Ther. Drug Carrier Syst., 8(3), 271–303.

Duchateau, G.S.M.J.E., Zuidema, J., and Merkus, F.W.H.M. (1986). Bile salts and intranasal drug absorption. Int. J. Pharm., 31, 193–197.

Gandhi, B.R., and Robinson, J.R. (1988). Bioadhesion in drug delivery. Ind. J. Pharm. Sci., 50, 145–152.

Gandhi, B.R., and Robinson, R.J. (1994). Oral cavity as a site for bioadhesive drug delivery. Adv. Drug Deliv. Rev., 13, 43–74.

Ganem-Quintanar, A., Quintanar-Guerrero, D., Falson-Rieg, F., and Buri, P. (1998). Ex vivo oral mucosal permeation of lidocaine hydrochloride with sucrose fatty acid esters as absorption enhancers. Int. J. Pharm., 173, 203–210.

Gordon, G.S., Moses, A.C., Silver, R.D., Flier, J.S., and Care, M.C. (1985). Nasal absorption of insulin: Enhancement by hydrophobic bile salts. Proc. Nat. Acad. Sci., USA., 82, 7419.

Gumbiner, B., and Simons, K. (1986). A functional assay for proteins involved in establishing an epithelial occluding barrier: Identification of a uvomorulin-like polypeptide. J. Cell Biol., 102(2), 457–468.

Harris, D., and Robinson, J.R. (1990). Bioadhesive polymers in peptide drug delivery. Biomaterials, 11(9), 652–658.

Harris, D., and Robinson, J.R. (1992). Drug delivery via the mucous membranes of the oral cavity. J. Pharm. Sci., 81, 1–10.

Hoogstraate, A.J., Verhoef, J.C., Pijpers, A., van Leengoed, L.A.M.G., Verheijden, J.H.M., Junginger, H.E., and Bodde, E.H., (1996a). In vivo buccal delivery of the peptide drug buserelin with glycodeoxycholate as an absorption enhancer in pigs. Pharm. Res., 13(8), 1233–1237.

Hoogstraate, J.A., Senel, S., Cullander, C., Verhoef, J., Junginger, E.H., and Bodde, H.E. (1996b). Effects of bile salts on transport rates and routes of FITC-labeled compounds across porcine buccal epithelium in vitro. J. Control. Release, 40(211–221).

Jasti, R.B., Zhou, S., Mehta, C.R., and Li, X. (2000). Permeability of antisense oligonucleotides through procine buccal mucosa. Int. J. Pharm., 208, 35–39.

Lee, L.H.V., and Yamamoto, A. (1990). Penetration and enzymatic to peptide and protein absorption. Adv. Drug Delivery Rev., 4, 171–207.

Li, X., Zhou, S., Mehta, R.C., and Jasti, B.R. (1997). Buccal permeation of a model antisense oligonucleotide. Int'l. Sym. Contr. Rel. Bioact. Mater., 24, 667–668.

Morgan, R.P. (2000). Oral histology and oral pathology. Http://Www.Kcl.Ac.Uk/Ip/Petermorgan/Oral_Mucosa.Html. Retrieved, from the World Wide Web.

Nakada, Y., Awata, N., Nakamichi, C., and Sugimoto, I. (1988). The effects of additives on the oral mucosal absorption of human calcitonin in rats. J. Pharmacobio-Dyn., 11(395–401).

Oh, C.K., and Ritschel, A.W. (1990). Absorption characteristics of insulin through the buccal mucosa, Methods Find. Exp. Clin. Pharmacol., 12, 275–279.

O'Hagan, D.T., and Illum, L. (1990). Absorption of peptides and proteins from the respiratory tract and the potential for development of locally administered vaccine. Crit. Rev. Ther. Drug Carrier Syst., 7(1), 35–97.

Portero, A., Remunan-Lopez, C., and Nielsen, H.M. (2002). The potential of chitosan in enhancing peptide and protein absorption across the Tr146 cell culture model—an in vitro model of the buccal epithelium. Pharm. Res., 19(2), 169–174.

Rathbone, J.M., and Hadgraft, J. (1991). Absorption of drugs from the human oral cavity. Int. J. Pharm., 74, 9–24.

Senel, S., Duchene, D., Hincal, A.A., Capan, Y., and Ponchel, G. (1998). In vitro studies on enhancing effect of sodium glycocholate on transbuccal permeation of morphine hydrochloride. J. Control. Release, 51(2-3), 107–113.

Senel, S., and Hincal, A.A. (2001). Drug permeation enhancement via buccal route: Possibilities and limitations. J. Control. Release, 72(1-3), 133–144.

Senel, S., Kremer, J.M., Kas, S.H., Wertz, W.P., Hincal, A.A., and Squier, A.C. (2000). Enhancing effect of chitosan on peptide drug delivery across buccal mucosa. Biomaterials, 21, 2067–2071.

Shojaei, A. (1998). Buccal mucosa as a route for systemic drug delivery: A review. J. Pharm. Pharmaceut. Sci., 1, 15–30.

Shojaei, A.H., Berner, B., and Li, X. (1998). Transbuccal delivery of acyclovir (I): In vitro determination of routes of buccal transport. Pharm. Res., 15(8), 1182–1188.

Squier, C.A. (1991). The permeability of oral mucosa. Crit. Rev. Oral Biol. Med., 2(1), 13–32.

Thorsteinn, and O'Fee, R. (2002). Cyclodextrin-based drug delivery. Business Briefing. Pharmatech.

Veuillez, F., Kalia, Y.N., Jacques, Y., Deshusses, J., and Buri, P. (2001). Factors and strategies for improving buccal absorption of peptides. Eur. J. Pharm. Biopharm., 51(2), 93–109.

Washington, N., Washington, C., and Wilson, C. (2001a). Physiological Pharmaceutics—Barriers to Drug Absorption (2d ed.), New York: Taylor and Francis.

Washington, N., Washington, C., and Wilson, G.C. (2001b). Drug delivery to the oral cavity or mouth. Biological Pharmaceutics: Barriers to Drug Absorption (pp. 37–58). New York: Taylor and Francis.

Xiang, J., Fang, X., and Li, X. (2002). Transbuccal delivery of 2′,3′-Dideoxycytidine: In vitro permeation study and histological investigation. Int. J. Pharm. 231(1), 57–66.

Yamamoto, A., Hayakawa, E., Kato, Y., Nishiura, A., and Lee, V.H. (1992). A mechanistic study on enhancement of rectal permeability to insulin in the albino rabbit. J. Pharmacol. Exp. Ther., 263(1), 25–31.

Zhang, H., and Robinson, R.J. (1996). Routes of Drug Transport across Oral Mucosa in Oral Mucosal Drug Delivery. New York: Marcel Dekker.

4

Oral Transmucosal Delivery of Protein and Peptide Therapeutics

Bhaskara R. Jasti and William Abraham

CONTENTS

I. INTRODUCTION

Search for nonparenteral routes of drug delivery is driven mainly by advances in the development of peptides and proteins as a major class of therapeutics. Therapeutic uses of peptides and proteins range over a variety of pharmacological classes including enzymes (i.e., tissue plasminogen activator), enzyme inhibitors (i.e., captopril used as ACE inhibitor), hormones (i.e., leutinizing hormone releasing hormone, LHRH), immunomodulators (i.e., interferons, vaccines), and antimicrobial agents (i.e., penicillins). Although the oral route is the most common route of drug administration, poor bioavailability of peptides and proteins administered orally makes the transmucosal and transdermal routes of delivery an attractive alternative to oral delivery.

The mucosal lining of the nasal, oral, vaginal, and rectal cavities and the skin offer potential routes for peptide and protein delivery that avoid first-pass clearance by the liver as well as presystemic metabolism in the gastrointestinal tract. These routes are noninvasive and may be better accepted by patients compared to the parenteral administration. For peptides, the nasal route has already achieved commercial status for the delivery of peptides such as LHRH and calcitonin. However, the advantage of high permeability of the nasal epithelium is offset by disadvantages such as the potential irritation and toxicity to ciliary cells associated with the chronic or prolonged use of the dosage form in the nasal cavity. Also intra- and intersubject variability in mucus secretion in the nasal mucosa can affect nasal delivery. Drug delivery across the vaginal and the rectal mucosae suffers from poor patient acceptance. These factors make the mucosa of the oral cavity such as the buccal, gingival, and sublingual regions attractive for systemic delivery of peptides and proteins.

In this chapter, delivery of peptide drugs via the buccal route is reviewed by describing the structure of oral mucosa and the epithelial barrier to transport, the experimental methods used to evaluate buccal delivery, and the drug delivery systems unique to buccal delivery. Recent reviews in this field describe the different aspects of buccal delivery (1–4).

II. BARRIER TO PEPTIDE DELIVERY

Drug delivery across the mucosal lining of the oral cavity can be classified as sublingual delivery wherein the drug is administered across the highly permeable, thin epithelium of the floor of the mouth, buccal delivery wherein the delivery is across the mucosal lining of the inside of the cheek, and local delivery into the oral cavity. When considering the barrier to peptide transport across the mucosa, one has to consider the passive diffusional barrier as well as the enzymatic metabolic barrier resulting from the enzymatic activity of the epithelium.

A. Passive Barrier

The diffusional barrier to transport has been identified by Squier et al. (5–7) to reside in the outer one-third of the epithelium. Interestingly, one of the earliest studies to identify the diffusional barrier in buccal transport used a water-soluble protein, horseradish peroxidase, with a molecular weight of 40 kDaltons and a size of approximately 5 to 6 nm (5). The enzymatic properties of this protein enabled visualization of this tracer by light and electron microscopy. It was shown that when applied either topically or subepithelially, to keratinized as well as nonkeratinized epithelia of monkey, dog, pig, rabbit, hamster, and rat, the protein did not penetrate the top 25 percent of the epithelium (5–7). Similar conclusions were reached from studies with an organ culture system (8).

In a different study using a smaller peptide, thyrotropin releasing hormone (TRH), transport across rabbit buccal mucosa showed the upper 50 μm of the epithelium as the barrier to transport (9). Permeability of horseradish peroxidase was shown to increase across sublingual mucosa whereas there was no change in permeability across the buccal tissue in experiments where the epithelium was subjected to tape stripping (10). This was explained in terms of the difference in the thickness between sublingual and buccal epithelia (11). The barrier layer was not completely removed by tape stripping in the case of the thicker buccal epithelium. The basal lamina was suggested to present a major barrier based on electrical resistance, morphology, and permeability data from dermatomed porcine buccal mucosa (12). The relative contribution of the epithelium and the basal lamina of the oral mucosa barrier depends on the molecular structure of the drug and on the physical integrity of the tissue.

The permeability differences among different regions of the oral mucosa were measured using tritiated water (13). The nonkeratinized epithelia of the buccal and the sublingual regions were more permeable than the keratinized epithelia of the gingiva and the palate (7,13). Furthermore, the floor of the mouth (e.g., sublingual mucosa) was shown to be more permeable than the buccal mucosa. Also, the thinner sublingual epithelium gives a faster onset of action than the thicker buccal epithelium. Thus, the sublingual route is ideal for situations where a rapid delivery is desired and has been used for the delivery of nitroglycerin from quick-dissolving tablets for management of acute pain associated with angina pectoris. However, delivery of less permeable drugs such as the peptides requires sustained-release delivery systems that would deliver the drugs long enough for the plasma level to reach steady state. The sublingual region is not suited for application of mucoadhesive systems due to the steady flow of saliva and a highly mobile environment due to the movement of the tongue. Alternatively, the buccal site offers a smooth, relatively immobile surface that has a permeable vascularized mucosa for the application of a sustained-release drug delivery system. The increased residence time enables the attainment of the steady-state plasma levels. Thus, buccal mucosa is better suited for peptide delivery compared to other sites within the oral cavity.

B. Pathways for Peptide Transport

There are two possible pathways for penetration of peptides across the buccal mucosa: (1) the intracellular or the transcellular pathway where the peptide takes the shortest route across the epithelial cells, and (2) the intercellular or the paracellular pathway where the peptide molecule diffuse through the intercellular matrix, taking a tortuous pathway. Although there is some evidence for a transcellular pathway for the penetration of small molecules (14,15), there is growing evidence in the literature that indicates that the intercellular route may be the predominant route for peptides and other drug molecules (16–18). Horseradish peroxidase was visualized in the intercellular space of the oral epithelium of a variety of animals (5–7).

Paracellular pathway was shown to be the major route for TRH diffusion in rabbit buccal epithelium based on autoradiographic grain density distribution in rabbit buccal transport studies using ^{3}H-TRH (9,17). Confocal laser scanning microscopy (CLSM) revealed the distribution of a 10 kDa molecular weight FITC-labeled dextran in the intercellular space of porcine buccal epithelium (18). Using an in situ method, wherein the permeation of fluorescent compounds was visualized by CLSM, it was shown that the diffusion of a hydrophilic marker such as FITC was slowed down at approximately 150 μm from the surface of the epithelium (4). This region corresponded to the cell layers where the membrane coating granules (MCGs) discharge their contents into the intercellular region (4,19).

It was shown that horseradish peroxidase penetrated the intercellular spaces of the entire junctional epithelium, which lacks MCGs (20). Permeability of horseradish peroxidase through the intercellular region increased upon enzymatic digestion of the intercellular barrier (21). Wertz and Squier suggested a relationship between the relative cytoplasmic volume occupied by the MCGs in different oral epithelia to the tissue permeability (22). An inverse relationship was apparent, suggesting that a greater volume of MCGs is associated with lower permeability. Thus, the contents of MCGs have been shown to form the barrier to passive diffusion in the extracellular space of the epithelium.

The paracellular pathway is quite often referred to as the polar pathway, which describes the passage of polar molecules through the junctional complex that binds the cells in a simple epithelium such as the nasal, intestinal, and the rectal mucosa. In the buccal mucosa, the paracellular route for hydrophilic peptide permeation was described as aqueous polar pathway, which describes the passage of polar molecules between cells through the junctional complex (23). The mechanism of transport of over 30 compounds ranging from simple amino acids to peptides and proteins such as TRH, vasopressin, oxytocin, calcitonin, and insulin was reviewed in terms of the penetration pathway for these hydrophilic molecules. Based on the transport studies of a homologous series of model peptides, it was shown that the charged amino acids and their t-butyl-oxycarbonyl derivatives have smaller buccal permeabilities than structurally similar charged nonamino acids. The zwitterionic peptides have lower permeability despite a comparable partition coefficient, compared to the model compounds. Thus, the n-octanol/water partition coefficients were not good indicators of buccal permeability. Results from TRH and insulin transport studies in vehicles having different pH values also led to similar conclusions (9,24).

C. Enzymatic Barrier

Compared to the oral route, not much is known about the enzymatic barrier to peptide absorption in the buccal mucosa. The enzymatic activity in mucosal tissue homogenates was determined from different mucosal tissue, as a first step towards understanding the nature of the enzymatic barrier (25–28). The enzymatic barrier of the buccal mucosa has been suggested to be comparable to other mucosal routes. This was based on the hydrolysis of methionine enkephalin and leucine enkephalin in the tissue homogenates of rectal, vaginal, buccal, and nasal mucosa from rabbits (25). The hydrolysis was shown to be predominantly due to aminopeptidases. The activities of esterase, aminopeptidases, carboxypeptidases, and endopeptidases were evaluated in the buccal and intestinal homogenates of the rat and the hamster (26). Carboxypeptidase activity was shown to be high in the buccal mucosa whereas endopeptidase activity was greater in the intestinal mucosa (27).

The proteolytic activities against small peptides did not differ much among the mucosal routes although the activities were different for proteins such as insulin and proinsulin, with the buccal route showing the least activity (28). The relative proportion of the proteases in the different mucosae was not characterized. It was shown that in nasal and ileal mucosae, the aminopeptidase activity was membrane-bound, whereas it was principally cytosolic in the buccal, rectal, and the vaginal mucosae. In light of increasing evidence for the paracellular pathway for peptide penetration in buccal epithelium as discussed earlier, the tissue homogenate studies may be an overestimation of the tissue enzyme activity compared to the actual pathway of penetration. Nevertheless, there could be some enzymatic activity present in the intercellular space, for example, resulting from the MCGs. During the paracellular transport, the membrane-bound enzymes of rabbit buccal epithelium were suggested to be responsible for the disappearance of intact oxytocin (29).

A more useful study examined the simultaneous diffusion and metabolism of the leucine-p-nitroanilide across excised hamster cheek pouch (30). Leucine-p-nitroanilide was completely hydrolyzed and the metabolic activity was substantially reduced in the presence of Bestatin, an aminopeptidase inhibitor. The permeation results were analyzed using a model for simultaneous diffusion and metabolism. Such a model was described in detail by Ho (31). In this model, the drug diffuses through the epithelium after partitioning from the donor medium and undergoes enzymatic degradation or metabolism in the tissue. This model treats the buccal epithelium as an homogeneous, metabolically active hydrogel. The diffusion of p-nitrophenol (PNP) across freshly excised monkey buccal tissue was evaluated by applying this model (31). The tissue metabolism of PNP to its sulfate and glucouronide conjugates was shown to be 98.3 percent.

However, a simple consideration of the epithelial thickness, which is 75 μm for the monkey buccal tissue versus 700 μm for the thickness of the excised full thickness mucosa that was used in this study, and the residence time in the epithelium versus the full thickness mucosa gives a whole different perspective. The residence time can be calculated from the thickness (δ) and the diffusion coefficient (D) in the tissue using the relation $t = \delta/D^2$. It was shown that t_{epi} = 56 sec and t_{muc} = 4900 sec. The extent of epithelial metabolism was estimated to be only 12 percent compared to the experimentally measured value of 98.3 percent. Such differences in the thickness of the barrier layer (i.e., the epithelium) should be put in perspective when extrapolating the results to the in vivo situation. Thus it is crucial to use appropriate in vitro and in vivo models while estimating the enzymatic barrier to buccal delivery of peptides.

Use of various protease inhibitors to suppress the proteolytic activity and enhance nasal and gastrointestinal (GI) absorption of peptides has been described (32). However, the use of these agents for suppressing any enzymatic activity in the buccal mucosa has not been explored extensively and the validity of such an approach remains to be tested.

III. BIOADHESIVE SYSTEMS

Bioadhesive systems are best suited for peptide delivery as these systems adhere readily upon application to the buccal mucosa and are unobtrusive and retained in place for several hours. This provides enough time for the drug to reach therapeutic levels in the plasma. Bioadhesion or mucoadhesion defines the interaction between a biological surface such as the oral mucosa and the polymer. Mucoadhesion is said to occur through the following steps:

1. Initial physical interaction or mutual wetting between the polymer and the mucus
2. Interpenetration of the polymer and the glycoproteins (from the mucus)
3. Electrostatic interactions, hydrogen bond formation, and van der Waal's forces between the polymer and the mucosal surface

This is in accordance with the adsorption–interdiffusion bonding theory (33). Mucoadhesive materials are generally hydrophilic polymers that are called wet adhesives as they adhere to the mucosal surface upon wetting. These mucoadhesive polymers fall under one of the following three categories: (1) anionic polymers such as polyacrylic acid, sodium carboxymethyl cellulose, and sodium alginate; (2) cationic polymers such as chitosan and diethyl aminoethyl dextrans; and (3) nonionic polymers such as cellulose derivatives (hydroxypropyl cellulose, hydroxypropylmethyl cellulose), polyvinylpyrrolidone, polyvinylalcohol, starch, and polyethyleneoxide. Recently, buccal tablets made by direct compression of a novel natural gum were shown to achieve adequate mucoadhesive strength (34). The process involved in mucoadhesion differs with different materials used for the fabrication of the dosage form as well as on the site within the oral mucosa. Some of the formulation variables that affect mucoadhesion are discussed below.

Swelling contributes to bioadhesion favorably (35). However, excessive water content leads to a drop in adhesion strength, suggesting disentanglement of the polymer and the mucus glycoprotein (36). Only polycarbophil and chitosan were shown to retain good adhesion upon full hydration among a wide range of materials tested (37). It has been reported that the bioadhesive strength increased as the molecular weight of the polymer increased, suggesting that the length of the polymer chain determined the extent of interpenetration and molecular entanglements (38). The adhesive strength was shown to decrease in highly concentrated systems, suggesting that the solvent-poor systems lower the number and extent of polymer chains that are available for interpenetration. In other words, the polymer is not plasticized enough to be an adhesive (36). Bioadhesion also decreases in extensively crosslinked systems for similar reasons (39).

The effect of polymer polarity and segmental mobility on bioadhesion was investigated in acrylic acid–butyl acrylate copolymer (40). As the content of butyl acrylate increased, the polarity decreased, while the segmental mobility increased and the T_g of the copolymer decreased. This led to increased mucoadhesion up to 20 percent butyl acrylate in the copolymer, beyond which the increased hydrophobicity counteracted the fluidity increase of the polymer chain, thereby decreasing mucoadhesion (40). In general, anionic polymers such as polyacrylic acid have shown superior bioadhesion compared to nonionic and cationic polymers, mainly due to extensive hydrogen bonding and surface energy matching of the hydrated polymer with the mucosal surface (41). Also, polymers containing carboxyl groups were shown to be better than those containing sulfate groups (42).

All these parameters should be taken into consideration as key formulation variables in the fabrication of mucoadhesive delivery systems. Quite often the selection of the formulation with a desired drug release profile leads to some compromise in the bioadhesive property of the system. Bioadhesive force can be measured by a variety of in vitro methods based on either shear or tensile stress measurements (43,44). Although these methods are useful for ranking the different polymer systems in terms of adhesion properties, the residence time and comfort of wear are defined more precisely by wear studies in humans.

Fabrication of laminated mucoadhesive patches for buccal peptide delivery has been described (45,46). Aqueous solutions of polymers with peptide drug and other adjuvants were cast onto a backing layer sheet and dried at a constant temperature of 38°C for ~2 hours. For hydroxyethyl cellulose, the adhesion duration versus viscosity grade showed a maximum, with lower and higher viscosity grades showing less adhesion. For PVA and PVP, increase in polymer viscosity resulted in prolonged adhesion. Similarly, oxytocin buccal patches consisting of carbopol 974P and silicone mucoadhesive polymers were prepared by solvent casting at ambient temperature and drying at 50°C for 30 minutes (29). At the end of 4 hours, the permeation of oxytocin from these buccal patches amounted to 0.012 percent of the total loaded drug. In vivo release of model compounds such as salicylate or protirelin was shown to be prolonged upon increasing the viscosity of the polymer as well as the polymer load in the patch, suggesting that the drug release was controlled by polymer dissolution (45). In a recent study, a buccal tablet made by direct compression of a novel natural gum from *Haea gibbosa* was found to sustain the release of salmon calcitonin (34). The serum concentrations of salmon calcitonin from such buccal tablets were found to be sufficient to achieve biological activity.

IV. PEPTIDE AND PROTEIN DELIVERY

Only a small number of peptides and proteins have been explored for systemic delivery across buccal mucosa. A representative list of peptides and polypeptides investigated for buccal delivery along with their molecular weights is shown in Table 4.1. Most of these studies have shown only modest bioavailability, which was increased by coadministration of penetration enhancer(s). In the remainder of this chapter, selected studies are reviewed with emphasis on factors such as penetration enhancement, formulation variables, and bioavailability.

Table 4.1 Representative Peptide-Based Drugs Studied for Buccal Delivery

Peptide	Molecular Weight	Ref.
Thyrotropin releasing hormone (TRH)	362	9, 17, 59, 63
ACE inhibitors	~370	69
Somatostatin analogue	~850	58
Oxytocin	1007	70–72
Vasopressin and its analogues	~1050	73
LHRH and its analogues	~1100	4, 66
Calcitonin	3600	67
Insulin	6000	24, 46–48, 51–52

A. Insulin

Due to the crucial therapeutic importance of insulin, it is not surprising that it is one of the most widely studied peptides for noninvasive delivery routes, including the buccal route of delivery. Insulin is a large protein with 51 amino acids, organized into two subunits, and with a molecular weight of 6000. One of the earliest studies using dogs showed a low bioavailability of 0.5 percent when insulin was administered with sodium glycocholate in a cocoa butter matrix (47). Later, more detailed studies in rats showed a higher bioavailability of up to 25 to 30 percent compared to intramuscular administration (24,47,48). Insulin solutions containing adjuvants at different pH were administered using a microfilter syringe at 0.2 ml/kg dosing volume in anesthetized rats. Blood samples were collected for up to four hours and insulin absorption efficacy was expressed as percent efficacy of buccal insulin relative to intramuscularly (IM) delivered insulin, by measuring plasma glucose concentrations (49).

Among the several adjuvants studied, CHAPS (3-3-Cholamidopropyl-dimethyl ammonio-1-propane sulfonate (Pierce)), was shown to be the most potent enhancer on a molar concentration basis. Nonionic surfactants with laurate hydrophobic groups were studied, as the 12-C long hydrophobic group has shown maximal absorption enhancement for nasal, rectal, and transdermal routes. Efficacy relative to IM administered insulin was shown to attain a maximum of 27 percent with laureth-9 ($CH_3(CH_2)_{10}$-CH_2-$(OCH_2CH_2)_9OH$). However, the absorption enhancement effect of the nonionic surfactants was not related to their hydrophilic–lipophilic balance (HLB). It was shown that the nonionic surfactants with an ether group linking the hydrophobic and the hydrophilic regions were more effective than those with ester linkages (24).

It was also shown that the buccal absorption of insulin at pH 3.4 and at pH 7.4 was higher than that at pH 5.4. For example, the efficacy relative to IM administered insulin was shown to be 32 and 27 percent at pH 3.4 and pH 7.4, respectively, compared to an efficacy of 15 percent at pH 5.4 with 5 percent laureth-9. This was attributed to increased solubility of insulin at pH 3.4 and 7.4, which increased the chemical potential across the buccal tissue. Insulin is not very soluble at pH 5.4 (the isoelectric point of the peptide) even with 5 percent surfactant. It is interesting to note that the higher solubility of the ionized form led to increased buccal absorption. The solubility of the adjuvant is also a key factor, as was shown for sodium laurate, which showed higher insulin absorption at pH 8.9 compared to pH 7.4. Bile salts, which are steroidal detergents with ionizable group(s), showed insulin absorption enhancement of 20 to 25 percent relative to IM administered insulin. Overall, it was shown that higher concentrations of up to 5 percent of bile salts were required for enhancement of buccal insulin absorption in these studies.

Other adjuvants such as polyacrylic acid and EDTA did not show any enhancement of buccal absorption, although these have been shown to enhance peptide absorption across simple columnar epithelium such as the nasal mucosa (50). Coadministration of a potential inhibitor of insulin metabolism, such as Z-Gly-Pro-Leu-Gly-Pro, had no effect on the buccal absorption of insulin, suggesting that a metabolic barrier does not play a significant role in buccal insulin absorption (51).

Absorption of insulin through rabbit buccal mucosa was investigated using a buccal absorption cell (52,53). The effects of application time, dose size, and pH of the insulin solution, absorption enhancers such as Brij-35, sodium taurocholate, sodium lauryl sulfate, deoxycholate, sodium methoxy salicylate, sodium dextransulfate, and EDTA were investigated. The hypoglycemic response was used to monitor buccal absorption. The various doses and solution pH had no effect on insulin absorption in the absence of absorption enhancers. Brij-35 was found to show maximum buccal absorption enhancement of insulin, with a bioavailability of 12 percent compared to a bioavailability of 4.3 percent with a tablet formulation. A marked increase in the hypoglycemic response was observed when the application time was increased from 0.5 to 1 hour, which was confirmed by the insulin blood levels. The t_{max} of hypoglycemia was shown to be about 1 hour after the removal of the buccal cell used for insulin administration, suggesting a tissue reservoir function of the buccal mucosa.

Buccal insulin absorption in beagle dogs was reported by Ritschel et al. (54). Insulin was administered using a specially designed buccal cell or in the form of gelatin films, with different absorption enhancers and at different pH. Pharmacological activity (PA) was calculated as the ratio of the area under the percent glucose reduction versus the time curve for buccal to intravenous administration. The solution at pH 7.5 with permeation enhancers Labrafil® M 1944 and Azone® gave a PA of 18.3 compared to a PA of 0.9 percent for the control solution in normal saline. Highest reduction in plasma glucose concentration was found from a taurocholate–linoleic acid combination. The PA for the buccal solution varied between 0.9 and 22.3 percent for the different formulations and between 11 and 15 percent for the gelatin films, much higher than the previously reported insulin bioavailability of 0.5 percent in dogs (47).

In the solution study, t_{max} occurred after an administration time of 0.5 hours. In some of the formulations, t_{max} ranged from 1 to 3 hours, and the application was terminated at 30 minutes. This suggested a reservoir effect for insulin in the dog buccal tissue, similar to the report from the rabbit study (52,53). These studies with beagle dogs and rabbits are more meaningful compared to the studies reported in rats as the rat buccal epithelium is fully keratinized and is not a relevant model for buccal absorption in humans. Also, in the rat study, the insulin solutions were maintained at the dosing site (i.e., the buccal region) but the absorption area was not clearly defined, whereas in the studies with the buccal cell or the gelatin film in dogs and rabbits, the absorption area was defined more precisely.

Alkylglycoside surfactants have been shown to enhance insulin absorption across rat buccal mucosa (55). Octyl and decyl glycoside at 5 percent concentrations were the most effective among the various adjuvants studied and were shown to enhance insulin absorption across buccal mucosa more than the nasal and rectal absorption. There was no consistent relationship between the alkyl chain lengths of the alkylglycosides or the critical micellar concentrations and their absorption-promoting effect. The alkylglycosides are nondenaturing and have been shown to stabilize insulin against aggregation and enzymatic degradation (56). It was also suggested that alkyl glycosides could have reduced irritation potential (55). The absorption enhancement was suggested to occur by alteration of the lipid/glycolipid intercellular matrix of the barrier layer (55). A more recent study showed that insulin associated with erythrocyte ghost membranes was absorbed across rat buccal mucosa (57,58). The authors concluded that the pharmacological effect did not result in a significant therapeutic effect. As in the earlier study (55), the absorption area was not well defined in this study (57). The buccal absorption of insulin seems to show promise despite the large size of the peptide and could well represent the upper limit in terms of the size of the peptides that can be delivered via the buccal route.

B. Somatostatin Analogue

Merkle and Wolany have made a systematic study of the permeation of octreotide, an enzymatically stable cyclic octapeptide somatostatin analogue (59). In this study, a porous patch soaked with the peptide solution containing adjuvants was placed on the buccal mucosa using a peripheral bioadhesive ring. Thus the absorption was restricted to a defined area of the buccal mucosa. Studies were performed in anesthetized rats and in conscious beagle dogs, and interesting differences in the pharmacokinetic profiles were observed between the two species (59).

Upon increasing the buccal dose from 50 to 1600 μg octreotide per rat (~300 g body weight), the plasma level increased from ~0.1 ng/ml to 2 ng/ml. The plasma level showed a hundred fold increase in C_{max} as the buccal dose was increased from 0.2 to 2 mg/kg in the dog. Buccal bioavailability ranged from 0.2 to 0.6 percent in rats, whereas it was 0.3 to 2 percent in dogs as measured by $AUC_{0\text{-}6}$, in comparison to the intravenous data. The permeation enhancement caused by the adjuvants, sodium glycocholate, taurocholate, and Azone®, was more pronounced in dog compared to that in rat, as shown in Table 4.2. This difference was attributed to the greater perturbation of the nonkeratinized epithelium of the dog by the adjuvants, compared to the fully keratinized epithelium of the rat.

Table 4.2 Bioavailability of Octreotide in Rat and in Dog[a]

Species	Bioavailability $AUC_{0\text{-}6}$	Enhancement Factor [b]
Rat	0.2–0.6	0.6–1.2
Dog	0.3–2	4–6

[a] From Reference (58).

[b] Enhancement factor = $AUC_{0\text{-}6(enh)}/AUC_{0\text{-}6(std)}$. Permeation enhancement caused by adjuvants such as sodium glycocholate is more pronounced in dog compared to that in rat.

Another difference between rat and dog is in the drug clearance after buccal administration, as shown in Table 4.3. The terminal half-life of the drug is significantly longer in rat when delivered buccally compared to intravenous, suggesting a depot effect for the keratinized tissue. In dog, there is presumably less binding and thus a lower terminal half-life. There is, however, an increase in the $t_{1/2}$ compared to intravenous in dog also. Thus there could be variations in the pharmacodynamic profiles of peptide drugs delivered buccally, compared to intravenous administration. In addition, there were differences reported in the t_{max}, based on the adjuvants used for absorption enhancement. For example, use of cholic acid derivatives resulted in a change of t_{max} from 1 to 0.5 hours whereas the use of Azone® resulted in a t_{max} of ~2 hours. Such differences in t_{max} were

Table 4.3 Terminal Half-Life of Octreotide in Different Species Following Buccal Delivery[a,b]

	Half-Life (hours)		
Species	Intravenous	Buccal	Buccal + Enhancer
Rat	0.61	2.34	8.68
Dog	0.5	1.1	—

[a] From Reference (58).
[b] Longer half-life resulting in slower drug clearance following buccal administration suggests a depot effect of the keratinized tissue.

explained in terms of differences in the type of interaction between the adjuvant and the constituents of the epithelial barrier layer.

This study exemplifies the need for the choice of relevant animal model and defining the absorption area precisely such that the absorption enhancement can be quantitated. Also, the use of animals such as rats with keratinized buccal epithelium versus dogs with nonkeratinized buccal epithelium could lead to drastically different bioavailability estimations and conclusions.

C. Thyrotropin Releasing Hormone (TRH)

Buccal absorption of TRH, a tripeptide (L-pyroglutamyl-L-histidyl-L-proline amide) with a molecular weight of 362, was studied in vitro in rabbit buccal mucosa (9) and in vivo in humans (60). The in vitro study using rabbit buccal tissue established TRH permeation pathway to be paracellular based on autoradiography grain distribution in the extracellular space, when the experiments were performed using ^{3}H-TRH. This is in agreement with the earlier results from horseradish peroxidase tracer studies (6,7). It was also shown that the steady-state permeability of TRH was lower under hypotonic and higher under hypertonic conditions relative to its permeability in isotonic conditions. It was suggested that under hypotonic conditions, using a nontransportable solute such as mannitol, the net water flux into the cells leads to close cell packing and smaller intercellular opening and therefore decreased permeability, whereas there is increased permeability under hypertonic conditions due to wider intercellular opening resulting from the net water flux out of the cells. TRH was shown to undergo tissue metabolism, with the rate and extent of metabolism being less when the peptide was applied to the mucosal side as compared to the application on the serosal side.

Tissue viability was monitored in this study by measuring tissue ATP levels and by electron microscopic (EM) examination (9). The tissue ATP levels were shown to decrease ~30 percent in the first hour and ~40 percent after 6 hours. However, the threshold tissue ATP level required for normal tissue function is not clearly known. More meaningful information regarding tissue integrity was obtained from EM and transport data. It was shown that freshly excised mucosa

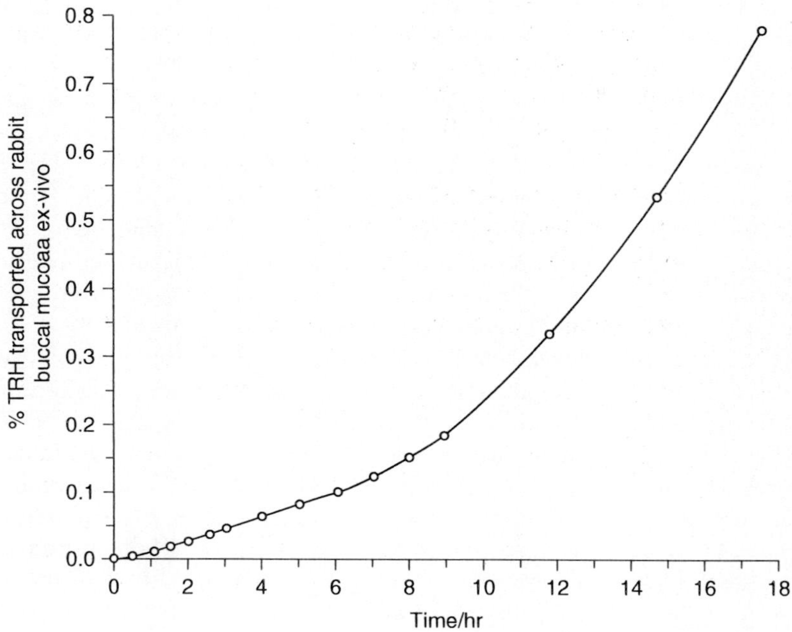

Figure 4.1 In vitro buccal TRH permeability as a function of time (reproduced with permision from Reference 9).

could not be distinguished from the 3-hour tissue, and the 6-hour tissue showed large intercellular vacuoles or cell–cell separation in the lower epithelial cell layers. However, the superficial and the intermediate epithelial cell layers were shown to be similar in freshly excised 3-hour and 6-hour tissue, suggesting that the barrier layer was intact up to 6 hours. TRH transport data showed a steady-state permeability from 2 to 8 hours, and then a progressively increasing permeability, indicating changes to the permeability barrier, as seen in Figure 4.1.

It is not clear if the change to the rate-limiting barrier is due to cell death per se or if it is related to the prolonged hydration such as that occurring in the side-by-side diffusion cell. Tissue permeability was shown to increase drastically upon prolonged hydration in the nonkeratinized tissue, and the keratinized tissue permeability was increased only slightly (5,7). This difference could be attributed to the difference in the intercellular lipids that constitute the barrier to passive diffusion. Thus, the major lipid components in the nonkeratinized epithelia are the glycolipids, which are of the polar swellable type, whereas the major components of the keratinized epithelia (i.e., the ceramides) are of the nonpolar nonswellable type (61). It should be interesting to perform studies in a Franz-type diffusion cell wherein the formulation in the form of a gel or a mucoadhesive patch could be applied on the mucosa for a fixed period of time and then removed to expose the mucosal side to air. Nevertheless, the 8-hour tissue viability could be utilized for in vitro studies to perform transport studies that reflect the buccal dosage administration time in vivo.

The permeability of TRH was shown to be independent of pH. This was explained in terms of the high solubility of TRH in both charged and uncharged states. Thus, the pH partition hypothesis relating to lipophilicity of ionizable solutes diffusing through biological membranes may not be applicable to the diffusion of TRH and other hydrophilic peptides in the buccal epithelium (62). It was shown that hydrogen bonding was more important than lipophilicity in peptide absorption across Caco-2 cells (63). It is not clear if this is also true in the case of stratified squamous epithelia such as the buccal epithelium.

TRH buccal absorption in humans from mucoadhesive patches was reported by Merkle et al. (60,64). Using both high- and low-viscosity grade hydroxyethyl cellulose, it was shown that thyrotropin and prolactin stimulation was independent of polymer viscosity, although the fraction of the peptide drug remaining after 30 min application was much lower, 0 to 15 percent, for the low-viscosity grade polymer compared to 0 to 50 percent for the high-viscosity grade polymer. Intravenous and nasal applications were more efficient than buccal administration, however, the drastic side effects such as nausea associated with the former two routes were absent when the peptide was administered by the buccal route. This suggested a moderate pituitary stimulation kinetics of buccal TRH compared to the intravenous and the nasal route.

Effects of penetration enhancers, sodium taurodihydrofusidate (STDHF), didecanoyl phosphatidyl choline (DDPC), and lysophosphatidylcholine (LPC) on in vitro transport of TRH across porcine buccal mucosa have been reported (65). A permeability coefficient of 2.03×10^{-7} cm sec^{-1} was reported for porcine buccal mucosa compared to a value of 1.3 to 1.5×10^{-7} cm sec^{-1} reported for rabbit buccal mucosa (9). The porcine tissue was shown to be viable for up to 8 hours as monitored by tissue ATP levels, EM, and electrical resistance measurements. Enhancement ratios of 2.4, 3.3, and 7.1 were reported in the presence of 0.5 percent w/v of LPC, DDPC, and STDHF, respectively.

D. LHRH

Buccal delivery of the nonapeptide buserelin, an LHRH analogue, was examined in a crossover study with intravenous bolus injection using six pigs (4). Buserelin was administered using a Hilltop chamber that was attached to the buccal mucosa using Orahesive® (Squibb). The peptide was applied as a 50 mg/ml solution in 0.9 percent saline with and without 10 mM glycodeoxycholate (GDC). Blood samples were collected up to 8 hours for a 4-hour application on the mucosa. The plasma concentration–time profile after buccal administration with and without GDC was reported by Hoogstraate (4). The elimination half-life of buserelin after cessation of the buccal administration at 4 hours was ~100 min, similar to that observed after intravenous injection (4). A depot effect observed earlier for peptides and other drugs (52,54,59,66) was not observed in this study. The coadministration of GDC at 10 mM concentration resulted in an enhanced buccal absorption with a bioavailability of 5.3 percent compared to 1 percent bioavailability from solutions without GDC (4).

Buccal delivery of LHRH from a two-layered mucoadhesive device was reported recently (67). The mucoadhesive consisted of a fast-release layer made of lower molecular weight PVP (K-30), and a sustained-release layer made of Carbopol 934 and PVP (K-90). Permeation studies using freshly excised pig buccal mucosa showed permeation enhancement with increased content of sodium cholate, reaching a maximum at 20 mg of sodium cholate in the fast-release layer (68). The authors claim that the formulation of the mucoadhesive device can be varied to achieve a desired rate of transmucosal penetration of LHRH. This claim remains to be confirmed in in-vivo studies.

E. Other Polypeptides

Buccal absorption of salmon calcitonin, a polypeptide composed of 32 amino acids, with a molecular weight of 3600, was reported in anesthetized mongrel dogs (68). Calcitonin plasma level profiles were shown to reach greater than the therapeutic levels of 0.1 to 0.4 ng/ml from both solution and buccal tablet formulations, although no details of the formulations were given in this study. Buccal absorption of human calcitonin was studied in anesthetized rats (69). The decrease in plasma calcium level 0 to 4 hours after dosing HCT solutions with different additives was determined from the area under the plasma calcium level versus time curves. The decrease in plasma calcium levels at pH 7.4 was greater than that at pH 3.6, but there was no significant difference between pH 7.4 and pH 11.3. Additives such as sodium deoxycholate, taurocholate, quillajasaponin, sodium lauryl sulfate, and sugar esters showed absorption enhancement, with the bile salts being the most effective. This is in contrast to the observation with octreotide, an octapeptide, where the enhancement of buccal absorption in rats was modest (69).

The permeation of a novel angiotensin converting enzyme (ACE) inhibitor, CGS 16617, across buccal and colonic mucosa in vitro was evaluated using rabbit, pig, and dog tissue (70). The compound was structurally similar to a tripeptide. Transport studies were carried out in Ussing chambers for up to five hours and the integrity of the epithelial barrier was monitored by tissue electrical resistance measurement as well as by histological examination. No significant histological changes were reported over the time course of the experiments. This is in contrast to the changes reported in the rabbit buccal studies of TRH transport (9), wherein vacuole formation in the lower epithelial layers was reported within six hours. In the latter study, the tissue was examined under an electron microscope and therefore could have picked up cell–cell separation at the ultrastructural level that is not visible in the histological sections used by Quadros et al. (70).

It was noted that the buccal tissue of all three species (i.e., the rabbit, the pig, and the dog) were nonkeratinized, and that the buccal epithelium of the pig was considerably thicker than that of the rabbit and the dog. The flux of CGS 16617 was inversely related to the thickness of the epithelial layer, with the highest flux being observed in the dog, followed by the rabbit, and then the pig. A thicker passive membrane would only increase the lag time and should not affect the steady-state flux. Thus, the difference in the permeability could be attributed to differences in the metabolic rates, which would depend on the thickness of the tissues. Unfortunately, this aspect of the barrier was not addressed (69).

Induction of labor by buccal delivery of oxytocin has been reported in humans (71–73). Buccal demoxytocin has been shown to be as effective as an oxytocin drip for induction of labor (72). Although the buccal oxytocin is not used in the United States, the nonapeptide with a molecular weight of 1007 has been administered via the buccal route under clinical settings for ~30 years in Europe and Asia (71–73). The in vitro release and permeation of oxytocin from a buccal patch was studied by Li et al. (29). The apparent permeability coefficient and diffusion coefficient of oxytocin in rabbit buccal mucosa was found to be 1.94×10^{-7} cm sec^{-1} and 9.2×10^{-8} cm sec^{-1}, respectively. Desglycinamide-arginine-vasopressin (DGAVP), a decapeptide, has been shown to permeate buccal mucosa (74). Using freshly excised porcine buccal mucosa, the authors measured a permeability of 1.1×10^{-8} cm sec^{-1} and 3.5×10^{-8} cm sec^{-1}, without and with 0.067 M EDTA as the permeation enhancer. EDTA has been shown to be an ineffective buccal permeation enhancer in some of the earlier studies (24).

Thus, a wide variety of peptides of different molecular weight range has been studied for buccal delivery. The permeability coefficient of these peptides in different animal species are summarized in Table 4.4. The general permeability of peptides is in the range of 10^{-7} to 10^{-8} cm sec^{-1}. Based on the permeability coefficient and the pharmacological half-life, steady-state plasma levels have been estimated for TRH, DDAVP, and insulin (75). Based on such estimations, the authors concluded that DDAVP with a relatively long plasma half-life may be better suited for buccal delivery than insulin or TRH. Although such estimations are a useful exercise before embarking upon transport studies, the significant progress reported in the literature, as reviewed in this and other chapters in this

Table 4.4 Buccal Permeability of Peptide-Based Drugs In Vitro

Peptide	Species	P (cm sec^{-1})	Ref.
TRH	Rabbit	1.5×10^{-7}	9
	Pig	2.03×10^{-7}	
ACE inhibitor	Rabbit	6.02×10^{-8}	69
	Pig	2.74×10^{-8}	
	Dog	1.0×10^{-7}	
DGAVP	Pig	$1.1–3.5 \times 10^{-8}$	73
LHRH	Pig	$0.92–5.2 \times 10^{-7}$	4

book, in terms of penetration enhancement should also be considered before eliminating any potential candidate for buccal delivery.

V. SUMMARY

Buccal delivery of peptides, polypeptides, and proteins requires coadministration of adjuvants to increase the permeation of these hydrophilic molecules. The search for penetration enhancers is an active area of research (refer to Chapter 3 in this book). The ability to increase peptide absorption without damage to the mucosa is crucial for the success of buccal delivery technology. The buccal mucosa is a robust tissue with a fast turnover and devoid of Langerhans cells, which makes it less prone to sensitization. The recovery of the permeability barrier after exposure to permeation enhancers was shown to occur within five to eight hours in vivo in dogs (76). Although only a few peptide-based therapeutics have been studied for buccal delivery (cf. Table 4.1), the list is bound to grow with the discovery of more potent molecules. In addition to providing a readily accessible site for application of the mucoadhesive system, the buccal route may provide an advantage in terms of pharmacodynamic response for potent peptide-based drugs. Because the pharmacodynamic response is determined by rate and extent of absorption, slower buccal absorption may avoid some of the side effects associated with intravenous administration, as shown for TRH.

REFERENCES

1. Harris D and Robinson JR. Drug delivery via the mucus membranes of the oral cavity. J Pharm Sci 1992; 81:1–10.
2. Merkle HP and Wolany GJM. Intraoral peptide absorption. In: Biological Barriers to Protein Delivery, Audus KL and Raub TJ, eds., Plenum, New York, 1993, pp. 131–160.
3. Merkle HP and Wolany G. Buccal delivery for peptide drugs. J Control Rel 1992; 21:155–164.
4. Hoogstraate J. Buccal Drug Delivery. Kinetics and Histology of Transport Enhancement *in vitro* and *in vivo*. Ph.D. Thesis. Leiden University, The Netherlands, 1995.
5. Squier CA. The permeability of keratinized and non-keratinized oral epithelium to horseradish peroxidase. J Ultrastruct Res 1973; 43:160–177.
6. Squier CA and Hall BK. The permeability of mammalian nonkeratinized oral epithelia to horseradish peroxidase applied *in vivo* and *in vitro*. Arch Oral Biol 1984; 29:45–50.
7. Squier CA and Hall BK. The permeability of skin and oral mucosa to water and horseradish peroxidase as related to the thickness of the permeability barrier. J Invest Dermatol 1985; 84:176–179.
8. Hill MW and Squier CA. The permeability of rat palatal mucosa maintained in organ culture. J Anat 1979; 128:169–178.
9. Dowty ME, Knuth KE, Irons BK, and Robinson JR. Transport of thyrotropin releasing hormone in rabbit buccal mucosa *in vitro*. Pharm Res 1992; 9:1113–1122.

10. Squier CA and Hall BK. In vitro permeability of porcine oral mucosa after epithelial separation, stripping, and hydration. Arch Oral Biol 1985; 30:485–491.
11. Squier CA and Hill MW. Oral mucosa. In: Oral Histology, Development, Structure, and Function. A.R. Ten Cate, ed., C.V. Mosby, St. Louis, MO, 1989, p. 341.
12. de Vries ME, Bodde HE, Verhoef HC, et al. Localization of the permeability barrier inside porcine buccal mucosa: A combined *in vitro* study of drug permeability, electrical resistance and tissue morphology. Int J Pharm 1991; 76:25–35.
13. Lesch CA, Squier CA, Cruchley A, et al. The permability of human oral mucosa and skin to water. J Dental Res 1989; 68:1345–1349.
14. Siegel IA, Izutsu KT, and Watson E. Mechanism of non-electrolyte penetration across dog and rabbit oral mucosa *in vitro*. Arch Oral Biol 1981; 26:357–361.
15. Siegel IA. Permeability of the rat oral mucosa to organic solutes measured *in vivo*. Arch Oral Biol 1984; 29:13–16.
16. Squier CA and Lesch CA. Penetration pathways of different compounds through epidermis and oral epithelia. J Oral Pathol 1988; 17:512–516.
17. Dowty ME, Knuth KE, and Robinson JR. Enzyme characterization studies on the rate-limiting barrier in rabbit buccal mucosa. Int J Pharm 1992; 88:293–302.
18. Hoogstraate AJ, Bodde HE, Cullander C, and Junginger JE. Diffusion rates and transport pathways of FITC-labelled model compounds through buccal epithelium. Pharm Res 1994; 11:83–89.
19. Lavker RM. Membrane coating granules: The fate of discharged lamellae. J Ultrastruct Res 1976; 55:79–86.
20. Romanowski AW, Squier CA, and Lesch CA. Permeability of rodent junctional epithelium to exogenous protein. J Periodont Res 1989; 23:81.
21. Squier CA. Effect of enzyme digestion on the permeability barrier in keratinizing and non-keratinizing epithelia. Br J Dermatol 1984; 111:253–264.
22. Wertz PW and Squier CA. Cellular and molecular basis of barrier function in oral epithelium. Crit Rev Ther Drug Carrier Syst 1991; 8:237–269.
23. Ho NFH, Barsuhn CL, Burton PS, and Merkle HP. Routes of delivery: Case studies. (3) Mechanistic insights to buccal delivery of proteinaceous substances. Adv Drug Del Rev 1992; 8:197–236.
24. Aungst BJ and Rogers NJ. Comparison of the effect of various transmucosal absorption promoters on buccal insulin delivery. Int J Pharm 1989; 53:227–235.
25. Dodda-Kashi S and Lee VHL. Enkephalin hydrolysis in homogenates of various absorptive mucosae of the albino rabbit: Similarities in rates and involvement of aminopeptidases. Life Sci 1986; 38:2019.
26. Garren KW and Repta AJ. Buccal drug absorption. I. Comparative levels of esterase and peptidase activities in rat and hamster buccal and intestinal homogenates. Int J Pharm 1988; 48:189–194.
27. Stratford RE and Lee VHL. Aminopeptidase activity in homogenates of various absorptive mucosae in the albino rabbit: Implications in peptide delivery. Int J Pharm 1986; 30:73–82.
28. Lee VHL and Yamamoto A. Penetration and enzymatic barriers to peptide and protein absorption. Adv Drug Del Rev 1990; 4:171–207.
29. Li C, Bhat PP, and Johnston TP. In vitro release and permeation of oxytocin from a mucoadhesive buccal patch. Pharm Dev Tech 1996; 1(4):357–364.
30. Garren KW, Topp EM, and Repta AJ. Buccal absorption. III. Simultaneous diffusion and metabolism of an aminopeptidase substrate in the hamster cheek pouch. Pharm Res 1989; 6:966.

31. Ho NFH. Biophysical kinetic modeling of buccal absorption. Adv Drug Del Rev 1993; 12:61–97.
32. Lee VHL. Protease inhibitors and penetration enhancers as approaches to modify peptide absorption. J Control Rel 1990; 13:213–223.
33. Kaelble DD, Ed. Physical Chemistry of Adhesion. Wiley Interscience, New York, 1971.
34. Alur HH, Beal JD, Pather SI, Mitra AK, and Johnston TP. Evaluation of a novel, natural oligosaccharide gum as a sustained-release and mucoadhesive component of calcitonin buccal tablets. J Pharm Sci 1999; 12:1313–1319.
35. Huntsberger JR. Surface energy, wetting and adhesion. J Adhesion 1981; 12:3–12.
36. Gurny R, Meyer JM, and Peppas NA. Bioadhesive intraoral release systems: design, testing and analysis. Biomaterials 1984; 5:336–340.
37. Lehr CM, Bowstra JA. Schacht EH, and Junginger HE. In vitro evaluation of mucoadhesive properties of chitosan, and some other natural polymers. Int J Pharm 1992; 78:43–48.
38. Chen JL and Cyr CY. Composition producing adhesion through hydration, In. Adhesion in Biological Systems, RS Manly, ed., Academic, New York, 1970, pp. 164–181.
39. Kammer HW. Adhesion between polymers. Acta Polymerica 1983; 34:112–118.
40. Bodde HE, de Vries ME, and Junginger HE. Mucoadhesive polymers for the buccal delivery of peptides. Structure-adhesiveness relationships. J Control Rel 1990; 13:225–231.
41. Lehr CM, Bouwstra JA, Bodde HE, and Junginger HE. A surface energy analysis of mucoadhesion: contact angle measurements on polycarbophil and pig intestinal mucosa in physiologically relevant fluids. Pharm Res 1992; 9:70–75.
42. Park K and Robinson JR. Bioadhesive polymers as platforms for controlled drug delivery: method to study bioadhesion. Int J Pharm 1984; 19:107–127.
43. Gu JM, Robinson JR, and Leung SHS. Binding of acrylic polymers to mucins/epithelial surfaces, structure property relationships. CRC Crit Rev Ther Drug Carrier Syst 1988; 5(1):21–67.
44. Peppas NA and Buri PA. Surface, interfacial and molecular aspects of polymer bioadhesion on soft tissue. J Control Rel 1985; 2:257–275.
45. Anders R and Merkle HP. Evaluation of laminated mucoadhesive patches for buccal drug delivery. Int J Pharm 1989; 49:231–240.
46. Merkle HP, Anders R, and Wermerskirchen A. Mucoadhesive buccal patches for peptide delivery. In: Bioadhesive Drug Delivery Systems, Lenaerts V and Gurny R, eds., CRC Press, Boca Raton, FL, 1990, pp. 105–136.
47. Ishida M, Machida Y, Nambu N, and Nagai T. New mucosal dosage form of insulin. Chem Pharm Bull 1981; 29:810–816.
48. Aungst BJ and Rogers NJ. Site-dependence of absorption promoting actions of Laureth-9, Na-salicylate, Na_2-EDTA, and aprotinin on rectal, nasal, and buccal insulin delivery. Pharm Res 1988; 5:305–308.
49. Aungst BJ, Rogers NJ, and Shefter E. Comparison of nasal, rectal, buccal, sublingual and intramuscular insulin efficacy and the effects of a bile salt absorption promoter. J Pharmacol Exp Ther 1988; 244:23–27.
50. Morimoto K, Iwamoto T, and Morisaka K. Possible mechanisms for the enhancement of rectal absorption of hydrophilic drugs and polypeptides by aqueous polyacrylic acid gel. J Pharmacobio-Dyn 1987; 10:85–91.

51. Hori R, Komada F, and Okumura K. Pharmaceutical approach to subcutaneous dosage forms of insulin. J Pharm Sci 1983; 72:435–439.
52. Oh CK and Ritschel WA. Absorption characteristics of insulin through the buccal mucosa. Methods Find Exp Clin Pharmacol 1990; 12:275–279.
53. Oh CK and Ritschel WA. Biopharmaceutic aspects of buccal absorption of insulin. Methods Find Exp Clin Pharmacol 1990; 12:205–212.
54. Ritschel WA, Ritschel GB, Forusz A, and Kraeling M. Buccal absorption of insulin in the dog. Res Commun Chem Pathol Pharmacol 1989; 63:53–67.
55. Aungst BJ. Site-dependence and structure-effect relationships for alkylglycosides as transmucosal absorption promoter for insulin. Int J Pharm 1994; 105:219–225.
56. Hovgaard L, Mack EJ, and Kim SW. Insulin stabilization and G.I. Absorption. J Control Rel 1992; 19:99–108.
57. Al-Achi A, Greenwood R, and Walker B. Buccal administration of erythrocyte-ghosts-insulin in rats. J Control Rel 1994; 28:267–268.
58. Al-Achi A and Greenwood R. Buccal administration of human insulin in Streptozotocin-diabetic rats. Res Commun Chem Pathol Pharm 1993; 82:297–306.
59. Merkle HP and Wolany G. Buccal delivery for peptide drugs. J Control Rel 1992; 21:155–164.
60. Schurr W, Knoll B, Ziegler R, Anders R, and Merkle HP. Comparative study of intravenous, nasal, oral, and buccal TRH administration among healthy subjects. J Endocrinol Invest 1985; 8:41–44.
61. Small D. The Physical Chemistry of Lipids. Plenum, New York, 1986, p.93.
62. Ho NFH and Higuchi WI. Quantitative interpretation of *in vivo* buccal absorption of n-alkanoic acids by the physical model approach. J Pharm Sci 1971; 69:537–541.
63. Conradi RA, Hilger AR, Ho NFH, and Burton PS. The influence of peptide structure on transport across Caco-2 cells. Pharm Res 1991; 8:1453–1460.
64. Merkle HP, Anders R, Wermerskierchen A, et al. Buccal route of peptide and protein drug delivery. In: Peptide and Protein Drug Delivery, VHL Lee, ed., Marcel Dekker, New York, 1990, pp. 545–578.
65. Vyas K, Kellaway IW, Hanu AC, and Dennis A. The effects of some penetration enhancers on the *in vitro* diffusion of TRH across porcine buccal mucosa. Eur. J Pharm Sci 1994; 2:161.
66. Cassidy JP, Landzert NM, and Quadros E. Controlled buccal delivery of buprenorphine. J Control Rel 1993; 25:21–29.
67. Lee Y and Chien YW. Oral mucosa controlled delivery of LHRH by bilayer mucoadhesive polymer systems. J Control Rel 1995; 37:251–261.
68. Heiber SJ, Ebert CD, Dave SC, Smith K, Kim SW, and Mix D. In-vivo buccal delivery of calcitonin. J Control Rel 1994; 28:269–271.
69. Nakada Y, Awata N, Nakamichi C, and Sugimoto I. The effect of additives in the oral mucosal absorption of human calcitonin in rats. J Pharmacobio-Dyn 1988; 11:395–401.
70. Quadros E, Cassidy J, Gniecko K, and LeRoy S. Buccal and colonic absorption of CGS 16617, a novel ACE inhibitor. J Control Rel 1991; 19:77–86.
71. Mehta AC. Buccal and oral drugs: Induction of labour. Acta Chirurgica Hungarica 1986; 27:157–163.
72. Muller K and Osler M. Induction of labour. A comparison of intravenous, intranasal, and transbuccal oxytocin. Acta Obstet Gynecol Scand (Suppl) 1967; 9:59–62.

73. Poulsen EF. The induction and stimulation of labour with buccal desamino oxytocin and oxytocin tablets. Acta Obstet Gynecol Scand 1973; 52:303–305.
74. de Vries ME, Bodde HE, Verhoef JC, and Junginger HE. Transport of the peptide desglycinamide-arginine-vasopressin across buccal mucosa. J Control Rel 1990; 13:316.
75. Harris D and Robinson JR. Bioadhesive polymers in peptide drug delivery. Biomaterials 1990; 11:652–658.
76. Zhang J, Niu S, Ebert C, and Stanley TH. An *in vivo* dog model for studying recovery kinetics of the buccal mucosa permeation barrier after exposure to permeation enhancers: apparent evidence of effective enhancement without tissue damage. Int J Pharm 1994; 101:15–22.

5

Sublingual Transmucosal Delivery of Organic Nitrates: Treatment of Angina Pectoris

John Anthony Bauer and Jennifer L. Seifert

CONTENTS

I. INTRODUCTION

In 1879 Sir William Murrell published a seminal report demonstrating the effectiveness of lingual nitroglycerin (NTG) for patients suffering from acute chest pain (angina pectoris) (1). Because bloodletting of patients in this era was not uncommon as "therapy" for these angina symptoms, this discovery of a reliable pharmacological approach was a significant therapeutic advance. Much later, in the 1960s, sublingual NTG tablets were made commercially available in the United States. Interestingly, despite numerous advances in cardiovascular medicine, sublingual NTG remains the drug of choice for the pharmacological treatment of acute angina episodes and for the prevention of angina (2,3).

As the first commercially marketed product employing oral transmucosal delivery for systemic action, NTG sublingual tablets helped to demonstrate the feasibility of this administration route and demonstrated important issues and principles for its success. In this chapter, we review currently available oral transmucosal NTG and other related products.

II. MEDICAL RATIONALE FOR SUBLINGUAL NITROGLYCERIN

A. Pharmacological Rationale

The primary symptom of ischemic heart disease is angina pectoris. This acute sensation of pain is the result of an inadequate oxygen supply versus demand relationship and may be caused by decreased myocardial oxygen supply due to changed or reduced coronary flow or increased myocardial oxygen demand due to exertion or both (4,5). In many patients a reproducible threshold of myocardial oxygen consumption or heart "work" exists for the onset of chest pain, and is often associated with only certain levels of exertion. This condition, termed stable angina, is usually the result of narrowed or blocked atherosclerotic coronary arteries (6). In contrast, variant or unstable angina is due to local or diffuse coronary vasospasm and/or transient platelet aggregates causing transient derangement of oxygen delivery even at rest (6,7). In general, the pharmacological agents used for angina therapy either decrease myocardial oxygen demand primarily by reducing heart rate, contractility, or systemic vascular resistance, or increase myocardial oxygen availability by increased or redistributed coronary blood flow.

The primary indications for sublingual NTG products are the immediate treatment or prevention of angina symptoms. This benefit is derived from the direct relaxant action of NTG on blood vessels, leading to improved myocardial oxygen delivery and in some cases reduced demand (8,9). The vasodilating properties of NTG and other organic nitrates are quite unique, a fact that probably explains the continued prevalence of NTG therapy for 120 years. Studies using isolated blood vessels and intact cardiac imaging have demonstrated that pharmacologically relevant concentrations of NTG preferentially relax large conduit arteries and veins rather than smaller resistance vessels (10,11), which are the primary controllers of systemic blood pressure. Thus, at low doses in normotensive patients systemic blood pressure is usually only slightly reduced or unchanged.

In the coronary circulation NTG's selectivity causes increased bulk blood flow and improved blood distribution to the ischemic myocardium without causing increased demand on myocardial performance (12). This benefit is also derived from reduced cardiac preload and cardiac pressure during diastolic filling. A further advantage is that NTG provides direct vasorelaxation of stenotic vessel regions (13,14) thus enhancing blood flow past and around partially occluded regions. Prevention of platelet aggregation may also play a role in NTG efficacy, particularly in the setting of unstable angina (15). These pharmacologic actions are also valuable in the management of congestive heart failure, particularly with long-acting nitrate products (2,3) such as isosorbide dinitrate and isosorbide mononitrate.

NTG and other organic nitrates elicit their vasorelaxant effects through local metabolic conversion to nitric oxide (NO, a simple but critical molecule in vascular tone regulation) (3,16). Endogenous vascular NO is derived from endothelial cells but metabolic conversion of organic nitrate occurs primarily in vascular smooth muscle cells. Thus, nitrates can elicit relaxation even in vessels with damaged or dysfunctional endothelium (13,14). The selectivity of nitrate action is probably derived from regional distribution of the enzyme(s) responsible for NO conversion in various vascular regions (17).

Pharmacological drawbacks of sublingual NTG use are primarily dose-related, associated with its vasoactivity, and readily reversible. Severe headaches are a common problem with acute dosing, a side effect that may discourage drug use (18). This side effect is so common and reproducible that NTG now serves as a tool for the study of cluster headache mechanisms (19). The headaches are probably an index of NTG effects on cranial blood flow and are usually rapidly reversible due to NTG's short duration of action. Postural hypotension, which can be severe in volume-depleted individuals or patients taking other vasodilator medications, is also a concern particularly with high doses (18). Methemoglobinemia has been rarely reported with very high doses, but this is not likely to be a clinical concern with acute sublingual use.

B. Pharmacokinetic Rationale

In pure form NTG is a nonionized oil at room temperature with extremely low water solubility. This lipophilic nature provides for several key aspects of NTG pharmacokinetics, including rapid transfer across mucosal barriers and extensive tissue distribution. This affords very rapid absorption of NTG from sublingual sites and an onset of action in one to three minutes (see Table 5.1). NTG can be further characterized as a very high-clearance drug whose metabolic removal exceeds liver blood flow and is correlated with cardiac output (20,21). Owing to extensive metabolism in nearly every body organ including blood, the elimination half-life of NTG is roughly two minutes in humans (22).

The "pharmacologically productive" vascular metabolism to NO probably only accounts for a few percent of total NTG removal. Other pathways include

Table 5.1 Drugs and Dosage Forms Used for Oral Transmucosal Cardiovascular Therapy[a]

Drug	Brand Name (Manufacturer)	Availability	Formulation/ Excipients	Dose/ Directions	Onset of Action (min)	Duration of Action	Labeled Indications
Nitroglycerin							
Sublingual tablet	Nitrostat®[b] (Parke-Davis)	0.3, 0.4, or 0.6 mg/tab	PEG 3350, lactose, sucrose	1 tab under tongue q5 min up to 3 in 15 min	1–3	30–60 min	Prophylaxis, treatment, and management of angina pectoris
Aerosol spray	Nitrolingual® (Rhone-Poulenc Rorer)	0.4 mg/spray	CFC-11, CFC-12, caprylic/capric/ diglyceryl succinate, flavors	1–2 sprays under tongue q5 min up to 3 in 15 min	1–3	30–60 min	Acute relief of an attack or prophylaxis of angina pectoris
Buccal tablet	Nitrogard® (Forest)	1, 2, and 3 mg/tab	Hydroxypropyl methylcellulose, silicone dioxide, lactose, stearic acid, flavoring agents	Place tablet under upper lip or between cheek and gum and allow to disolve over 3–5 hours	2–5	1–3 hr	Angina prophylaxis
Isosorbide dinitrate							
Sublingual tablet	Isordil® (Wyeth-Ayerst)	2.5, 5, and 10 mg/tab	Cellulose, lactose, magnesium stearate, starch, dyes	Place tablet under tongue 15 minutes prior to activity expected to cause angina—can be used for acute attack but not DOFC	3–5	Approx 2 hr	Prevention and treatment of angina pectoris Not drug of first choice for angina relief

Table 5.1 Drugs and Dosage Forms Used for Oral Transmucosal Cardiovascular Therapy[a] (continued)

Drug	Brand Name (Manufacturer)	Availability	Formulation/ Excipients	Dose/ Directions	Onset of Action (min)	Duration of Action	Labeled Indications
Nifedipine**							
Gel-filled capsule	Procardia® (Pratt) Adalat® (Bayer)	10 and 20 mg/capsule	Glycerin, peppermint oil, PEG, soft gelatin capsule	Swallow whole, bite and swallow, or bite and hold sublingually	10	Approx 2 hr	**Not labeled for acute angina relief Management only of vasospastic and chronic stable angina pectoris

[a] Portions of this table were adapted from References 2 and 8.

[b] Other brand names exist.

the production of nitrate and nitrite ions, both inactive metabolites (23), as well as dinitrate metabolites. These dinitrates can be further metabolized to NO and are also vasorelaxants, but are roughly 10- to 100-fold less potent than NTG at the vascular level (24). Thus, in the setting of acute sublingual NTG the dinitrate metabolites probably do not contribute to efficacy. However, continuous therapy, using transdermal patches, for example, may lead to their accumulation and achievement of pharmacologically active levels.

A critical requirement of acute angina therapy includes a rapid onset of action in order to abort the chest pain. The combination of a very high potency (blood concentrations of NTG for antianginal effects are in the 20 to 400 pg/ml range) and rapid absorption from the oral mucosa makes sublingual NTG an effective approach. The sublingual route of NTG administration has the additional advantage of avoidance of first-pass hepatic metabolism, thus allowing a lower total dose required for efficacy. Although NTG's short biological half-life limits the duration of action for angina relief, it also affords easy titration of effects and rapid dissipation of most side effects.

III. ORAL TRANSMUCOSAL PRODUCTS FOR ANGINA THERAPY

NTG tablets were first employed as sublingual angina therapy, however, other nitrate and non-nitrate products are also currently available in the United States. Figure 5.1 shows chemical structures and attributes of the most commonly used

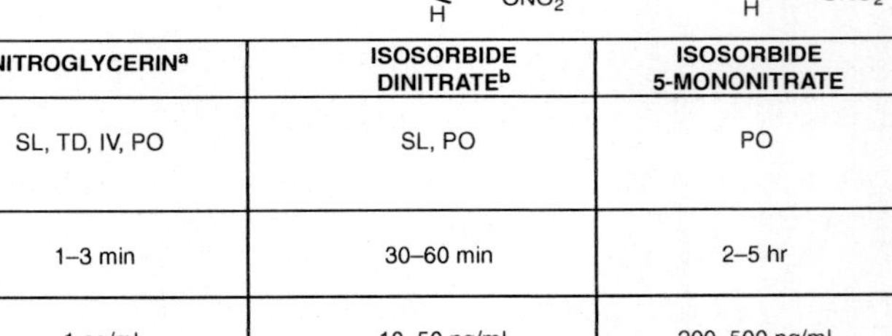

Organic nitrate	NITROGLYCERIN[a]	ISOSORBIDE DINITRATE[b]	ISOSORBIDE 5-MONONITRATE
Dosage forms available	SL, TD, IV, PO	SL, PO	PO
Elimination $T_{1/2}$	1–3 min	30–60 min	2–5 hr
Active plasma concentrations (with acute dosing)	<1 ng/ml	10–50 ng/ml	200–500 ng/ml

[a] SL, sublingual (oral transmucosal); TD, transdermal; PO, oral
[b] Pharmacologically active metabolites are produced

Figure 5.1 Organic nitrates used for anginal therapy in the United States.

organic nitrates in the United States. Additionally, descriptions of the various oral transmucosal products used for angina therapy are provided in Table 5.1 and described in detail below.

A. Organic Nitrate Products

1. Nitroglycerin

Available formulations of NTG include sublingual tablets, a lingual/sublingual spray, and buccal tablets (see Figure 5.1). As a small and highly lipophilic molecule, NTG rapidly penetrates the oral mucosa producing effects within minutes of application. These physicochemical properties combined with high potency provide for generally simple oral transmucosal formulations that differ primarily in their convenience and ease of use rather than therapeutic efficacy.

Nearly all angina patients respond to sublingual NTG therapy at adequate doses. Sublingual NTG tablets are the oldest of these products and remain so commonly prescribed that Nitrostat tablets are routinely in the top 100 U.S. prescription drugs (e.g., #86 in a 1997 review) (25). Placement of one tablet (0.15 to 0.6 mg/tablet) under the tongue usually provides relief of angina symptoms and is also useful for short-term prophylaxis prior to physical exertion. Onset of action is within 1 to 3 minutes, and the patient may feel a "tingling" sensation at the application site as evidence of local NTG vasodilation and tablet activity (26). Additional tablets (up to total of three tablets) may be employed if symptoms are not alleviated rapidly; a physician should be contacted if symptoms continue. This immediate release formulation, a simple lactose tablet with polyethylene glycol stabilizer, provides efficacy for 30 to 60 minutes (27). As such, sublingual nitroglycerin tablets are useful only for acute relief of symptoms or prophylaxis and are not convenient for sustained anti-angina protection. Because of avoidance of NTG first-pass hepatic metabolism and rapid oral mucosal penetration, bioavailability after sublingual dosing is nearly 100 percent (28).

One drawback of these tablets is related to the physicochemical properties of NTG. As an oil at room temperature, NTG is volatile and to a lesser extent unstable and has a tendency to interact with various plastics. For these reasons, NTG tablets need to be stored in tightly sealed original glass containers and should not be stored with other tablets, cotton filler, and so on. Migration and/or loss of NTG from tablets is a common limitation, but is easily overcome by proper storage and frequent replenishment of this inexpensive product.

NTG lingual spray has indications and efficacy virtually identical to sublingual tablets but has some advantages in convenience and product stability (albeit at a higher cost). One metered dose sprayed onto or under the tongue provides 0.4 mg NTG. The lingual spray may be better absorbed than sublingual tablets by patients with dry mucosal membranes (29,30). Onset of action, duration of efficacy, and side effects are not significantly different from those following sublingual tablets (29,30). When compared to the tablets, this product may be more convenient for patients with poor manual dexterity. Additionally, the sealed pressurized aerosol container also provides prevention of NTG volatile loss and provides more reliable dosing after prolonged storage.

In addition to the rapid-release sublingual products described above, NTG is also available in a tablet to be placed under the lip or in the buccal cavity. After application, these methylcellulose-based tablets adhere to the mucosa and slowly erode over 3 to 5 hours. This provides a slower NTG delivery with peak blood levels at approximately 10 to 15 minutes and a longer duration of activity than sublingual products (see Table 5.1). The vasodilatory effect persists only if the tablet remains intact and in contact with the mucosa (2). For these reasons this product is more appropriate for prevention of angina development rather than acute pain relief. The popularity of this product is not great, primarily due to the wider patient acceptance of transdermal NTG devices for prolonged delivery (i.e., 12 to 14 hours) and the utility of sublingual tablets for acute needs.

2. Other Organic Nitrates

Although NTG is the most popular treatment for angina pectoris, other organic nitrates used for the management of angina include isosorbide dinitrate (ISDN) and its major metabolite isosorbide 5-mononitrate (IS5MN). These agents have greater water solubility than NTG and improved bioavailability after oral (gastrointestinal) administration, lesser first-pass metabolic clearance, and longer elimination half-lives. The potency of these drugs is lower than NTG in vivo and in vitro, apparently due to reduced rates of vascular enzymatic NO conversion, but the general pharmacological properties are identical at active concentrations.

ISDN is formulated to be used sublingually or to be swallowed. Sublingual ISDN tablets provide a slightly slower onset and longer duration of action relative to immediate-release NTG preparations (see Table 5.1). These attributes make ISDN more appropriate for angina prevention rather than acute relief (ISDN is not the drug of first chioce for angina relief). They are consistent with a slower or less efficient penetration of the oral mucosa (less lipophilic compared to NTG), and a longer biological half-life (e.g., one to three hours) for ISDN. Bioavailability of sublingual ISDN is approximately 40 to 50 percent, probably due to slower mucosal penetration relative to NTG and swallowing a portion of the dose, which is subsequently metabolized. IS5MN is a major metabolite of ISDN and is not formulated for transmucosal delivery (Figure 5.1).

B. Organic Nitrate Tolerance

A well-established clinical problem associated with organic nitrate therapy is tolerance development to the therapeutic effects of the drug. This phenomenon is associated with a loss of anti-anginal efficacy during chronic and continuous nitrate use and is a general problem for all nitrates when administered in this manner (32,33). Interestingly, convincing evidence for the nitrate tolerance problem coincided with the advent of transdermal devices that provided stable and continuous blood concentrations. The mechanism for this pharmacodynamic tolerance has not been completely defined but probably is due to more than one factor. Currently postulated mechanisms include decreased vascular metabolism to NO due to changes in enzyme properties, increased presence of other free radicals reducing the efficacy of NO, or increased activity or sensitivity to endogenous vasoconstrictor pathways (32,33). Although recent studies suggest that co-administration of antioxidants may reduce tolerance development, the only currently established clinical approach to reduce or avoid tolerance development is the provision of a "drug-free" interval. This is achieved by 12 hour on/off patch use or "eccentric" dosing of oral nitrates (32).

The persistent use of long-acting nitrates has also been shown to decrease the hemodynamic response of acute sublingual NTG, demonstrating the phenomenon of cross-tolerance (34). Because a loss of NTG efficacy can develop during just 12 hours of continuous patch application (35), frequent or continuous use of the extended-release NTG buccal tablet or ISDN tablets may lead to reduced efficacy as well.

C. Non-Nitrate Vasodilators for Oral Transmucosal Delivery

The organic nitrates described above (NTG and ISDN) are the only currently available products designed, approved, and labeled for sublingual or buccal administration and indicated for the treatment of angina in the United States. In addition, gel-filled nifedipine capsules (Procardia®, immediate release) have been used as a sublingual dosage to treat hypertensive emergencies and to a lesser extent for acute angina episodes (36). More recently, several lines of evidence suggest that this is inappropriate therapy that should be discouraged (see below).

The pharmacological properties of nifedipine are distinctly different from organic nitrates. Unlike nitrates, nifedipine and other calcium ion channel antagonists cause direct vasodilation of large blood vessels as well as smaller arterioles, resulting in more extensive reductions in systemic vascular resistance and cardiac afterload (37). At lower plasma concentrations reduced cardiac contractility is also observed, thus anti-anginal effects are due to both reduced cardiac oxygen demand and improved oxygen availability (38).

Sublingual nifedipine dosing is achieved by biting the gel-filled capsule and holding it under the tongue or by puncturing the capsule and squeezing the contents under the tongue. The rationale for this administration has been to provide more rapid absorption and thus shorten the onset of action. No evidence is available, however, to demonstrate this advantage, and it appears that sublingual absorption of nifedipine is negligible under these conditions (39).

Recent studies have also suggested that acute nifedipine dosing may cause more harm than good in some settings (39). Owing to the potent action on resistance vessels, acute nifedipine can cause rapid and sometimes drastic hypotension. This effect can lead to a "steal" phenomenon, for example, shifting the blood away from ischemic injury or infarction (40). These complications are apparently due to rapidly fluctuating nifedipine concentrations, because they have not been associated with extended-release oral products used for the chronic management of hypertension or angina. Recent FDA warnings have discouraged the use of immediate-release nifedipine products for angina or hypertension due to reports of increased risk for cardiovascular events. For all of these reasons, the use of sublingual nifedipine for cardiovascular indications is not advised. This generally inappropriate and unsafe use of sublingual nifedipine, which is an appropriate agent in other settings, further underscores the clinical value and apparent safety of NTG products for acute angina therapy.

IV. REGULATORY ASPECTS

The Food and Drug Administration (Division of Cardio-Renal Drug Products) currently considers anti-anginal therapy as symptomatic therapy (41). Therefore, therapeutic benefit is judged by simple demonstration of improved anginal symptom relief relative to placebo treatment, and proven anti-ischemic effects. Thus, approval for the treatment of angina requires demonstration of efficacy using randomized, placebo-controlled, dose-ranging clinical trials (41). In these studies patients are typically asked to walk on a treadmill or exercise on a bicycle until they experience chest pain. Drug efficacy is demonstrated when treatment causes increased time or distance to elicit the same pain level. A statistically significant improvement in exercise duration or distance (relative to placebo) is considered a direct measure of symptom relief. Other useful endpoints to demonstrate anti-anginal efficacy include documented reduction of angina attack frequency in normal daily life. In addition to symptom relief, a new anti-anginal agent must be shown to reduce myocardial ischemia. This may be demonstrated by cardiac imaging approaches (e.g., wall motion abnormalities, regional perfusion) or electrocardiogram monitoring (41).

As mentioned above, NTG and ISDN are the primary agents specially formulated and approved for sublingual cardiovascular drug therapy. Given the long-standing acceptance, established efficacy, safety, and low cost of these existing products (especially sublingual NTG tablets), newer agents or formulations may be hard to justify. Other organic nitrate derivatives, such as pentaerythritol tetranitrate (Peritrate®) and erythrityl tetranitrate (Cardilate®) are also available, but not widely used in the United States.

One fundamental aspect of NTG therapy is a high degree of variability in both pharmacokinetics and pharmacodynamics. This variability is an important issue with respect to long-acting dosage forms, particularly regarding bioequivalence of NTG transdermal systems, and is also illustrated in wide ranges of onset and duration following sublingual dosing. This variability may be related to differences in the clinical study designs employed (i.e., angina relief during exercise challenge vs. relief at rest) (42).

VI. SUMMARY

In many ways the sublingual NTG tablet is the "gold standard" for acute anti-anginal therapy, and only marginal improvements have been made with newer and more expensive therapies. The success of this initial product demonstrated the potential value of the oral mucosal adminstration route for appropriate agents (i.e., those with high potency, low irritation, and adequate penetration). Further insight has been gained by the reasonable success of NTG lingual spray and other organic nitrate products. Thus, the development and use of these products have served an important clinical need and have provided valuable experience for the design and development of other oral transmucosal products for systemic therapies.

ACKNOWLEDGMENTS

Supported by grants from the American Diabetes Association, American Heart Association (Ohio–West Virginia Affiliates), and the National Institutes of Health (HL059791, DK55053). The technical assistance of Ms. Yuhua Ma is greatly appreciated.

REFERENCES

1. Murrell W. Nitroglycerin as a remedy for angina pectoris. Lancet 1879; 1:80,113,225,284,642–646.
2. Abrams J. The role of nitrates in coronary heart disease. Arch Intern Med 1995; 155:357–364.
3. Bauer JA, Booth BP, Fung H-L. Nitric oxide donors: Biochemical pharmacology and therapeutics. In: Ignarro L, Murad F, eds. Advances in Pharmacology Series: Nitric Oxide: Biochemistry, Molecular Biology, and Therapeutic Implications. San Diego, Academic Press 1995; 361–381.
4. Sonnenblick EH et al. Oxygen consumption of the heart: Physiologic principles and clinical implications. Mod Concepts Cardiovasc Dis 1971; 40:9–14.
5. Braunwald E. Control of myocardial oxygen consumption: Physiologic and clinical considerations Am J Cardiol 1971; 27:416–419.
6. Ross RS. Ischemic heart disease: An overview. Am J Cardiol 1975;36:496–501.
7. Conti CR. Use of nitrates in unstable angina pectoris. Am J Cardiol 1987; 60:31H–34H.
8. Thadani U. Role of nitrates in angina pectoris. Am J Cardiol 1992; 70:43B–53B.
9. Greenberg H. Effects of nitroglycerin on the major determinants of myocardial oxygen consumption: An angiographic and hemodynamic assessment. Am J Cardiol 1975; 36:426–431.
10. Winbury MM. Redistribution of left ventricular blood flow produced by nitroglycerin. Circ Res 1971; 28:I14–I147.

11. Harrison DG, Bates JN. The nitrovasodilators. New ideas about old drugs. Circulation 1993; 83:1461–1467.
12. Cohn PF, Maddox D, Holman BL, Markis JE, Adams DF, See JR, Idione J. Effect of sublingually administered nitroglycerin on regional myocardial blood flow in patients with coronary artery disease. Am J Cardiol 1977; 39:672–676.
13. Gorman MW, Sparks HV Jr. Nitroglycerin causes vasodilation within ischemic myocardium. Cardiovasc Res 1980;14:515–521.
14. Brown BG, Bolson E, Petersen RB, Pierce CD, Dodge HT. The mechanisms of nitroglycerin action; stenosis vasodilation is a major component of drug response. Circulation 1981; 64:1089–1097.
15. Loscalzo J. Antiplatelet and antithrombotic effects of organic nitrates. Am J Cardiol 1992; 70:18B–22B.
16. Moncada S, Higgs A. The l-arginine nitric oxide pathway. New Engl J Med 1993; 329:2003–2012.
17. Bauer JA, Fung H.-L. Arterial *versus* venous metabolism of nitroglycerin to nitric oxide: A possible explanation of nitrate venoselectivity. J Cardiovasc Pharmacol 1996; 28:371–375.
18. Abrams J. Therapy of angina pectoris with long-acting nitrates: Which agent and when? Can J Cardiol 1996; 12:9C–16C.
19. Nordin M, Fagius J, Waldenlind E. Sympathetic vasoconstrictor outflow to extremity muscles in cluster headache. Recordings during spontaneous and nitroglycerin-induced attacks. Headache 1997; 37:358–367.
20. Armstrong PW, Armstrong JA, Marks GS. Blood levels after sublingual nitroglycerin. Circulation 1979; 59:585–589.
21. Bogaert MG. Clinical pharmacokinetics of nitrates. Cardiovasc Drugs Ther 1994; 8:693–699.
22. Thadani U, Whitsett T. Relationship of pharmacokinetic and pharmacodynamic properties of the organic nitrates. Clin Pharmacokin 1988; 15:32–42.
23. Bauer JA, Fung H-L. Specific binding of nitroglycerin to bovine coronary artery microsomes: Evidence of a vascular nitrate binding site. Biochem Pharmacol 1996; 52:619–625.
24. Feelisch M. Biotransformation to nitric oxide of organic nitrates in comparison to other nitrovasodilators. Eur Heart J 1993; 14:I123–I132.
25. Scott-Levin Associates Audit Services. The top 200 drugs of 1997. Am Druggist, Feb. 1998:46–53.
26. Copelan HW. Burning sensation and potency of nitroglycerin sublingually. JAMA 1972; 219:176.
27. Kattus AA. Comparison of placebo, nitroglycerin, and isosorbide dinitrate for effectiveness of relief of angina and duration of action. Chest 1979; 75:17–22.
28. Jensen KM, Mikkelsen S. Studies on the bioavailability of glyceryl trinitrate after sublingual administration of spray and tablet. Arzneimittel-Forschung 1997; 47:716–718.
29. Parker JO, Vankoughett KA, Farrell B. Nitroglycerin lingual spray: Clinical efficacy and dose response relation. Am J Cardiol 1986; 57:1–5.
30. Sato, Koretsune Y, Taniguchi T, Fukui S, Shimazu T, Sugii M, Matsuyama T, Karita M, Hori M. Studies on the response of nitroglycerin oral spray compared with sublingual tablets for angina pectoris patients with dry mouth. A multicenter trial. Arzneimittel-Forschung 1997; 47:128–11.

31. Klaus AP. Comparative evaluation of sublingual long-acting nitrates. Circulation 1973; 48:519–522.
32. Fung H-L, Bauer JA. Mechanisms of nitrate tolerance. Cardiovasc Drugs Ther 1994; 8:265–275.
33. Thadani U. Nitrate tolerance, rebound, and their clinical relevance in stable angina pectoris, unstable angina, and heart failure. Cardiovasc Drugs Ther 1997; 10:735–742.
34. Thadani U, Manyari D, Parker JO, Fung H-L. Tolerance to the circulatory effects of oral isosorbide dinitrate. Rate of development and cross tolerance to glyceryl trinitrate. Circulation 1980; 61:526–535.
35. DeMots H, Glasser SP. Intermittent transdermal nitroglycerin therapy in the treatment of chronic stable angina. J Am Coll Cardiol 1989; 13:786–793.
36. Opie LH. Calcium channel antagonist in the management of anginal syndromes: Changing concepts in relation to the role of coronary vasospasm. Prog Cardiovasc Dis 1996; 4:291–314.
37. Rutherford JD. Pharmacologic management of angina and acute myocardial infarction. Am J Cardiol 1993; 72:16C–20C.
38. Ferlinz J. Nifedipine in myocardial ischemia, systemic hyertension, and other cardiovascular disorders. Ann Int Med 1986; 105:714–719.
39. Grossman E, Messerli FH, Grodzicki T, Kowey P. Should a moratorium be placed on sublingual nifedipine capsules given for hypertensive emergencies and pseudoemergencies? JAMA 1996; 276:1328–1331.
40. Cheng TO. Adverse effects of sublingual nifedipine in acute myocardial infarction. Crit Care Med 1989; 17:1364.
41. Lipicky RJ. Natural history endpoints and angina trials. In: Cardiovascular Drug Development: Protocol Design and Methodology. JS Borer, JC Somberg, eds., New York, Marcel Dekker 1999:185–189.
42. Tremblay G. Quantitation pitfalls in comparative studies of antianginal therapies. Am Heart J 1975; 89:521–524.

6

Oral Transmucosal Delivery of Melatonin

Roy L. McQuinn, Luce Bénès, and Françoise Horrière

CONTENTS

I. INTRODUCTION

A. Biosynthesis and Chronokinetics of Secretion

Melatonin {N-[2-(5-Methoxy-1H-indo-3-yl)ethyl]acetamide}or N-acetyl-5-methoxytryptamine (Figure 6.1) is a low molecular weight indolic compound that appears to be synthesized in almost all vertebrates. It is a neuroendocrine hormone with a purported wide range of biological activities including immunomodulation (1–4), reproductive regulatory activity (5–7), antitumor activity (8–11), antioxidant activity (12–14), and most notably the regulation or modulation of circadian rhythms (15).

The synthesis and secretion of melatonin occurs primarily in the pineal gland. However some synthesis of the hormone is thought to occur in other tissues such as the retina, iris-ciliary body, Harderian gland, and the lacrimal glands (15). The biosynthesis of melatonin originates with L-tryptophan and then proceeds via 5-hydroxytryptophan, serotonin, and N-acetylserotonin to the final product melatonin. Melatonin is only synthesized and secreted at night, regardless of whether the animal is nocturnal or diurnal. Thus in both nocturnal and diurnal species, melatonin concentrations are high during the night hours and low during the day (Figure 6.2). It is interesting to note that the peak of neuronal activity of the brain's master clock, the suprachiasmatic nucleus (SCN), which is also reported to be the site where melatonin receptors are densest, occurs during the daylight hours in nocturnal and diurnal species.

B. Metabolic and Pharmacokinetic Considerations

Once melatonin is synthesized, it is rapidly released from the pineal gland into the circulatory system. Thus, plasma concentrations of melatonin are a reasonably accurate reflection of the amount of melatonin being synthesized. Plasma melatonin levels typically begin to increase at approximately 2100 hours, peak between 0200 and 0400, and return to baseline at 0700 to 0900 hours. Systemic concentrations of melatonin during the secretory period, as judged by serum levels, vary greatly among individuals, ranging from as low as 15 pg/ml to as much as 100 pg/ml or more in normal healthy adult humans. However, an individual's plasma profile of melatonin is normally very consistent from day to day, with the exception that melatonin levels usually show a steady decline with advancing age (16).

Melatonin is rapidly eliminated from human blood. Its biological half-life is estimated to be approximately 45 minutes in healthy human adults (17). Melatonin appears to be eliminated from the body almost entirely by metabolic processes. It was generally thought that melatonin was primarily metabolized in

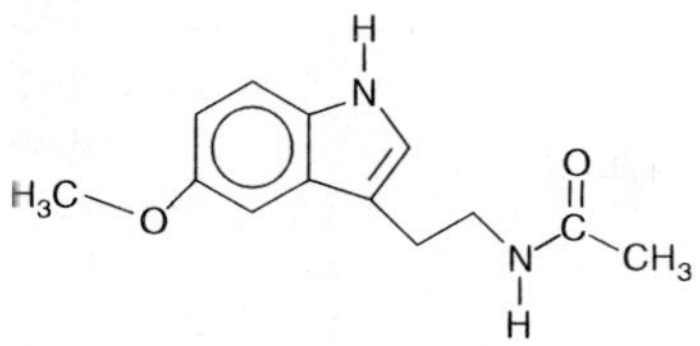

Figure 6.1 Chemical structure of melatonin.

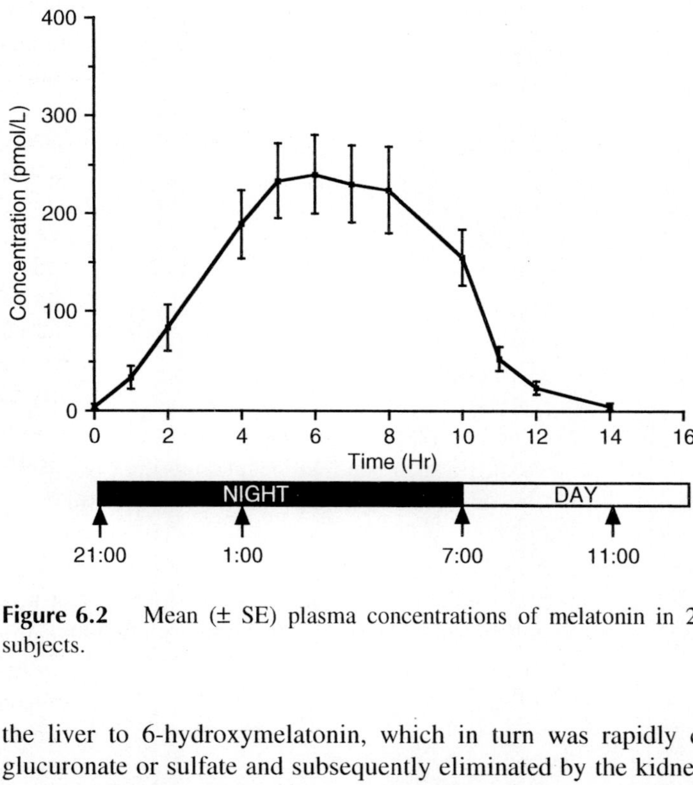

Figure 6.2 Mean (± SE) plasma concentrations of melatonin in 20 healthy human subjects.

the liver to 6-hydroxymelatonin, which in turn was rapidly conjugated as a glucuronate or sulfate and subsequently eliminated by the kidney into the urine. In fact, urinary concentrations of the sulfate conjugate of 6-hydroxymelatonin, 6-sulphatoxymelatonin have been used as a noninvasive index of pineal function (18,19). More recent findings indicate that melatonin also undergoes a substantial amount of extrahepatic metabolism leading to the formation of a large number of metabolites including such compounds as N-acetylserotonin, 5-methoxytryptamine, methoxytryptophol, and several kynurenamines, including N-acetyl-5-methoxy-kynurenamine and N-acetyl-2-formyl-5-methoxy-kynurenamine (13,15,20). Several of these metabolites have now been shown to have potent biological activities of their own (13,20).

C. Potential Therapeutic Indications and Safety Considerations

Melatonin has been widely publicized throughout the world as a completely safe, totally "natural" hormone with therapeutic activity against a wide variety of human ailments including insomnia, migraine headache, sarcoidosis, rheumatoid arthritis, and several types of cancer. In addition, it is said to blunt the effects of jet lag and possibly to slow the process of aging (12,13,15).

Melatonin is readily available throughout much of the world. In the United States and in many other countries, it is available over the counter without prescription. In Europe melatonin has now been classified as a neurohormone and can no longer be sold over the counter. It has been estimated that millions of people are now taking this hormone, almost exclusively by the oral route (21), and usually at very high doses that are likely to produce supraphysiological systemic concentrations of melatonin and of its metabolites.

Despite its widespread use, very little is known about the toxicology and safety of melatonin and its metabolites (20). No careful, well-designed studies to investigate its safe use in humans at any dose or by any route of administration have been reported. In spite of the lack of supportive evidence, it has generally been thought that the compound is safe, even at the very large doses that are often used.

Nevertheless, anecdotal reports have begun to appear in the literature that suggest melatonin may not be as safe as previously thought. Hong and colleagues reported that one patient developed autoimmune hepatitis shortly after taking 3 mg of melatonin daily for about two weeks (22). The authors suggest that there may be a causative relationship between the exogenous melatonin and the development of hepatitis in that patient. Bardazzi and colleagues reported two cases of "fixed drug eruption" of the genitalia, which they ascribed to exogenous melatonin, because rechallenge with melatonin caused a reoccurrence of the condition (23). Nagtegaal and colleagues reported on 97 patients taking exogenous melatonin for 2 to 12 months for the treatment of circadian rhythm disorders, and who spontaneously reported any adverse events after initiating therapy (24). Twenty-five of these patients reported a total of 35 adverse drug-related events including fever, hyperkinesia, dizziness, gastrointestinal disorders, headaches, hemorrhages, pigmentation, ankle edema, flushing, diplopia, hepatic pain, thrombosis, and hyperglycemia (in patients with diabetes).

Given this information and the wide range of physiological activities that appear to be regulated by nanomolar concentrations of melatonin, it seems unwise to overload membrane receptors with micromolar and millimolar concentrations of melatonin and its active metabolites inasmuch as only microgram quantities of melatonin are normally synthesized and secreted each night. The massive uncontrolled assault on the circadian clock and the endocrine system by large doses of melatonin must be contemplated for the possibility of unintended adverse consequences.

Taking into account the possible benefit of maintaining physiological levels of melatonin in the body and the unknown risks associated with large oral doses of the hormone, we sought to discover a more effective way to deliver the drug that would more closely mimic the body's natural physiological profile. For melatonin, the normal physiological profile in blood (see Figure 6.2) is best characterized as a rapid rise in concentrations upon initiation of pineal secretion, achievement of a semisteady-state (plateau) level lasting for several hours, and then a rapid fall in blood concentrations as pineal secretion stops (so-called square-wave profile). In our opinion, the best delivery device for melatonin would be one that gives a plasma level profile most closely resembling the natural square-wave profile of the hormone.

A controlled-release oral formulation of melatonin might be able to produce such a profile. However, the oral bioavailability of melatonin is reported to be low (<20 percent; presumably due to extensive first-pass metabolism), highly variable between subjects, and, in the case of controlled-release preparations, subject to variable absorption due to nutritional status (17,25,26). In addition, first-pass metabolism would likely result in decidedly nonphysiological systemic concentrations of some of the metabolites of melatonin. Transdermal administration of melatonin may be feasible in some cases. The physicochemical properties of melatonin, low molecular weight (232.27), relatively low melting point (116 to 118 °C), and a moderately high octanol/water partition coefficient (calculated log $P = 0.98$), are consistent with the properties normally associated with good candidates for transdermal delivery. Transdermal delivery devices for melatonin, which should avoid first-pass metabolism and may have the potential to more closely mimic the normal endogenous plasma melatonin profile, have previously been described by several laboratories (27–30). However, the lag time and depot effect often associated with transdermal delivery might be expected to limit the usefulness of this route of delivery for melatonin. Yates and colleagues claimed to overcome many of these limitations to transdermal delivery of melatonin by incorporating permeation-enhancing agents into their patch and by use of rate-controlling membranes (28). However, the long-term safety of many of the most effective permeation-enhancing agents is still uncertain.

Based upon our own experience with the development of controlled oral (CR), transdermal (TDD), and oral transmucosal drug delivery (TMD) systems, we concluded that TMD would most likely give the best approximation of the normal physiological profile for systemic melatonin. With TMD, first-pass metabolism is largely avoided and physiological ratios of plasma melatonin to metabolite are more likely to be obtained than with oral melatonin. Because of the much thinner layer of keratinized tissue making up the outer layer of the oral mucosa when compared with the skin, TMD melatonin could reasonably be expected to have a quicker onset of action, less potential for a depot effect, and a faster decline in plasma concentrations after patch removal than might be anticipated with TDD melatonin. For these reasons and others we sought to

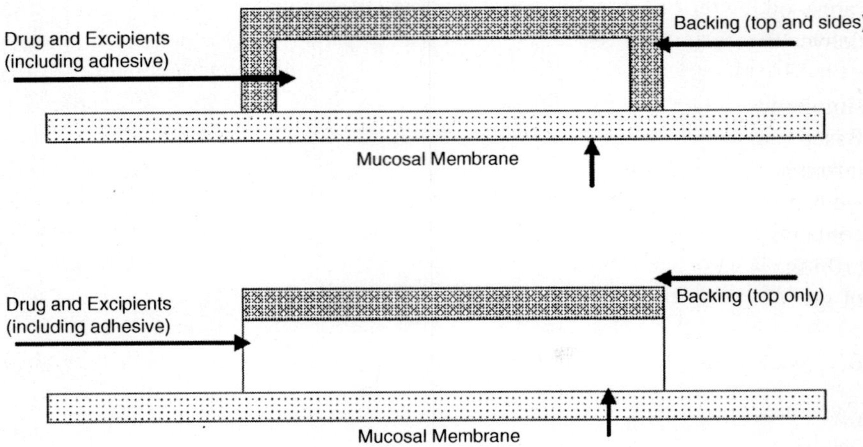

Figure 6.3 Cross-sectional view of matrix TMD system with and without protective backing extending around the edges of the patch.

develop a melatonin TMD patch utilizing TMD technology currently being developed in our laboratory.

II. TMD DEVICE DESIGN CONSIDERATIONS AND IN VITRO TESTING

A. Materials and Geometry

Our laboratory has been active in the development of TMD systems for the delivery of a variety of small and large organic compounds. To cover a wide range of compounds and their specific requirements for effective oral TMD delivery, we have employed several different patch designs. For melatonin, we chose a simple drug dispersed within a bioadhesive matrix design. The matrix design, in its most basic form, is a thin flexible disk composed of the active drug and a polymeric mucoadhesive resin dispersed within a mixture of elastomeric compounds (Figure 6.3). A preferred mucoadhesive and elastomer are polyacrylic acid (PAA) and polyisobutylene (PIB), respectively, although other mucoadhesives and elastomers have been used. In some cases a protective backing is laminated to one side of the disk and may even extend around the edges of the disk in order to shield it from saliva. This reduces erosion of the patch and limits the amount of drug that is swallowed. Without the protective backing, drug delivery from the patch can be bidirectional, allowing for both systemic and local delivery simultaneously. A typical matrix patch will have a surface area of 0.5 to 1.0 cm^2 (one side) and will be about 1 to 3 mm thick. However, both smaller and larger patches have been used and are usually well tolerated. To cover a wide range of compounds and their specific requirements for effective oral TMD delivery, we have also developed a reservoir patch.

For our initial efforts to develop a melatonin TMD system, we chose the simpler matrix patch approach (drug evenly dispersed within an adhesive matrix). Based upon the known rapid systemic clearance of melatonin and the lack of information regarding its absorption rate and absolute bioavailability from TMD, we constructed patches with surface areas ranging from about 0.5 to 1.0 cm^2 and containing 0.25 to 10 mg of melatonin. This relatively high dose of melatonin (10 mg) was chosen in order to be reasonably assured of having doses capable of giving physiologically relevant blood levels of melatonin.

B. Manufacturing Processes

Two fundamentally different manufacturing approaches were used to construct patches. In one approach (solvent-cast), all patch excipients including the melatonin were codispersed in an organic solvent and coated onto a sheet of release liner. After solvent evaporation, a thin layer of the protective backing material is laminated onto the sheet of coated release liner to form a laminate that is die-cut to form TMD patches of the desired size and geometry. With the other approach, patches are manufactured without the use of solvents (solvent-free). Drug and excipients are mechanically mixed by direct milling or by kneading, usually without the presence of any liquids. After the mixing process, the resultant material is rolled on a release liner until the desired thickness is achieved. The backing material is then laminated as previously described. While there are only minor or even no differences in patch performance between patches fabricated with the two processes, the solvent-free process is preferred because there is no possibility of residual solvents and no associated solvent-related health issues.

C. In Vitro Testing: Adhesion and Delivery

Although there are many reports describing the use of excised animal tissue and cultured cell preparations for in vitro assessment of bioadhesion and drug delivery from TMD devices (31,32), we have relied more on the use of synthetic membranes for most of our in vitro bioadhesion and drug delivery screening procedures. We have found that measurements of adhesion and drug delivery using a proprietary hydrogel film, composed of a blend of polyvinylpyrrolidone and cellulose (33), is helpful in understanding which formulation parameters most affect adhesion and drug delivery. This information can be a very useful tool for selecting the most promising formulations to assess in vivo. In vitro adhesion is assessed by measuring the adherence of a hydrated patch to metallic plate after 10 minutes of contact. The experiment is performed on an Adamel Lhomargy tensile machine (Model DY34) at a constant extension rate of 5 mm/min at 25°C and 55 percent relative humidity. This test has been found to be reliably predictive of the initial tack of the adhesive.

Using in vitro techniques, we were able to show that a common form of polyacrylic acid (PAA; Carbopol 934P; B.F. Goodrich), when formulated into a preferred matrix of elastomers, had superior bioadhesive strength compared with other known mucoadhesive materials such as hydroxypropyl methylcellulose, chitosan, and acacia (34). The initial strength of adhesion, as represented by the peel strength of the PAA patches, was approximately 3 times greater (0.021 Kg/mm) than the patches made with hydroxypropyl methylcellulose or chitosan, and at least 20 times greater than the acacia patches. These results were in agreement with the findings of other researchers and thus provided an initial basis for the selection of Carbopol 934P as the preferred bioadhesive (35).

Because several hours of adhesion to the oral mucosa would be required for an effective melatonin patch and for a number of other therapeutic agents, we studied the effects of several factors on bioadhesion, including PAA load in the patch, elastomer content, initial hydration state of the PAA, and backing properties. As expected, adhesion was highly dependent upon the amount of PAA in the patch (36). The relationship between the concentration of PAA in the patch and the maximal strength of adhesion was determined in vitro and is illustrated in Figure 6.4. The duration of adhesion was found to increase (Figure 6.5)

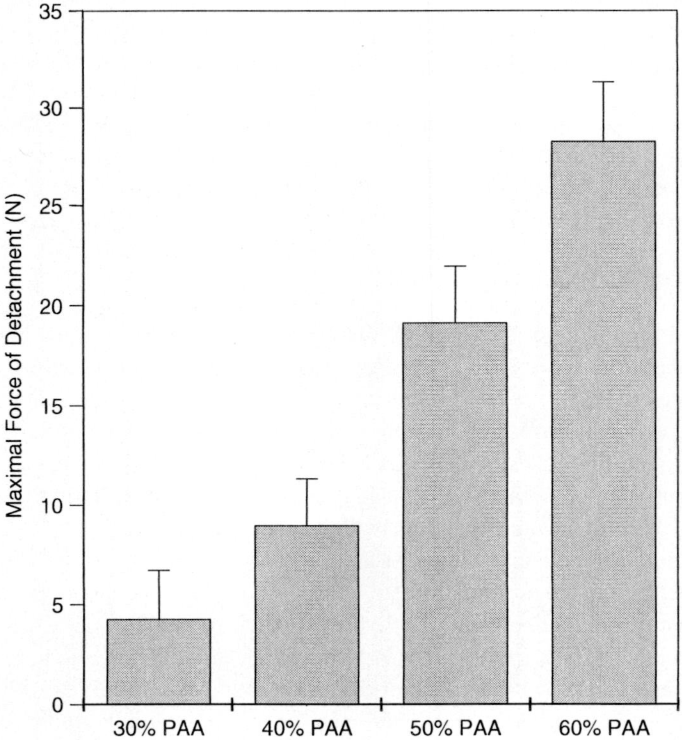

Figure 6.4 Effect of PAA concentration on the maximum detachment force (mean ± SE).

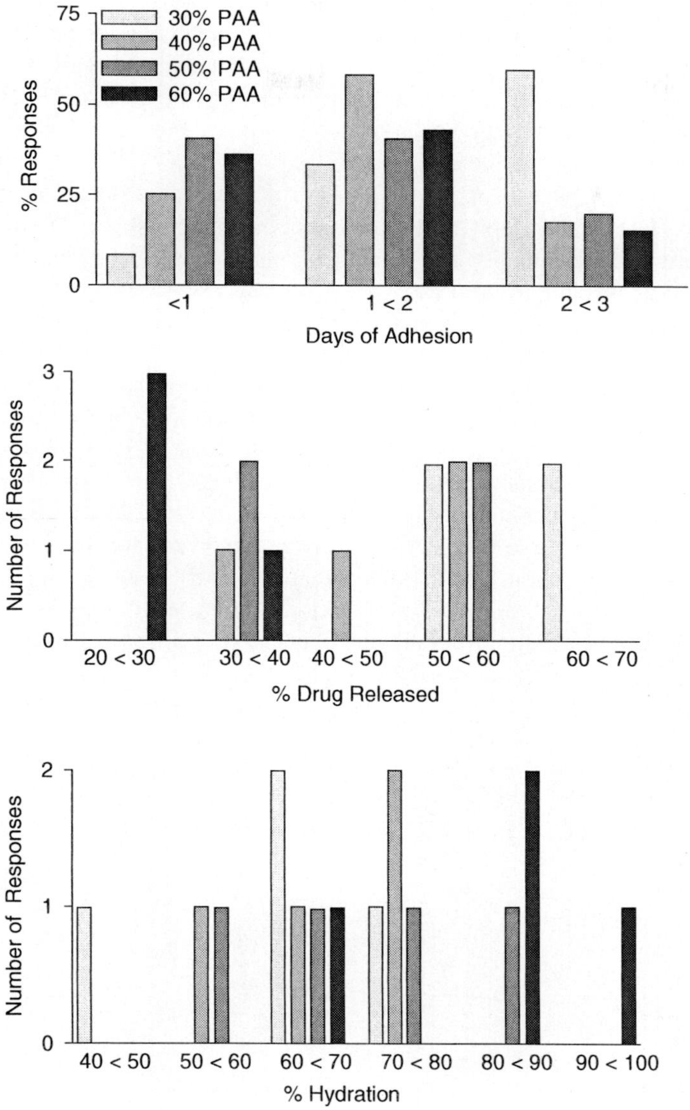

Figure 6.5 In vivo performance factors for TMD Patches as a function of total PAA content.

inversely with increasing percent PAA (w/w) in the patch (36). However, when the PAA content exceeded approximately 65 percent by weight, the amount of elastomer as a percentage of the total patch became too low to effectively maintain patch integrity. Not unexpectedly, the extent of patch swelling also increased as the percentage of PAA content in the patches increased (both patch weight and patch diameter were measured at frequent intervals after immersing the patches in phosphate buffer pH 2.6 or after adhering to a fully hydrated polyvinylpyrrolidone/cellulose membrane). This relationship is illustrated in Figure 6.6. When one side of the patch was covered by a protective backing, the rate of swelling was slower.

In general we rely primarily on in vivo data to study rate and extent of drug release from our TMD patches. However, some useful information relating to the effect on drug release by formulation and manufacturing changes can be obtained from in vitro studies. For the in vitro assessment of drug release, two different methods were used. The first method, which is a dissolution method, is described in the USP (37). The flow-through cells are immersed in a water bath and the temperature is maintained at 37 ± 0.5°C. The dosage form is placed on top of the beads and aliquots of the solution are taken periodically for drug concentration analysis. In the second method, which is a diffusion method, patches are adhered to a hydrated polyvinylpyrrolidone/cellulose hydrogel film, which is stretched across a modified Franz cell containing buffered solution at 37°C with stirring; aliquots of this solution are also taken periodically and analyzed for drug concentration. The top of the Franz cell is capped in order to ensure a humid environment around the patch and thus more closely mimic conditions in the mouth. The patch absorbs solution from the hydrogel and then releases dissolved drug through the hydrogel and into the receptor phase.

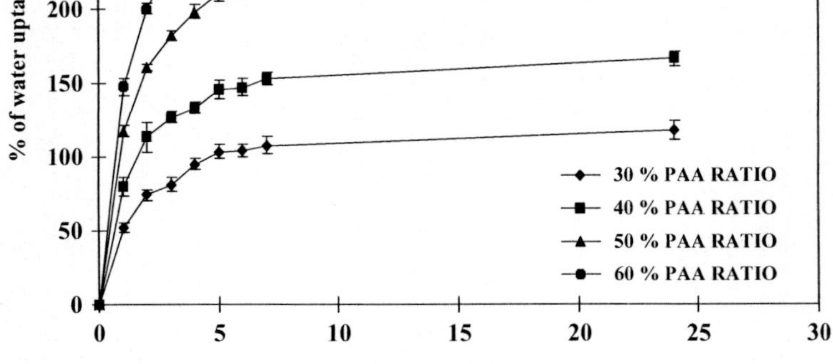

Figure 6.6 Effect of PAA concentration on patch hydration as measured by water uptake (mean ± SE).

For melatonin, in vitro release data using both the dissolution method and the diffusion method are contrasted in Figure 6.7. With patches made by the same manufacturing process, approximately 50 percent of the drug in the patches was released in about one hour and five hours when determined with the dissolution and the diffusion methods, respectively. The initial amount of drug in the patches did not seem to alter the rate of release. However, rate and extent of melatonin release from patches manufactured by the solvent-cast method and the solvent-free methods and containing 0.5 mg of melatonin indicated little or no effect of the manufacturing process on melatonin release (Figure 6.8).

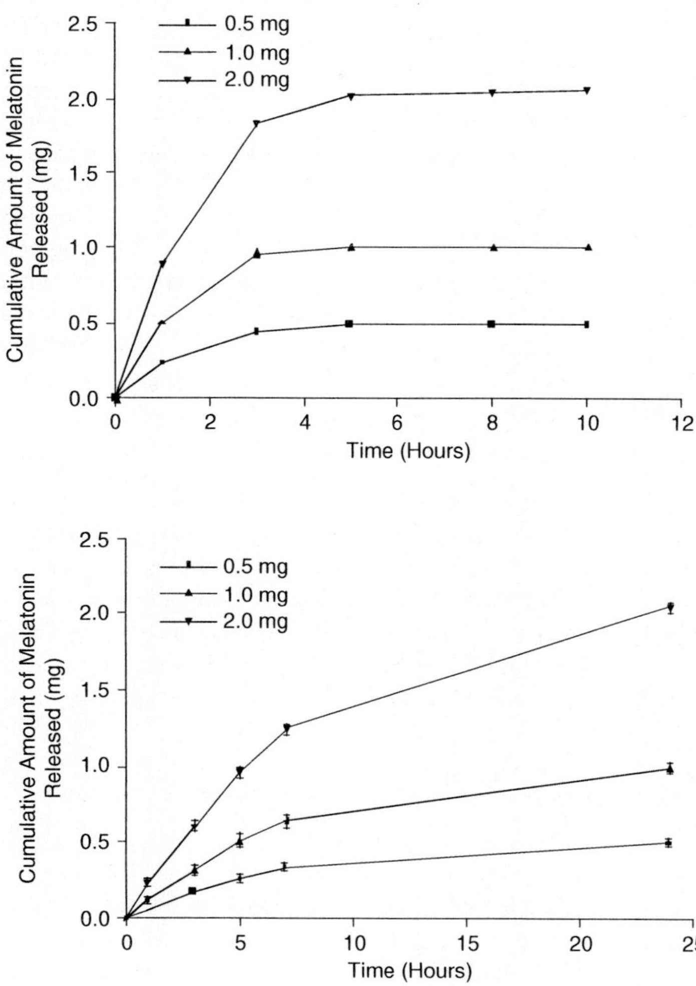

Figure 6.7 Melatonin release from TMD patches containing different amounts of drug: dissolution method (top panel) and diffusion method (bottom panel).

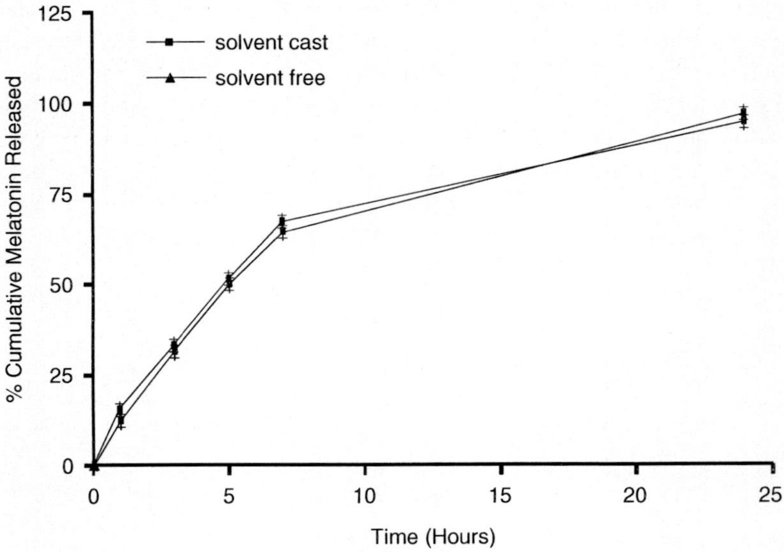

Figure 6.8 Release of melatonin from TMD patches manufactured by two different processes.

III. ANIMAL TESTING

A. Choice of Models

We have, as many other investigators, typically relied upon the beagle dog for the majority of in vivo testing of TMD systems. The beagle dog is well suited for TMD studies. Its large size, the availability of a large amount of mucosal tissue, and docile behavior facilitate frequent blood sampling and oral manipulation. Its oral mucosal tissue is generally nonkeratinized and highly lipoidal, much like that of humans. In general, we have found that TMD drug permeabilities determined in the beagle dog have usually been very indicative of results later obtained in humans.

B. Correlation between Buccal Absorption in Dogs and in Humans

As an initial assessment of the in vivo blood-level profile of exogenous melatonin when delivered via a TMD patch, we applied one patch each to the gingiva of six beagle dogs and then collected venous blood samples over the following 24 hours. Each patch contained 10 mg of melatonin (10 percent w/w) in a matrix of polyisobutylene with approximately 50 percent (w/w) of polyacrylic acid, had a surface area of 1 cm^2, had an impermeable backing protecting both the surface opposite to the contact side of the mucosa and the edges of the patch, and was manufactured using the solvent-cast process. The relatively large dose of 10 mg was chosen in order to maximize the probability of quantitating plasma melatonin concentrations during the wear period. The dogs were dosed at approximately 0800 hours in order to avoid most of the secreted melatonin during the majority of the wear period of the TMD. Patches were applied to the upper gum and removed after 10 hours of wear.

Plasma concentrations of melatonin were determined with a radioimmunoassay method (38,39). Plasma melatonin concentrations increased rapidly after patch application and achieved approximately 65 percent of peak by 3 hours postdose. Average peak plasma concentration was 11,900 pg/ml and was attained by 8 hours; the average area under the plasma level versus time curve for the 0 to 12 hour time period ($AUC_{0\text{-}12}$) was 68,020 pg*hr/ml. After the patches were removed at 10 hours postdose, plasma melatonin concentrations fell rapidly and were near baseline levels by 12 hours postdose. These patches adhered well, were well tolerated, and did not result in any obvious irritation to the dogs' mouths or gums. Qualitatively, the plasma-level profile for melatonin in these dogs with the TMD patches was very much like that reported for endogenous plasma melatonin levels in humans.

To confirm that results obtained in dogs with melatonin TMD patches were predictive of human results, we dosed six healthy human volunteers with single patches of the same design used in the dog study. As before, dosing was performed in the morning with patches applied to the upper gum just above the lateral incisor-canine juncture. Venous blood samples were collected over the next 24 hours for analysis of plasma melatonin concentrations. Mean plasma concentrations from this study as well as those from the previous dog study are depicted in Figure 6.9. There was a remarkable similarity in the plasma melatonin profiles between the two species. When corrected for the differences in dose on a mg/kg basis (the mean applied dose was 0.6 mg/kg and 0.14 mg/kg for dogs and humans, respectively), the maximum melatonin concentrations achieved in plasma and the plasma $AUCs_{0\text{-}12}$ were essentially identical for the two species.

C. Effect of Formulation Factors on Bioadhesion and Delivery

In addition to TMD drug delivery, intensity and duration of the TMD patch's adhesion as well as TMD patch tolerance was assessed in the dog. The average duration of bioadhesion of patches on the dog's gingiva was found to be approximately 30 hours. When we investigated the bioadhesion of a number of patch formulations in the dog, we discovered that in order for the formulation to adhere for a minimum of 7 hours to the gingiva, it must have an initial maximum detachment force of at least 5*N*. These studies were very useful in assisting with the process of selecting a specific formulation to be used for human testing.

Additional studies were conducted in the dog in order to assess the effects of various patch design factors, formulation changes, and manufacturing processes on drug delivery, bioadhesion, and patch tolerance. Some of the findings from these studies indicated that thinner patches delivered more melatonin,

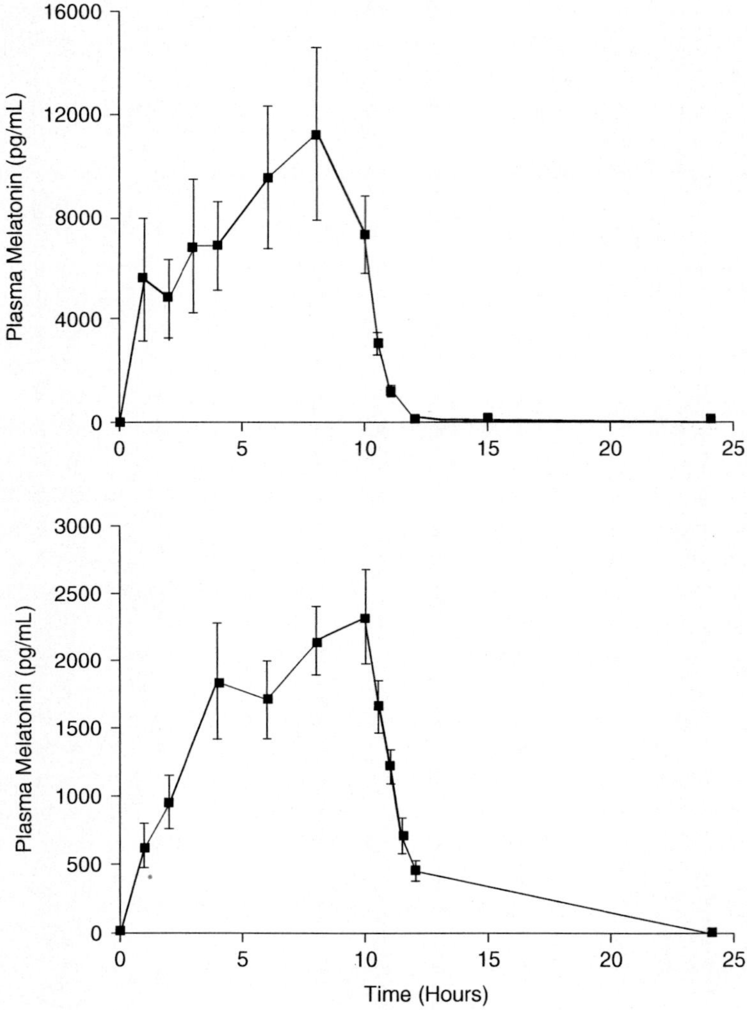

Figure 6.9 Plasma concentrations (mean ± SE) in dogs (upper panel) and human subjects (lower panel) versus time after application to the gingiva of a single TMD patch containing 10 mg of melatonin.

percentwise, than thicker ones, even though the melatonin content of the patch was a constant. They also indicated that the extent of melatonin delivery from patches of the same design was directly proportional to the melatonin concentration in the patch matrix, and that removing the protective backing from around the edges of the patches significantly accelerated the rate of melatonin delivery by allowing for extraction of the drug into saliva, and thus increasing the mucosal surface area exposed to the drug as well as allowing for some oral absorption.

D. Safety Assessment in Dogs

Because it is anticipated that use of this product in humans will entail multiple applications of the patch in individual patients, we sought to study the potential for local irritation of the gingiva with repeat applications of the patches in beagle dogs. Eighteen dogs (nine of each sex) were divided into three test groups with each group containing three males and three females. Group 1 received a single TMD patch containing 1 mg of melatonin applied daily at one of six different sites on the gum over a four-week period. Group 2 was treated in a similar fashion except that the TMD patch was a matched placebo. Group 3, a control group, received no patch. Clinical examinations and estimates of food consumption were performed daily. At the end of the study, one animal of each gender was selected from each test group and kept for a four-week recovery period. The remaining animals were humanely euthanized, and the patch application sites were removed, examined grossly, and submitted to a full microscopic inspection.

All animals (except those killed) survived the study, gained weight, and had no apparent systemic response to either the active or placebo patches. There were no apparent differences in body weight gain or food consumption between the test and control groups. No gross changes were noted in the gingiva of animals from any group. Microscopic examination of application sites from active and placebo treated dogs revealed a small degree of mononuclear cell aggregation. Mononuclear cell aggregation was not detected in samples taken from the gingiva of control animals. Overall, results from this study indicated that daily application of a TMD patch to the gums of beagle dogs over a four-week period induced slight inflammatory reactions of similar intensity and severity in animals receiving either the melatonin or the placebo patches. This finding was completely reversible as evidenced by its absence in the recovery animals. These data were supportive of further studies in humans.

IV. HUMAN TESTING

A. Initial Pharmacokinetics and Dose Selection

Our first trial in humans with a melatonin TMD patch indicated that it might be possible to produce a blood-level profile (square-wave profile) over an 8- to 12-hour time period that closely resembles the profile seen in normal healthy humans. It also indicated that onset of systemic melatonin was relatively prompt, and that melatonin plasma levels fall rapidly after removal of the patch, consistent with the absence of any depot effect in the gingival tissue. However, the magnitude of the plasma melatonin levels measured in that study with the single application of a TMD patch containing 10 mg of melatonin was severalfold higher than is normally seen with secreted melatonin. Therefore, our next study investigated the plasma-level profile of melatonin and of a principal metabolite, 6-sulphatoxymelatonin, in humans after application of TMD patches containing different doses of melatonin. The doses of melatonin were chosen based upon results from the previous study and with the intent to explore doses that were more likely to yield plasma levels of a similar magnitude to those of secreted melatonin.

Utilizing a randomized, placebo-controlled crossover study design, we investigated the relationship between TMD melatonin doses and plasma melatonin concentrations in 12 healthy human volunteers (40). All patches had a surface area of 0.5 cm^2, a thickness of 1.4 mm with an impermeable backing membrane, and contained 0, 0.5, 1, or 2 mg of melatonin. Patches were applied at 0800 hours on the gum, just above the lateral incisor–canine juncture, and were removed after 10 hours of wear. A one-week washout was allowed between dosing periods. Blood samples for determination of melatonin and the metabolite 6-sulphatoxymelatonin in plasma (38,39) were obtained over a 24-hour period.

Plasma concentrations versus time for melatonin at all doses and for 6-sulphatoxymelatonin at the 1 mg dose are depicted in Figure 6.10. As previously noted with the 10 mg TMD patch, the onset of absorption from the patches occurred quickly after dosing, and the rate of absorption was relatively rapid. Near peak plasma levels of melatonin were obtained within two to four hours and then plasma levels tended towards a gradually ascending plateau. After patch removal, plasma levels fell rapidly, characteristic of the short plasma half-life of melatonin in humans. Peak plasma melatonin levels and plasma melatonin $AUC_{0\text{-}12}$ were directly proportional to the melatonin dose. Plasma levels of 6-sulphatoxymelatonin showed the same qualitative profile as the parent drug, but were approximately two to three times greater in magnitude. This indicates that some of the melatonin in the patches may have dissolved in saliva and a portion of that swallowed. The ratio of 6-sulphatoxymelatonin to secreted melatonin is usually about one to two, whereas the same ratio of metabolite to parent drug after oral melatonin is typically on the order of ten or more.

B. Chronokinetics

To assess the potential effect of time of day of patch application on the pharmacokinetics of TMD melatonin, we studied the pharmacokinetics of melatonin in patients after nocturnal application of the patches. We administered at 2100 hours, single TMD patches containing 0, 0.25, 0.5, 1, and 2 mg of melatonin in crossover fashion to 20 healthy human volunteers. Sequential blood samples were collected over a 24-hour period for measurement of plasma melatonin and 6-sulphatoxymelatonin concentrations. Results from this study were compared with the data obtained from the previous study in which subjects were dosed with TMD melatonin at 0800 hours.

The plasma concentration versus time profiles for melatonin (Figure 6.11) from this study were very similar to those obtained from the subjects receiving comparable doses of TMD melatonin, but dosed in the daylight hours (Figure 6.10). Peak plasma melatonin concentrations, time to reach peak levels, and plasma melatonin $AUC_{0\text{-}12}$ were all very similar between the two studies at comparable doses. However, two important differences were noted.

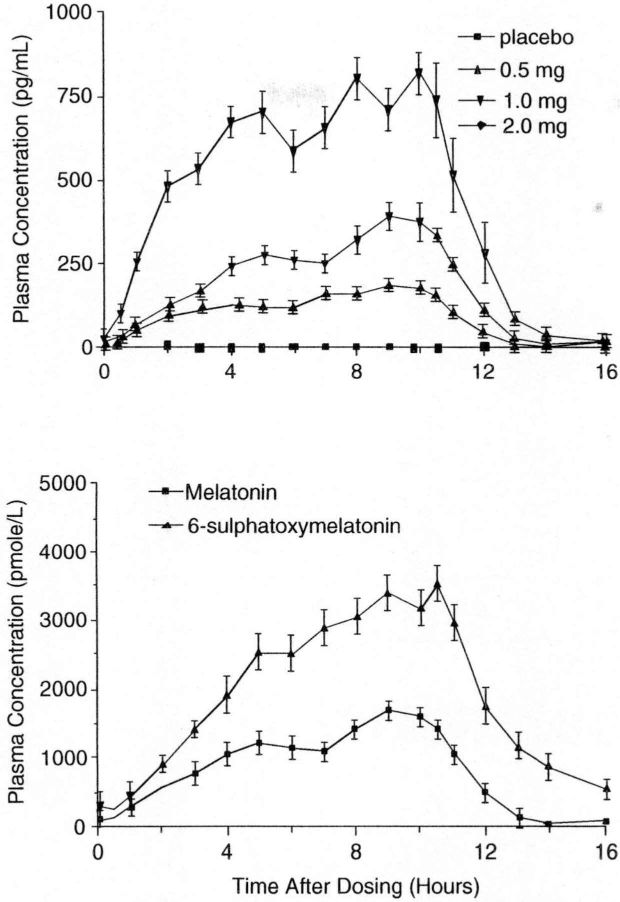

Figure 6.10 Plasma melatonin and plasma 6-sulphatoxymelatonin concentrations (mean ± SEM) in 12 healthy volunteers after application of a single TMD patch to the upper gum; patches were applied at 8:00 AM and removed at 6:00 PM.

For the nocturnally dosed subjects, the ratio of plasma 6-sulphatoxymelatonin to the parent compound melatonin was approximately one, which is essentially the same as that normally seen for secreted melatonin. For the subjects dosed at 0800 hours, the same ratio was approximately three, consistent with a small amount of melatonin being absorbed from the gastrointestinal tract. Another important difference between night and day dosed subjects was in the total amount of melatonin delivered. Measurement of residual melatonin in patches recovered from subjects after the ten-hour wear period indicated that about 50 percent of the dose was delivered from patches dosed in the daytime and about 26 percent was delivered from the nocturnally dosed patches. Also, the nocturnally applied

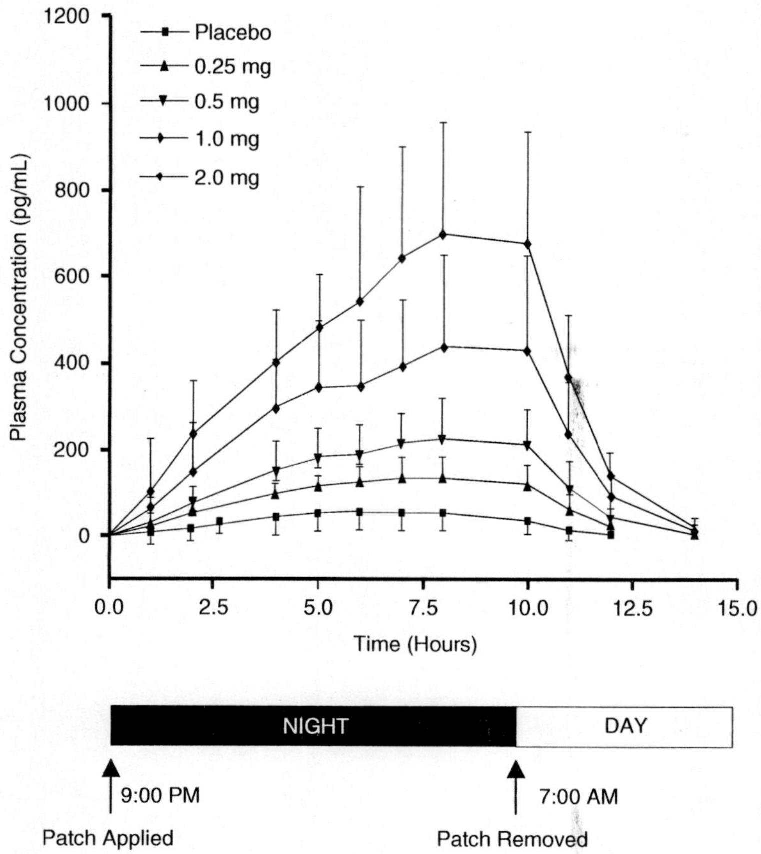

Figure 6.11 Plasma melatonin concentrations (mean ± SD) in 20 healthy volunteers after nocturnal application to the gum of a single TMD patch.

patches were less hydrated than those applied during the day. Taken together, these findings are consistent with little or no gastrointestinal absorption of melatonin from the nocturnally applied patches, but suggest that a portion of the melatonin that is delivered from patches applied during the daytime is swallowed and absorbed from the gastrointestinal tract. This would account for the difference in patch residual melatonin but lack of a difference in plasma melatonin concentrations.

C. Comparison of TMD Melatonin versus Oral Controlled Release and TDD

Although a physiological-like plasma-level profile for melatonin when delivered via TMD appeared to be possible, it might also be possible to produce a similar profile with other commonly used routes of administration. Two other drug delivery routes that looked promising to us were oral controlled-release formulations (CR) and transdermal delivery (TDD). To compare melatonin delivery from these other routes of administration, we compared plasma melatonin and 6-sulphatoxymelatonin concentrations versus time profiles in 12 healthy volunteers after administration of melatonin from each of three routes: TMD, TDD, and CR (41). This was an open label, single-dose crossover study with a one-week washout period between doses. All doses of melatonin were administered at 0800 hours and sequential blood samples for plasma melatonin and 6-sulphatoxymelatonin determination were collected over the subsequent 24 hours after dosing. Both the TMD and the TDD patches were removed after 10 hours of wear.

Mean plasma melatonin concentrations versus time profiles for each of the three routes of administration are depicted in Figure 6.12. As can be seen, only TMD gave the physiological-like square-wave profile for plasma melatonin (rapid onset after dosing, gradually ascending plateau, and a rapid offset after patch removal). Both TMD and TDD showed peak plasma 6-sulphatoxymelatonin to melatonin ratios of about one and the same ratio after CR was approximately nine. This indicates that TMD and TDD routes of administration largely avoid the first-pass metabolism that occurs with oral dosing of melatonin. Moreover, subject variability was found to be lower with the TMD route (41). Overall, the data from this study pointed to TMD as the preferred route of administration of

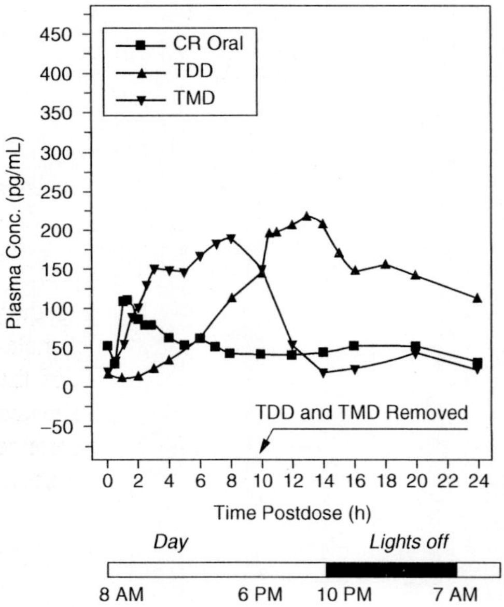

Figure 6.12 Plasma melatonin concentrations (mean ± SD) in 12 healthy volunteers after single doses by oral controlled-release (0.76 mg), TDD (8.0 mg), or TMD (0.5 mg).

melatonin in order to most closely mimic normal secreted melatonin concentrations in blood.

D. Patient Wear and Comfort

Systemic drug delivery via a TMD patch is a relatively novel concept. Patient acceptance of this new approach is by no means assured. And for an innovator seeking to develop a product using this novel drug delivery platform, very little data exists that systematically deals with the issue of patient acceptance. No matter how well the device performs as a delivery vehicle for the intended drug, it will not be widely used if patients perceive it to be irritating, excessively uncomfortable, or unreliable.

Therefore, we investigated the comfort and acceptance of the melatonin TMD patch in over 200 subjects and under a variety of conditions (i.e., with food and drink, without food, daytime wear, nighttime wear, application on the inside upper lip, application on the gum, etc.). One such study investigated the comfort and acceptance of our melatonin TMD patch in 130 patients (42). Patients wore the patches on their gums for 12-hour periods on three consecutive nights. In addition, the sequence of three consecutive applications was repeated up to five

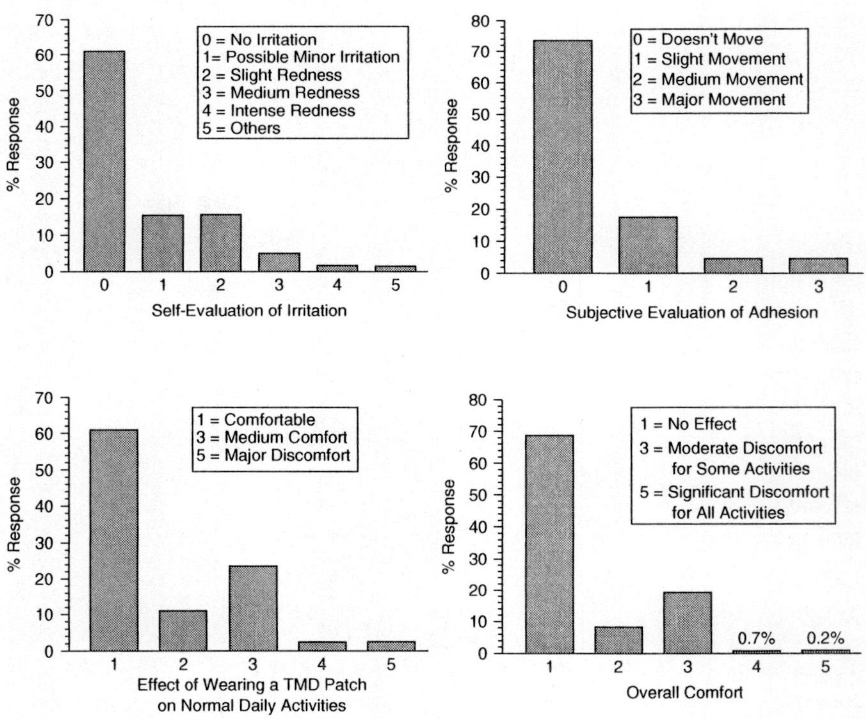

Figure 6.13 Summary of wear-panel results.

times in some patients. Comfort and acceptance were evaluated by the patient and by clinical personnel using nine different criteria, which included such things as bioadhesion, irritation, taste, overall sense of comfort by the patient, and effect of the patch on daily life or normal activities.

The results from this large trial were consistent with data obtained from several smaller previously conducted trials. Overall, the patches received high scores for comfort and bioadhesion, produced minimal irritation, and generally did not have a significant effect on the patient's daily life or normal activities (Figure 6.13). In addition, the patches were not difficult to remove, had little perceptible taste, and showed only a slight loss of patch integrity while worn in the mouth. These data suggest that a melatonin TMD patch has an excellent chance of being well accepted by patients undergoing therapy with melatonin as an alternative to conventional routes of delivery, particularly if the patches are to be worn during the nocturnal hours.

V. CONCLUSIONS

We have developed a small, thin, flexible circular disk that adheres well to the oral mucosa, is well tolerated by patients, has minimal irritation even with repeat application, and can produce a plasma-level profile of melatonin that closely resembles that seen with normal endogenously secreted melatonin. The device largely avoids absorption via the gastrointestinal tract, and thus does not result in abnormally large amounts of systemic melatonin metabolites that may have harmful or other unforeseen consequences. Melatonin delivery from the device is directly proportional to the amount of drug in the patch and follows predictable pharmacokinetics. Once the patch is removed from its application site in the mouth, blood melatonin levels begin to fall almost immediately. This gives an additional margin of safety which is not available to oral melatonin and possibly not available with TDD melatonin due to a possible depot effect with this route of administration. The device uses materials that have previously been well characterized in humans and judged to be safe. Manufacturing the patches on a commercial scale, although not a simple process, does seem possible. We have identified two manufacturing processes, one of which does not involve the use of any organic solvents, that may be amenable to scaling up to commercial processes. In summary, the melatonin TMD patch meets all the criteria of an effective, safe, and efficient delivery vehicle for melatonin and should be evaluated in clinical efficacy trials.

ACKNOWLEDGMENTS

The authors acknowledge Robert Scherrer, Sophie Burgaud, and Isabelle Seyler for careful review of this manuscript and Nicole Ducloux, Joelle Dupeyron, Pascal Jean, and Thierry Mathieu for technical assistance in conducting many of the experiments described in the manuscript.

REFERENCES

1. Liebmann PM, Wolfler A, Felsner P, Hofer D, Schauenstein K. Melatonin and the Immune System. Int. Arch. Allergy Immunol. 1997; 112:203–211.
2. Pieri C, Recchioni R, Moroni F, Marcheselli F, Marra M, Marinoni S, Primio RD. Melatonin Regulates the Respiratory Burst of Human Neutrophils and Their Depolarization. J. Pineal Res. 1998; 24:43–49.
3. Morrey KM, McLachlan JA, Serkin CD, Bakouche O. Activation of Human Monocytes by the Pineal Hormone Melatonin. J. Immunology 1994; 153:2671–2680.
4. Maestroni GJM. The Immunoneuroendocrine Role of Melatonin. J. Pineal Res. 1993; 14:1–10.
5. Hamed MY, Mostafa EM, Tour SS. Antifertility Effect of Orally Formulated Melatonin Tablets in Mice. Int. J. Pharma. 1991; 69:93–102.
6. Malpaux B, Daveau A, Maurice F, Gayrard V, Thiery JC. Sites of Action of Melatonin Involved in the Seasonal Control of Reproduction in the Ewe. In: Touitou Y, Arendt J, Pevet P, eds. Melatonin and the Pineal Gland—From Basic Science to Clinical Application. Amsterdam, The Netherlands: Elsevier, 1993:243–246.
7. Chemineau P, Maurice F, Daveau A. Re-initiation of Ovulatory Activity by Melatonin Given as a Constant-Release Implant in Long-Day Treated Ile-De-France Ewes, Depends on Endogenous Secretion of Melatonin. In: Touitou Y, Arendt J, Pevet P, eds. Melatonin and the Pineal Gland—From Basic Science to Clinical Application. Amsterdam, The Netherlands: Elsevier, 1993:247–250.
8. Gonzalez R, Sanchez A, Ferguson JA, Balmer C, Daniel C, Cohn A, Robinson WA. Melatonin Therapy of Advanced Human Malignant Melanoma. Melanoma Res. 1991; 1:237–243.
9. Grin W, Grunberger W. A Significant Correlation between Melatonin Deficiency and Endometrial Cancer. Gynecol. Obstet. Invest. 1998; 45:62–65.
10. Lissoni P, Ardizzoia A, Barni S, Brivio F, Tisi E, Rovelli F, Tancini G, Maestroni G, Fumagalli L. Efficacy and Tolerability of Cancer Neuroimmunotherapy with Subcutaneous Low-Dose Interleukin-2 and the Pineal Hormone Melatonin: A Progress Report of 200 Patients with Advanced Solid Neoplasms. Oncology Reports 1995; 2:1063–1068.
11. Shellard SA, Whelan RDH, Hill BT. Growth Inhibitory and Cytotoxic Effects of Melatonin and its Metabolites on Human Tumour Cell Lines in vitro. Br. J. Cancer 1989; 60:288–290.
12. Poeggeler B, Reiter RJ, Tan D-X, Chen L-D, Manchester LC. Melatonin, Hydroxyl Radical-Mediated Oxidative Damage, and Aging: A Hypothesis. J. Pineal Res. 1993; 14:151–168.
13. Hardeland R, Reiter RJ, Poeggeler B, Tan D-X. The Significance of the Metabolism of the Neurohormone Melatonin: Antioxidative Protection and Formation of Bioactive Substances. Neuroscience and Biobehavioral Reviews 1993; 17:347–357.
14. Gilad E, Cuzzocrea S, Zingarelli B, Salzman AL, Szabo C. Melatonin is a Scavenger of Peroxynitrite. Life Sciences 1997; 60 (10):169–174.
15. Yu H-S, Reiter RJ. Melatonin Biosynthesis, Physiological Effects, and Clinical Applications. Boca Raton, FL: CRC Press, 1993.
16. Reiter RJ. Pineal Function During Aging: Attenuation of the Melatonin Rhythm and Its Neurological Consequences. Acta. Neurobio. Exp. 1994; 54 (Suppl): 31–39.

17. Waldhauser F, Waldhauser M, Lieberman HR, Deng Mei-Hua, Lynch HF, Wurtman RJ. Bioavailability of Oral Melatonin in Humans. Neuroendocrinology 1984; 39:307–313.
18. Brown GM, Bar-Or A, Grossi D, Kashur S, Johannson E, Yie SM. Urinary 6-Sulphatoxymelatonin, an Index of Pineal Function in the Rat. J. Pineal Res. 1991; 10:141.
19. Aldous ME, Arendt J. Radioimmunoassay for 6-Sulphatoxymelatonin in Urine Using an Iodinated Tracer. Am. Clin. Biochem. 1988; 25:298.
20. Guardiola-Lemaître B. Toxicology of Melatonin. J. Biol. Rhythms 1997; 12 (6):697–706.
21. Turek FW. Melatonin Hype Hard to Swallow (Editorial). Nature 1996; 379:295–296.
22. Hong YG, Riegler JL. Is Melatonin Associated with the Development of Autoimmune Hepatitis? J. Clin. Gastroenterol. 1997; 25 (1):376–378.
23. Bardazzi F, Placucci F, Neri I, D'Antuono A, Patrizi A. Fixed Drug Eruption Due to Melatonin. Acta. Derm. Venerol. (Stockh.) 1997; 78:69–70.
24. Nagtegaal JE, Smits MG, Van Der Meer YG, Fischer-Steenvoorden MGJ. Melatonin: A Survery of Suspected Adverse Drug Reactions. Sleep-Wake Res. Netherl. 1997; 7:115–118.
25. Aldhous M, Franey C, Wright J, Arendt J. Plasma Concentrations of Melatonin in Man Following Oral Absorption of Different Preparations. Br. J. Clin. Pharmac. 1985; 19:517–521.
26. Lee B, Parrott KA, Ayres JW, Sack RL. Design and Evaluation of an Oral Controlled Release Dellivery System for Melatonin in Human Subjects. Int. J. Pharm. 1995; 124:119–127.
27. Lee B-J, Parrott KA, Ayres JW, Sack RL. Preliminary Evaluation of Transdermal Delivery of Melatonin in Human Subjects. Res. Communications in Mol. Pathol. and Pharmacol. 1994; 85 (3), 337–346.
28. Yates FE, Taskovich LT, Yum SI, Crisologo NM. Controlled Transdermal Administration of Melatonin. US Patent Number: 5,508,039; 1996.
29. Kappes JK, Hart R. Transdermal Melatonin Delivery System. US Patent Number: 5,385,736; 1995.
30. Konsil J, Parrott KA, Ayres JW. Development of a Transdermal Delivery Device for Melatonin in vitro Study. Drug Devel. and Indust. Pharm. 1995; 21 (12):1377–1387.
31. Zhang H, Robinson JR. In Vitro Methods for Measuring Permeability of the Oral Mucosa. In: Rathbone MJ, ed. Oral Mucosal Drug Delivery. New York: Marcel Dekker, 1996:85–100.
32. Audus, KL. Buccal Epithelial Cell Cultures as a Model to Study Oral Mucosal Drug Transport and Metabolism. In: Rathbone MJ, ed. Oral Mucosal Drug Delivery. New York: Marcel Dekker, 1996:101–120.
33. Ofstead RF. Polymer Blends with High Water Absorption. US Patent Number 4,771,089.
34. Guo J-H. Bioadhesive Polymer Buccal Patches for Buprenorphine Controlled Delivery: Formulation, In-Vitro Adhesion and Release Properties. Drug. Develop. Indust. Pharm. 1994; 20 (18):2809–2821.

35. Smart JD, Kellaway IW, Worthington HEC. An In Vitro Investigation of Mucosa Adhesive Materials for Use in Controlled Drug Delivery. J. Pharm. Pharmacol. 1984; 36:295–299.
36. Benes L, DeGrande G, Horrière F, Land M. Effect of Carbopol Concentration on In Vitro and In Vivo Characteristics of the 3M Buccal Delivery System (Cydot™). American Association of Pharmaceutical Scientists Meeting, Miami, FL, Nov 6–8, 1995.
37. USP 23, NF18, USP Pharmacopeia National Formulary, 1995.
38. Brun J, Claustrat B, Harthe C, Vitte PA, Cohen R, Chazot G. Melatonin RIA — Analytical and Physiological Criteria of Validity. Adv. in Biosciences. 1985; 53:41–45.
39. Harthe C, Claustrat B, Brun J, Chazot G. Direct Radio-immunoassay of 6-Sulfatoxymelatonin in Plasma with Use of an Iodinated Tracer. Clin. Chem. 1991; 37 (4):536–539.
40. Benes L, Brun J, Claustrat B, DeGrande G, Ducloux N, Geoffriau M, Horrière F, Karsenty H, Lagain D. Plasma Melatonin (M) and Sulfatoxymelatonin (aMT6s) Kinetics after Transmucosal Administration to Humans. In: Touitou Y, Arendt J, Pevet P, eds. Melatonin and the Pineal Gland—From Basic Science to Clinical Application. Amsterdam, The Netherlands: Elsevier, 1993; 347–350.
41. Benes L, Claustrat B, Horrière F, Geoffriau M, Konsil J, Parrott KA, DeGrande G, McQuinn RL, Ayres JW. Transmucosal, Oral Controlled-Release, and Transdermal Administration in Human Subjects: A Crossover Study with Melatonin. J. Pharm. Sci. (1997); 86 (10):1115–1119.
42. Benes L, Chau I, DeGrande G, Horrière F, Lacoste C. Evaluation of Patient Comfort and Acceptance of the 3M Transmucosal Delivery System. 7th International Symposium on Recent Advances in Drug Delivery Systems, Salt Lake City, UT; Feb. 28, 1995.

7

Oral Transmucosal Delivery of Nicotine: Smoking Cessation Therapy

William R. Pfister

CONTENTS

I. INTRODUCTION

Nicotine is one of the most addicting drugs of the twenty-first century and has had a long history of use and abuse. Nicotine is a natural plant-derived alkaloid and the addictive active chemical found in tobacco. Tobacco in the form of smoking cigars, cigarettes, and smokeless forms such as snuff, sachet, and chewing tobacco has been used by man as a central nervous system (CNS) stimulant for over 300 years. Introduced into Western civilization by the American Indian, tobacco use has spread around the globe. There are now over 500 million users of cigarettes worldwide and the number of users continues to increase despite the negative health consequences of smoking.[1] Smokers use the healthcare system 50 percent more than nonsmokers and generate an annual healthcare burden in the United States alone of over 100 billion dollars annually.[2] Chronic smoking leads to a number of diseases such as emphysema, lung cancer, and cardiovascular disease, which results in over 500,000 deaths annually.[3] The health risks of smoking are directly proportional to the number of cigarettes smoked per day, the number of years of cigarette smoking, and the amount of smoke inhaled, and the risk of sudden cardiac death is two to four times greater than that of nonsmokers.[4] The dangers in using smokeless tobacco products follow a close second to the long-term lethal effects of smoking tobacco and include risks associated with mouth and throat cancer, chronic mouth irritation, tooth loss, gum disease, and high blood pressure. It is now well recognized that use of tobacco products (i.e., smoking and use of smokeless tobacco) becomes an addictive habit with continued chronic use due to nicotine.[5] Nicotine, however, does not contribute to the negative health consequences of tobacco use, and therefore this advantage has been exploited over the past three decades to develop nicotine replacement therapy (NRT) products, which can deliver nicotine as an aid to break the habit of tobacco use (i.e., smoking and smokeless forms), addiction, and withdrawal.

It is now well established that the rapid onset of CNS effects including behavioral gratification from tobacco use (i.e., smoking and chewing tobacco) are attributed to nicotine. The short-term effects of tobacco use provide perceived pleasure, gratification, and reward, and continued use leads to establishment of the daily smoking "ritual." Nicotine is rapidly absorbed through skin, the tissues of the oral cavity (i.e., buccal, gingival, sublingual, lingual, and palatal mucosal), the nasal mucosa, as well as the lungs following inhalation. It is interesting, however, that nicotine is poorly absorbed following ingestion.[11] These properties have led to the exploitation of nicotine delivery by the transdermal, oral transmucosal (OT), nasal, and pulmonary routes, and introduction of numerous dosage forms for NRT.

This chapter provides a review of the development and commercialization of nicotine replacement (NR) products with a focus on OT dosage forms, and their evolution and use in smoking cessation therapy. The physicochemical and biopharmaceutical properties of nicotine are reviewed. The smoking cessation market, medical rationale for development of NR products, and various transdermal and OT nicotine dosage forms are discussed. The advantages and disadvantages of commercial NR products including nicotine transdermal patches for daily maintenance therapy are contrasted with the intraoral immediate-release nicotine gums, sprays, sublingual tablets, and lozenges, and other immediate-release dosage forms. In addition, OT nicotine products in the R&D pipeline for smoking cessation therapy are discussed. A summary of the patent domain encompassing nicotine intraoral delivery technology for smoking cessation therapy is provided. Finally, a glimpse of what the future may hold for new NR products that may eventually add to the effective armamentarium of nicotine smoking cessation regimens are discussed.

II. NICOTINE SMOKING CESSATION

A. Nicotine Physicochemical Properties

The IUPAC name for nicotine is (S)-3-(1-methyl-2-pyrrolidinyl) pyridine (CAS No. 54-11-5) and there are two enantiomeric isomers of nicotine, (R)-(+)-nicotine and (S)-(-)-nicotine. The (S)-(-)-enantiomer of nicotine is the natural active form found in tobacco products. Nicotine is a heterocyclic nitrogen containing pyridine alkaloid.[6] The physicochemical properties of nicotine base, the predominate form in NR products (i.e., transdermal systems, oral inhalers, and nasal solutions), and the pharmacokinetic parameters of nicotine in man are summarized in Table 7.1.[7]

Nicotine is also available as a stabilized complex (nicotine polacrilex USP, CAS NO. 96055-45-7) in most other intraoral NR products (i.e., chewing gum, lozenges, sublingual tablets, etc.). Nicotine present in NRT products is a highly purified extract obtained from the dried leaves of the tobacco plant (*Nicotiana tabacum*). Nicotine is also available in more stable salt forms (i.e., hydrochloride, dihydrochloride, sulfate, tartrate, bitartarate, and salicylate).[11–13]

Nicotine is poorly absorbed orally with a bioavailability (BA) of only 30 percent, however, it has physiochemical properties that result in high absorption by both skin and mucosal tissue without the need for permeation enhancers, solubilizers, or enzyme inhibitors to achieve therapeutic plasma levels (i.e., transdermal, sublingual, buccal, and nasal). Nicotine can be categorized as a highly soluble and highly permeable drug according to the Biopharmaceutical Classification System (BCS).[14] Nicotine also meets Lipinski's rule of five mnemonic as a drug ideally suited for absorption by mucosal tissue because it has high water solubility and ideal lipophilicity (i.e., octanol/water partition coefficient, log $P_{o/w}$ value of 1.1).[15]

The chemical structure of nicotine is shown in Figure 7.1. Nicotine bears structural similarity to the neurotransmitter acetylcholine, which explains its cholinergic agonist activity and stimulation of nicotinic cholinergic receptors in the central nervous system.

B. Nicotine Pharmacology

Nicotine is an agonist to one type of acetylcholine (ACh) receptor, whereas muscarine is an agonist to the other type of ACh receptor. Atropine is an antagonist

Table 7.1 Typical Properties of Nicotine

Physicochemical Data[a]	
Chemical name:	3-(1-methyl-2-pyrrolidinyl)pyridine
Empirical formula:	$C_{10}H_{14}N_2$
Molecular weight:	162.23
Description:	Colorless to pale yellow, oily liquid
Odor:	Slight fish odor when warm
Flash point (°C):	101
Boiling point (°C):	123–125
Freezing/melting point (°C):	–80
Vapor pressure (1 mm Hg/°C):	61.8
Vapor density (air =1):	5.61
Density (d_4^{20}):	1.0097
Refractive index (n_d^{20}):	1.5282
Optical rotation $[\alpha]_d^{20}$:	–169
PH:	10.2
pK (15°C):	$pK_1 = 6.16$, $pK_2 = 10.96$ (Reference 14) $pK_1 = 3.15$, $pK_2 = 7.87$ (Reference 8)
Log $P_{o/w}$:	1.2
Specific gravity/density (g/cm^3):	1.0100
Solubility in water (25°C):	Very high, miscible below 60 °C
Pharmacokinetic Data[b]	
$T_{1/2,}$ Initial (min)	2–3
$T_{1/2,}$ Terminal (hours)	2.0 (range: 1–4)
Volume of distribution (l/hr):	2-3
Clearance (Cl, l/hour)	78
Oral bioavailability (F_{oral},%)	30
Steady-state blood level (C_{ss}, ng/ml)	10–30

[a] Data taken from References 6, 8, 9, and 11–13.

[b] Data taken from References 7, and 10.

primarily to the muscarine and partial antagonist to the nicotinic ACh receptors respectively, and an antidote for nicotine toxicity (i.e., bradycardia, salivation, and diarrhea).[???] Nicotine is primarily a ganglionic (nicotinic) cholinergic-receptor agonist, and acts on the ACh receptors and their various subtypes in the autonomic ganglia and in the brain to provide its reinforcing and CNS effects, which also mediates its acute toxicity.[16] Stimulation of ACh receptors is also responsible for its pharmacologic reinforcing activity (i.e., addiction) as well as its acute toxicity (i.e., symptoms of cholinergic stimulation, including nausea, vomiting, and cardiovascular and respiratory arrest). Tolerance to the effects of nicotine use may be caused by desensitization of nicotinic receptors at both the central and peripheral synapses.[17]

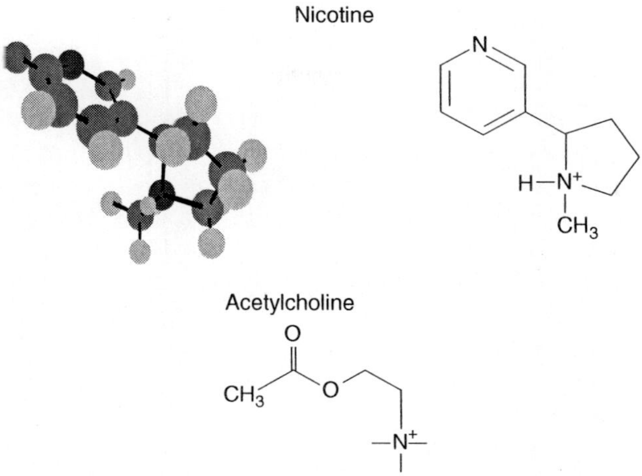

Figure 7.1 Chemical structure of nicotine: (a) space filling model and chemical structure of nicotine (upper panel); (b) chemical structure of acetylcholine (lower panel).

C. Nicotine Toxicity

Nicotine has been used as a pesticide and is one of the most toxic of all poisons.[10] Nicotine has a low LD_{50} in mice (0.3 mg/kg iv; 9.5 mg/kg i.p.; and 230 mg/kg orally) and other species.[1,9] It is also highly absorbed through the skin to the extent that it can lead to serious adverse effects in humans associated with its nicotinic cholinergic agonist activity (i.e., nausea, vomiting, sweating, and diarrhea). The minimum acute lethal dose of nicotine in adults is 40 to 60 mg orally (less than 1 mg/kg), which compares to a dose of 1 mg of nicotine delivered from smoking a cigarette.[10] Thus, nicotine has a relatively low therapeutic index (human oral lethal dose/effective dose: 60 mg/4 mg = 15). Nicotine poisoning can cause the following symptoms in increasing order of toxicity: nausea, salivation, abdominal pain, vomiting, diarrhea, cold sweat, dizziness, mental confusion, weakness, drop in blood pressure and pulse, convulsions, and paralysis of respiration.[13] Tolerance to the acute toxic effects of nicotine develops rapidly over a period of days in smokers. The degree of diminution in acute toxicity is a function of the length and frequency of use of nicotine products (i.e., smoking or use of smokeless tobacco). This is one reason why smokers easily tolerate NR products and nonsmokers can become very ill and exhibit cholinergic side effects (i.e., nausea, vomiting, etc.).

Although cigarette smoking leads to a number of diseases including cancer, there is no evidence that nicotine is carcinogenic in humans. Furthermore, low doses (e.g., 1 to 4 mg nicotine) administered acutely (i.e., within an hour or so) are well tolerated in smokers without adverse systemic toxicity. Thus, the low doses and slower rate of absorption of nicotine used in NRT products (e.g., lozenges, gums, and sublingual tablets which contain 1 to 4 mg per dose) result in lower toxicity and abuse liability compared to smoking cigarettes.[74]

D. Nicotine Intraoral Absorption

It is evident that drug penetration through skin and other tissues is primarily through the lipophilic regions of the stratum corneum (SC) and it is generally accepted that drugs in their nonionized form have better transport properties through biological membranes as compared to their more hydrophilic ionized forms.[18] The permeability of ionizable drugs across the buccal mucosa follows the pH-partitioning theory characteristic of passive diffusion.[19] For example, increasing the nonionized fraction of an ionizable drug will favor drug penetration through the transcellular route (i.e., across cell membranes). Nicotine is a diacidic base with 2 pKa values. Nonionized nicotine partitions into the lipid regions and permeates mainly via the transcellular pathway whereas the mono- and di-protonized forms partition into the aqueous regions and permeate via the paracellular pathway (i.e., a torturous path around cells).[20,21] Thus, variations in the pH of saliva can influence nicotine absorption and serum levels after administration of nicotine chewing gum and other IODs, where increasing pH increases the non-ionized fraction of nicotine, and increases its absorption and physiological effects.[22,23] For example, nicotine absorption is directly related to the pH when nicotine is delivered in either tobacco smoke or from nicotine polacrilex gum, and drinking coffee and carbonated beverages substantially reduces salivary pH and nicotine absorption.[22]

The permeability of nicotine across buccal mucosa increases significantly with increasing pH of the saliva, and thus a strategy to optimize buccal absorption of nicotine from an IOD is to maintain the pH of saliva on the alkaline side by use of appropriate buffers.[20]

The secretion of saliva is another factor that can alter the absorption of nicotine by the buccal mucosa and secretion of saliva can be promoted by acidic excipients, which can reduce absorption of nicotine due to the drop in pH as well as increase clearance of nicotine into the GIT along with swallowing of saliva.[24] The daily output of saliva in humans is between 750 and 1000 ml and the pH may range from 5.8 to 7.6 when salivary fluid production is stimulated.[25] Thus, based on the pKa of nicotine, a fraction will normally be in the ionized and nonionized forms over this pH range. Due to the continuous production of saliva and outflow from the mouth into the GIT, nicotine can be easily cleared from the mouth and thus provides a challenge for designing intraoral mucosal drug delivery systems intended to prolong delivery of nicotine. Even so, the permeability of nicotine across buccal mucosa is greater compared to skin and this advantage has spawned the development of a variety of IODs.[20,26]

E. Nicotine Addiction and Smoking Cessation

When a cigarette is inhaled, the physiologic response to nicotine is almost immediate. Within seconds of inhaling the smoke, nicotine enters the pulmonary venous circulation and is taken to the heart where it is subsequently delivered through the internal carotid and anterior cerebral vessels to the brain.[27] Typically, smoking one cigarette will deliver 1 mg of nicotine and a pack of cigarettes will deliver about 20 mg of nicotine over the course of several hours. The positive reinforcing effects that lead to addiction are reversed and amplified during the withdrawal process with a multiplex of symptoms including irritability, depression, insomnia, restlessness, anxiety, inability to concentrate, and so on. Few individuals are successful in their first attempt to quit smoking; in fact, smoking cessation is regarded as a cycle in which the smoker has to try several times to stop before achieving that goal permanently. Craving for cigarettes is one of the most difficult forms of addiction to alleviate, with the most consistent, most severe, and earliest withdrawal symptoms experienced by those who are attempting to quit, and it is the trigger for smokers to have another cigarette.[28] The degree of severity of craving and withdrawal symptoms varies widely among smokers, and depends on the length of exposure, dose, and other individual factors, but generally peaks over the first 24 to 72 hours of abstinence and then declines, although craving has been reported even after five years of abstinence.[29]

Thus, the objective of NRT is to substitute nicotine in a pharmaceutical product for other forms of tobacco (cigarettes, cigars, and smokeless forms). Most promising smoking cessation medications currently available deliver lower amounts of nicotine to the brain to mimic the effect of smoking. Smoking cessation therapy has a relatively short duration of one to several months in order to relieve the craving and withdrawal symptoms that make it difficult to stop smoking.

Nicotine transdermal delivery systems maintain baseline levels of nicotine in the blood. Although the low constant blood levels of nicotine provided by transdermal patches, and to a lesser extent by gums, relieve some of the symptoms of nicotine withdrawal, they do not, however, effectively address the breakthrough periods of craving.[30] This may be due to the fact that cigarette smoking provides an initial sharp rise in nicotine blood levels, which is a missing aspect of transdermal NRT. However, many nicotine products for NRT are now available that specifically address the acute craving component of nicotine addiction (i.e., lozenges, sublingual tablets, inhalators, nasal delivery, and to a lesser extent gums). Figure 7.2 illustrates the kinetic time profile of craving and withdrawal comprising a baseline period of craving, and superimposed on this are the intermittent periods of acute breakthrough craving, which must be controlled for smoking cessation therapy to be effective.

F. Medical Rationale: Nicotine Replacement Therapy

It is well established that nicotine meets all the criteria of a highly addictive drug, as recognized by the U.S. surgeon general.[31] Although tobacco use often escalates to compulsive use accompanied by tolerance and dependence, this is not usually observed with NR products that provide lower doses of nicotine.[31] Nicotine replacement therapy has become established as a successful aid to smoking and tobacco cessation programs. However, the pharmacological doses of nicotine

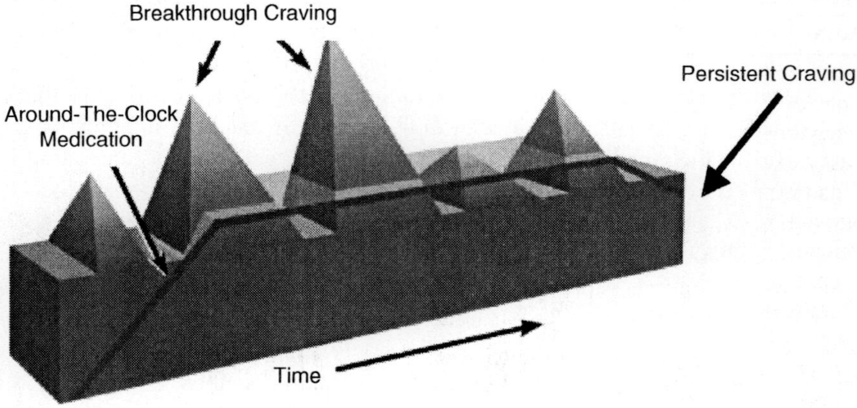

Figure 7.2 Illustration of the kinetics of persistent and breakthrough episodes of craving.

from these products usually do not relieve withdrawal symptoms to the same degree as smoked nicotine and smokers are often disappointed that they do not receive the same "smoking effect" provided by these products.[32] Increasing the availability and number of nicotine replacement products will provide a public health benefit, as no single therapy is effective for all smokers desiring to quit. Transdermal delivery of nicotine is effective to establish low-maintenance blood levels of nicotine; however, the onset time is slower compared to absorption through mucosal tissues of the oral cavity and respiratory tract (i.e., nasal and lung tissues). Therefore, intraoral delivery offers a unique opportunity to develop products with a rapid onset of action to address the acute periods of craving, as well as the period of breakthrough craving in patients on transdermal therapy. Thus, there is a compelling benefit that greatly outweighs the risks for NR products and a regulatory incentive for approval of this class of NRT.[33] The various advantages and disadvantages of delivery of nicotine by the intraoral route are summarized in Table 7.2.

Although NRT can double the success rates of quitting compared to placebo, particularly when combined with behavioral intervention such as aversion therapy, counseling, educational programs, hypnosis, and self-help literature, other pharmacological treatment strategies for further improving the efficacy of smoking cessation therapy have been proposed and evaluated.[27] The merits of combination therapy with transdermal delivery of nicotine from patches along with other IODs that permit ad libitum nicotine delivery (i.e., gums, lozenges, nasal sprays, inhalers, etc.) offer the opportunity for more effective control of nicotine withdrawal and craving symptoms.[34] Several studies report that combined therapy is more effective than monotherapy.[10] However, there are regulatory and commercial obstacles that must be considered and addressed for approvals of combination NR products in the future. Nonetheless, off-label use of combined

Table 7.2 Advantages and Disadvantages of Intraoral Nicotine Replacement Products

Advantages	Disadvantages
Solid dosage form to deliver 1 to 4 mg	Dose limitations are about 5 mg, higher acute doses are absorbed quickly and produce systemic and local toxicity
Convenience (dosing anywhere, anytime, without water)	Absorption is rapid and has a higher abuse liability vs. transdermal delivery
Less expense vs. nasal, transdermal, and inhalation devices	Dose limitation for sublingual tablets is about 4 mg
No water required, dissolution in saliva	Shorter duration of action vs. transdermal delivery
Potential to promote a more rapid onset of action vs. transdermal delivery	Frequent dosing required for tablets and lozenges
Rapid absorption to address acute craving and breakthrough period of craving not satisfied by transdermal therapy	Local mucosal irritation and toxicity with high concentrations of nicotine

OTC (over-the-counter) smoking cessation products (i.e., patches and more rapid-onset forms to address the period of acute craving) may provide a more effective treatment for smokers who are dependent on tobacco. The rationale for combination nicotine replacement therapy for smoking cessation has been evaluated over the past several years and has been patented as a method for smoking cessation therapy, but is not yet commercialized. For example, combined use of NTS and chewing gum may increase the success rate in quitting for more patients.[35]

G. Nicotine Dosage Form Considerations

The physiochemical properties of nicotine are ideally suited for delivery via a number of routes (i.e., transdermal, nasal, oral transmucosal, inhalation, etc.). The high absorption of nicotine through skin has led to the development of several transdermal nicotine smoking cessation products in the form of transdermal systems (i.e., skin patches) for NRT. Nicotine transdermal systems (NTS) were originally introduced as prescription products and are now available as over-the-counter products as well.[36] Nicotine base (nicotine USP) is used in all of the transdermal dosage forms on the market (i.e., Nicotrol® CQ, Nicoderm®, Prostep®, Habitrol®, and generic nicotine patches). Nicotine base is also used in nicotine nasal spray (i.e., Nicorette® Nasal Spray) and inhalation products (Nicorette® Inhalator) as well. Nicotine delivered by the oral transmucosal route is available in the form of a gum (i.e., Nicorette® gum), which contains nicotine absorbed onto a polacrylate resin (i.e., nicotine polacrilex USP).[37] The base form of nicotine is also used in lozenges (e.g., Nicorette® Microtab, Stompers®, and Commit™) designed for rapid absorption of nicotine in the oral cavity and tissues of the buccal mucosal. Nicotine salicylate is a more stable form of nicotine and was used in lollipops, which are no longer available in the United States, and the

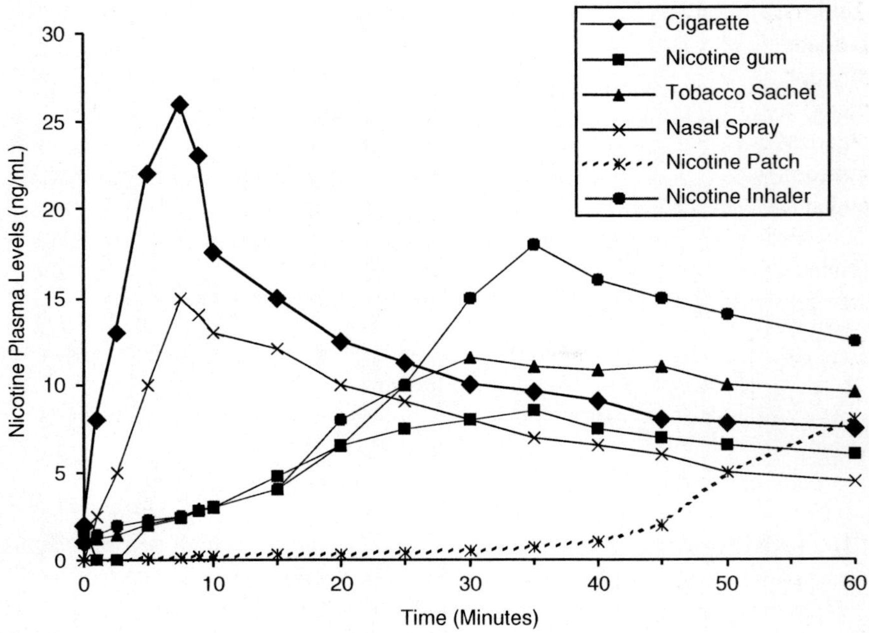

Figure 7.3 Comparison of nicotine plasma levels following smoking a cigarette and dosing with intraoral nicotine products (adapted from References 31, 40–42, and 107).

bitartarate form is used in sublingual tablets. Each of these products has its advantages and disadvantages and these aspects are discussed for each type of NR product.

H. Nicotine Pharmacokinetics

Smoking a high-nicotine content cigarette over a period of 10 minutes results in a rapid rise in nicotine plasma levels reaching a maximum concentration within 5 minutes and then a rapid decline over a period of 60 minutes. The nicotine plasma level time profiles produced by smoking a cigarette, nicotine gum, tobacco sachet, nicotine inhaler, nasal spray, and transdermal system are illustrated in Figure 7.3. The plasma levels of nicotine achieved after smoking a cigarette are both higher (having a C_{max} between 25 to 40 ng/ml) and sharper (having a T_{max} of between 5 to 10 minutes) compared to the blunted (lower C_{max} and longer T_{max}) and prolonged steady-state plasma levels provided by transdermal nicotine patches. The optimal steady-state plasma levels for nicotine replacement are between 10 to 15 ng/ml; however, a quick rise is probably necessary for more complete relief of craving in the early stages of cigarette withdrawal.[30] A rapid rise in nicotine plasma levels of at least 10 ng/ml within 10 minutes is required to produce the postsynaptic effects at the nicotinic cholinergic receptors in the

CNS, which may be responsible for the druglike "high" feelings such as lightheadedness or dizziness produced by smoking cigarettes.[30] In contrast, use of a nicotine sachet for 10 minutes results in a slower onset and lower maximum nicotine plasma level, which approaches the level comparable to smoking a cigarette after about 30 minutes. The objective of NRT is to mimic the behavioral reinforcing effects of smoking or use of smokeless tobacco products (i.e., sachets and snuff, etc.) and match the nicotine plasma levels.[39]

Following use of a 1 mg single dose of nicotine from a nasal spray (Nicotrol® NS), nicotine plasma levels are less than those observed following smoking and reach a steady-state level within about 20 minutes and provide effective short-term maintenance therapy. Similarly, after chewing a 2 mg nicotine gum (Nicorette® gum), nicotine plasma levels slowly increase to reach a C_{max} of approximately 8 ng/ml, less than that produced by smoking, after about 20 to 30 minutes and then plasma levels are maintained comparable to the levels observed following smoking for up to 60 minutes.[43]

III. COMMERCIAL NICOTINE REPLACEMENT PRODUCTS

A. Nicotine Transdermal Systems

Delivery of nicotine using transdermal patches was one of the first commercially available NRTs.[36] During late 1991 and early 1992, four forms of prescription nicotine NTS were approved by the FDA for smoking cessation therapy, and were generally approved for OTC use in other countries.[27] The advantage of the patch for NRT is transdermal absorption of nicotine, an effective means of bypassing hepatic first-pass metabolism, and poor absorption of nicotine from the GIT; however, nicotine transdermal systems are unable to provide the rapid increase and peaks in nicotine blood level produced by smoking.[122] The nicotine patches on the market are summarized in Table 7.3.

Many of these nicotine patches are now available as OTC products, due to their safe use history, ability of patients to self-diagnose their addiction and self-medicate, which attests to the effectiveness of these products. Nicotine transdermal delivery systems are effective in establishing a steady-state plasma level of nicotine over a 16- or 24-hour period. On April 19, 1996, an FDA advisory panel voted that Nicotrol® and Nicoderm® CQ should be available for nonprescription use, and today they are available without a prescription, as OTC products.[27] Two other brands, Habitrol® and Prostep®, are available only by prescription. Due to the diffusion barrier of the skin there is a lag time of up to an hour before steady-state nicotine plasma levels are achieved. The optimal length of treatment using NTS depends upon the individual and degree of nicotine dependence, and may vary from six to eight weeks.

These products have been developed in several dosage strengths to provide continuous delivery of nicotine over a period of 24 hours (i.e., Nicoderm®, Habitrol®, and Prostep®) or 16 hours (e.g., Nicoderm® CQ). The medical rationale for these

Table 7.3 Comparison of Nicotine Transdermal Dosage Forms

Product	Marketer	Dosage Strength[a] (mg/day)	Active Surface Area (cm^2)	Total Surface Area (cm^2)	Nicotine Content (mg)	Delivery Efficiency (%)[b]
Habitrol®[c]	Ciba-Geigy	7	10	10	17.5	40
		14	20	20	35.0	40
		21	30	30	52.5	40
Nicoderm®	Marion Merrell Dow	7	7	7	36.0	18
		14	15	15	78.0	18
		21	22	22	114.0	18
Nicoderm® CQ[c]	GlaxoSmithKline	7	7	7	360	18
Nicoderm® CQ		14	15	15	780	18
Clear Patch[c]		21	22	22	114	18
Nicotrol®	McNeil	5	10	10	8.3	60
		10	20	20	16.6	60
		15	30	30	24.9	60
Prostep®	Lederle	11	3.5	32	15	73
		22	7	32	30	73

[a] All patches are intended for 24-hour wear except Nicotrol®, which is worn for 16 hours.

[b] Delivery efficiency is the percent of label claim of drug delivered over the prescribed dosing period.

[c] Available as OTC products. Information adapted from Reference 36.

types of products is to provide for a "step-down" smoking cessation program beginning with the high-dose strength product which is applied daily for four weeks, followed by the next lower dose strength for a period of four weeks, and finally the lowest dose strength until the habit is "kicked." Unfortunately, NTS do not simulate the pharmacokinetics of nicotine plasma levels following smoking, and there are significant differences in the pharmacokinetics of nicotine delivered by many of the NTS on the market.[44] Due to the rather low constant nicotine plasma levels following transdermal delivery there is a therapeutic gap that cannot be controlled by the NTS, such as periods of breakthrough craving.

B. Nicotine Oral Transmucosal Products

Several OT products have been developed that may be classified as fast-dissolving or immediate-release (i.e., within 1 to 10 minutes), slow-dissolving and slow-release (i.e., within 10 to 30 minutes), and nondissolving and controlled-release delivery systems (delivery greater than 30 minutes up to 24 hours). Many of these dosage forms provide a more rapid onset and higher nicotine plasma level than can be achieved from nicotine patches. A number of tobacco alkaloids have been investigated as an aid in smoking cessation including nicotine, nicotine-like alkaloids such as nor-nicotine, and lobeline, in the base or pharmaceutically acceptable acid addition salts and resin complexes.

Loboline is a natural nicotinic cholinergic alkaloid structurally related to nicotine, which has similar effects and has been used in smoking cessation products as well.[16] Loboline is the main alkaloidal constituent isolated from Indian tobacco (*Lobelia inflata,* also known as Lobeliaceae), and its hydrochloride and sulfate salts have been formulated into NR products and given by mouth as a smoking deterrent.[45] Review of antismoking therapy generally considers lobeline to have little benefit compared to placebo.[46,47] However, lobeline is available in several countries as an aid to smoking cessation (Australia: Cig-Ridettes; Spain: Nofum, Smokeless; UK: AntiSmoking Tablets).[45] The FDA classified lobeline (Bantron®) as a category III agent, safe but no proven effectiveness, and ordered the removal of lobeline sulfate OTC smoking deterrent products (i.e., Cigarrest®, Bantron®, Tabmit®, Nikoban®, and others) from the U.S. market in 1993.[48,49] Nicotine gum, lozenges, sublingual tablets, intraoral sprays, and other nicotine delivery devices have been developed and are either on the market or in various stages of commercialization as aids for smoking cessation. Intraoral forms of nicotine such as lozenges have a low abuse liability in young and old adults and the subjective effects of lozenges have utility as a smoking cessation aid and are comparable to that of nicotine gum.[50]

The various NR products on the market, which deliver nicotine to the oral cavity, are summarized in Table 7.4. These OT-nicotine products are intended to release nicotine into the oral cavity where it is absorbed predominately through the tissues of the buccal mucosa, with a small fractional absorption by the gastrointestinal tract (GIT), to provide systemic blood levels sufficient to cross

Table 7.4 Nicotine Replacement Products on the U.S. Market

Product	Manufacturer (Distributor)	Dose Strength (mg)	Onset Time (min)
Nicorette® Gum[a] (original, mint, orange)	Pharmacia AB (GlaxoSmithKline)	2.0, 4.0	30
Nicotine Polacrilex Gum[a]	Watson Lab	2.0, 4.0	30
Commit™ Lozenge[a]	GlaxoSmithKline	2.0, 4.0	30
Nicotrol® NS Spray[b]	Pharmacia & Upjohn (McNeil Consumer HealthCare)	0.5 mg/spray (max. 40 mg/day)	10–30
Nicotrol® Inhaler[b, c, d]	Pharmacia & Upjohn (McNeil Consumer HealthCare)	10 mg cartridge (4 mg delivered)	30

[a] Over-the-counter product (OTC).
[b] Prescription product (Rx).
[c] Smokeless artificial cigarette system.
[d] Withdrawn from the market.

the blood–brain barrier and produce pharmacodynamic and behavioral reinforcing CNS effects. Buccal administration refers to any dosage form or device for oral administration of the drug that is held or placed in the mouth and is used to deliver the drug through the buccal mucosa, in whole or part, and into the systemic circulation and includes lozenges, capsules, tablets, gums, inhalators, and the like. The advantages and disadvantages of the various NR products are contrasted in Table 7.5.

A number of nicotine chewing gums, lozenges, and sublingual tablets are on the market or in various stages of research and development and their product characteristics are summarized in Table 7.6. All of these products are more or less effective in reducing craving and withdrawal symptoms associated with smoking.[51]

1. Chewing Gums

Chewing gums have been used for several decades as consumer products and represent over a $2.2 billion market in the United States alone.[53] The concept of using chewing gum as a matrix for drug delivery is not new: Aspergum® was introduced in 1928 as the first medicated gum containing acetylsalicylic acid, and later in the 1970s gums containing nicotine were introduced as drug delivery systems.[53] Therapy with buccal nicotine polacrilex chewing gum is useful as a temporary substitute for smoking and when used in conjunction with other therapies is beneficial in overcoming nicotine withdrawal.[10] Chewing gum technology was used to develop Nicorette® in the 1970s, a gum-base containing nicotine polacrylate for delivery of nicotine as an aid for smoking cessation. Chewing gums offer several advantages as a drug delivery system including local delivery,

Table 7.5 Advantages and Disadvantages of Nicotine Replacement Products

Product	Advantage	Disadvantage
Transdermal delivery systems	There are no local oral adverse effects and few systemic side effects Ease of administration Dosing is once a day, and discreet Maintain steady-state plasma levels	Some local skin irritation Dosage is less flexible than other products Nicotine is delivered relatively slowly Less than optimal to control acute period of breakthrough cravings Sleep disturbance side effects
Nicotine gums	Faster onset of action than patches A more suitable surrogate for smoking vs. patches Delivery via the fast oral mucosal route Can titrate to the level of craving	Must chew to release the drug Muscles of the mouth become "tired" Must dispose of the gum after delivery Time of effective dosing not well controlled
Nicotine nasal spray	Faster onset of action vs. patch and gum Effectively address the acute craving periods Required dose delivered quicker for acute craving vs. patch Less effective for dose titration	May be more irritating vs. patch and gum Not as effective over 24 hours as patches Multiple dose regime is needed Less discreet vs. patch and gum More prone to overdose
Nicotine inhaler	Faster onset of action vs. patch and gum Effectively address the acute craving periods and provide dose titration Dose delivered quicker for acute craving vs. patch	May be more irritating vs. patch and gum Not as effective over 24 hours as patches Multiple dose regime is needed Less discreet vs. patch and gum More prone to overdose
Nicotine lozenge and sublingual tablet	Faster onset of action than patches A more suitable surrogate for smoking vs. patches No chewing necessary compared to gums Delivery via the oral mucosal route Can titrate to the level of craving	Must dissolve to release the drug Short duration of delivery Time of effective dosing not well controlled Multiple daily dosing required Less effective in maintaining C_{maxs} compared to patches, gums
Artificial cigarettes	Satisfies a portion of the psychological component of smoking Is a surrogate for a cigarette	Less effective in addressing acute craving vs. other methods Less discreet than other methods

rapid buccal absorption, rapid onset, dose titration, dose termination, and product line extension, and other products are expected soon. Some of the ideal properties of a chewing gum for smoking cessation and delivery of nicotine include:[115]

Table 7.6 Comparison of Nicotine Replacement Products

	Nicotine Replacement Dosage Form[a]					
Characteristics	Patch	Gum	Nasal Spray	Inhalator	Sublingual tablet	Lozenge
Administration	Apply to the skin on the trunk or upper arm and place a new patch on a different site each day	Chew one piece slowly for approx. 30 min when the urge to smoke occurs	Use one spray in each nostril when the urge to smoke occurs	Inhale when the urge to smoke occurs	Place tablet under the tongue when the urge occurs	Suck on one lozenge when the urge to smoke occurs
Type of Dosage	Nondissolving Controlled release	Nondissolving Slow release	Liquid Fast release	Aerosol Fast release	Slow dissolving Moderate release	Slow dissolving Moderate release
Onset of action	Very slow	Moderate	Fast	Fast	Moderate	Fast to moderate
Duration	Long (16–24 hr)	Moderate (30–60 min)	Short (20 min)	Short (5 min)	Moderate (30 min)	Moderate (30 min)
Dose strengths	21, 14, 7 mg/24 hours 15, 10, 5 mg/16 hours	2 and 4 mg	500 mcg/spray	10 mg/cartridge 4 mg delivered	1, 2, and 4 mg	2 and 4 mg
Advantages	Easy to use, 24 hour patch can help with early morning craving	Easy to regulate dose	Provides fast relief for craving, easy to adjust dose	Keeps hands busy, easy to regulate dose	Discreet, easy to adjust dose, few side effects	Discreet and easy to use

Table 7.6 Comparison of Nicotine Replacement Products (continued)

	Nicotine Replacement Dosage Form[a]					
Characteristics	Patch	Gum	Nasal Spray	Inhalator	Sublingual tablet	Lozenge
Disadvantages	May irritate skin, 24 hr patches may disturb sleep	Sticks to dentures, must be chewed to work, buccal irritation, tired jaw muscles	May cause nasal irritation, may cause dependence	Not very effective for heavy smokers, not discreet	Must be used correctly and does not work if swallowed	May cause throat irritation and indigestion
Product examples	16 hr patches: Nicorette®, Nicotrol® 24 hr patches: Nicotinell®, NiQuitin® CQ, Habitrol®, Nicoderm®, Nicoderm® CQ, Nicoderm® CQ Clear, Prostep®	Nicotinell® gum Nicorette® gum	Nicorette® nasal spray	Nicorette® Inhalator	Nicorette® Microtab	Nicorette® Lozenge: Regular, mint, orange Nicotinell®

Source: Adapted from Reference 52.

- Controlled release of drug over time
- Drug release rate proportional to the drug loading dose
- Drug release rate a function of chewing rate (i.e., chew/min) and extent (i.e., degree of gum deformation), and amount of saliva (i.e., required for gum hydration, drug dissolution, and absorption)
- Drug release and absorption activated and controlled by chewing (i.e., easy dose titration)
- Satisfy the psychological component of smoking as a suitable smoking substitute
- Easy to incorporate drug and manufacture

Continuous mastication is needed for drug release from medicated chewing gums because the insoluble matrix acts as a controlled slow-release matrix in the static state. In the past, drug release from gums was evaluated in "chew out" studies using volunteers that chewed the gum and then measured the amount of drug remaining after various times. Now, in vitro methods have been developed using an apparatus that simulates the process of mastication in the presence of artificial saliva.[54]

Chewing gum-based delivery systems are unique dosage forms that have rather specialized manufacturing processes involving mixing, rolling, cutting, coating, and packaging unit operations. Unlike the products manufactured for consumer markets, all production must be in accordance with Current Good Manufacturing Practices (cGMPs). The key manufacturing unit operations for the manufacture of chewing gum-based products are summarized below.

Mixing: All chewing gum raw materials are mixed in high-shear blenders typically used in the rubber and polymer gum industry.

Rolling: The chewing gum mass is fed through colanders and is rolled into a continuous plate or ribbon.

Cutting: Automatic cutters divide the plate or ribbon into rectangular tablets.

Coating: Coating takes place in a specialized coating tumbler. The coating solution is sprayed over a long period of time in a rotating tumbler containing thousands of chewing gum tablets. Each tablet accumulates a smooth and even glossy coating or glaze.

Packaging: The coated tablets are transferred to the packaging line where specially designed machines pack the tablets into a variety of package types such as blister packs, boxes, or foil packs. Generally, inline weight control checks ensure that no pack leaves the packaging line that is not perfect in appearance and content or count. Finally, individual packs are then placed in secondary cartons for distribution.

Gums have been formulated with both nicotine base and resin complex (nicotine polacrilex) as a dosage form for the buccal delivery of nicotine. Nicotine is primarily released from gums during chewing, and delivered to the buccal mucosa where it is absorbed. However, a substantial portion of the nicotine is retained in the gum because of incomplete chewing, or is swallowed and not absorbed due to hepatic first-pass metabolism, resulting in a low systemic BA of between 30 to 40 percent. Chewing nicotine gum is an improvement over transdermal delivery, but is still a slower and less efficient method of delivery of nicotine compared to smoking. Several nicotine chewing gum products on the market are highlighted below.

Nicorette® Chewing Gum: The first medicated gum approved for prescription use in the United States was Nicorette® nicotine gum.[27] The gum contains nicotine polacrylate as a controlled-release complex and the gum was originally developed by Pharmacia AB. Prior to marketing in the United States the FDA required a rigorous abuse liability assessment to examine whether enhanced palatability of nicotine gum would increase its abuse liability.[55] In that study mint- and orange-flavored gum did not increase abuse liability, but was effective in reducing craving. Nicorette® 2 and 4 mg (nicotine polacrilex) gum was approved by the FDA (NDA 18-612 and 20-006) as a prescription drug on February 9, 1996 and launched by SmithKline Beecham Consumer Healthcare in the United States.[56] The inactive ingredients in the original flavor gum include flavors, glycerine, gum base, sodium carbonate, sorbitol, and D&C yellow #10. The inactive ingredients in the mint-flavored gum include: gum base, maganesium oxide, menthol, peppermint oil, sodium carbonate, xylitol, and D&C yellow #10 Al. lake. The FDA approved the prescription product (NDA 20-066) on September 25, 2000 and SmithKline Beecham Consumer Healthcare launched Nicorette® Orange in the United States to reduce the withdrawal symptoms, including nicotine craving, associated with smoking cessation therapy.[57] Nicorette® Orange is a sugar-free gum and is available in two strengths, 2 and 4 mg. The nicotine gum is now widely available in the United States as an OTC product and was launched in Japan on September 10, 2001 by Pharmacia KK and Takeda Chemical Industries, Ltd.[58] Nicorette® gum contains nicotine polacrylate resin, which releases nicotine from a gum by the process of chewing, uptake of saliva, and hydration and liberation of the drug, which is then delivered to the buccal cavity for systemic absorption. The course of therapy when using Nicorette® or other nicotine gums involves placing the gum in the mouth and chewing it slowly. After 15 bites, a peppery taste and slight tingling may be noticed which decreases over time; the gum is chewed a few more times, and then the gum is "parked" in another area of the mouth, and the cycle is repeated over 30 minutes as shown in Figure 7.4.[60]

The course of therapy begins by chewing one piece of gum every one to two hours over the first 6 weeks, then one piece every two to four hours for the next 3 weeks and finally one piece every four to eight hours up to the last 12 weeks of use.[60,62]

Generic Nicotine Chewing Gum: More recently, a generic nicotine polacrilex 2 and 4 mg nicotine gum was approved by the FDA on March 15, 1999 (ANDA 74-707), and March 19, 1999 (ANDA 74-707), respectively, filed by Circa Pharmaceuticals and marketed by Watson Labs.[63]

Figure 7.4 Nicotine gum cycle of chewing.

Nicotinell® Chewing Gum: A number of other nicotine chewing gum products are on the market outside the United States. For example, Nicotinell® fruit- and mint-flavored 2 mg nicotine gum is sold in the United Kingdom. The Nicotinell® 2 mg nicotine resin gum is manufactured by Fertin A/S and marketed by Novartis and contains the following inactive ingredients: gum base, sorbitol, calcium carbonate, glycerol, sodium carbonate, sodium bicarbonate, amberlite, levomenthol, butylated hydroxytoluene, acesulfame potassium, saccharin, saccharin sodium, carnuba wax, talc, and water.[64]

2. Lozenges

Commit™ Lozenge: A more recent NR product is an orally dissolving nicotine lozenge (Commit™ lozenge) containing nicotine polacrylate and marketed by GlaxoSmithKline. The FDA approved the Commit™ 2 and 4 mg nicotine lozenges (NDA 21-330) on October 31, 2002 as an OTC product, for adults 18 years of age and older, to reduce the withdrawal symptoms, including nicotine craving, associated with smoking cessation therapy.[66] The product is available as 2 and 4 mg dose strengths and the primary packaging for this OTC product is shown in Figure 7.5. The 2 and 4 mg Commit™ nicotine lozenges contain 1 and 2 percent nicotine polacrilex, respectively,[67] and the following active ingredients: aspartame, calcium polycarbophil, flavor, magnesium stearate, mannitol, potassium bicarbonate, sodium alginate, sodium carbonate, and xanthan gum.

The Commit™ lozenge is protected by U.S. patent 5,110,605 issued in May of 1992, subsequently licensed to GlaxoSmithKline, and based on the pioneering work of Oramed, Inc. in the late 1980s on developing compositions comprising a reaction complex formed by the interaction of a polycarbophil component with alginic acid or a salt thereof in the presence of a divalent cation and an active medicinal agent, including breath-freshening agents.[68,69] This patent claims the method for the controlled release of nicotine from the polymeric

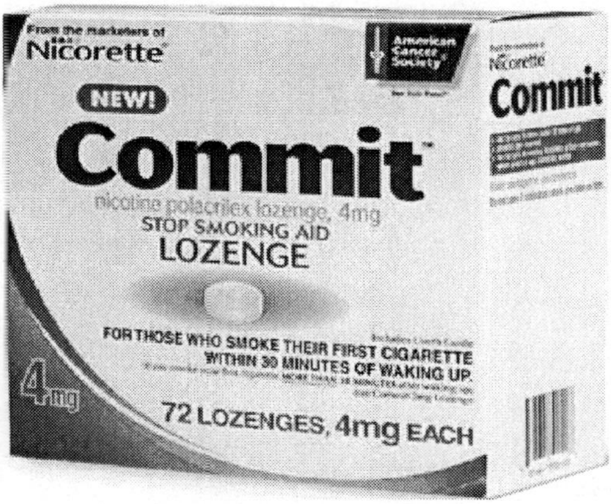

Figure 7.5 Commit™ lozenge, 4 mg dose strength.

complex in a matrix to the mucosal tissues of the mouth. In this case nicotine polacrylate resin is the polymeric complex with nicotine, which is released from the lozenge by a sucking action where the resin complex hydrates in the presence of saliva with liberation of nicotine, which is then absorbed by the buccal tissues. A companion patent was subsequently issued in 1992 where the controlled-release complex composition comprised calcium polycarbophil and a medicinal or breath-freshening agent.[70] The 2 and 4 mg nicotine lozenges are effective in significantly increasing abstinence after six weeks of use compared to placebo and is a safe and effective treatment for reducing craving and withdrawal symptoms in low- and high-dependence smokers.[71]

Nicotinell® Nicotine Lozenge: The Nicotinell® mint-flavored 1 mg nicotine lozenge is marketed by Novartis in the United Kingdom. Each lozenge contains 1 mg of nicotine bitartarate and the following inactive ingredients: maltitol, sodium carbonate, sodium bicarbonate, polyacrylate dispersion, xanthan gum, colloidal anhydrous silica, levomenthol, peppermint oil, aspartame, and magnesium stearate.[72] The lozenge can alleviate some of the unpleasant withdrawal effects that frequently occur when quitting smoking, such as craving and irritability. The dosing regimen includes taking 1 lozenge every one to two hours and 8 to 12 per day, with a maximum daily dose of 25 lozenges (i.e., 25 mg nicotine). Also, a 1 mg nicotine bitartarate salt lozenge (Nicotinell®) has been introduced in the United Kingdom and Sweden.[71]

3. Sublingual Tablets

Nicotine sublingual tablets have been evaluated as a dosage form for smoking cessation therapy. Sublingual tablets are generally smaller in size compared to lozenges and the tablets are intended to be placed under the tongue and dissolve without the need for sucking as in the case of lozenges. The sublingual tablets are smaller and generally dissolve faster than lozenges and therefore, may be able to achieve a somewhat faster onset of action owing to the more rapid dissolution and faster absorption of nicotine by the rich vascular supply of the sublingual mucosal tissue. The safety and efficacy of a 2 mg nicotine sublingual tablet has shown promise as a viable dosage form for smoking cessation therapy. Several nicotine sublingual tablets have been evaluated recently and some introduced into the marketplace. A 2 mg nicotine bitartarate sublingual tablet is available in some countries (e.g., United Kingdom, Sweden, and Denmark), however, the tablet delivers less nicotine than nicotine gum, and falls short of dose proportionality when two tablets are used to simultaneously mimic a 4 mg dose.[73] The following provides a summary of some of these commercial products, and those that are in the R&D pipeline.

Nicorette® Microtab: A 2 mg nicotine sublingual tablet has been evaluated compared to 2 mg Nicorette® nicotine gum in a double-blind, placebo-controlled trial in 247 adults who had smoked >10 cigarettes per day for at least three years. The dosage varied according to baseline nicotine dependence. Highly dependent smokers used up to 2 tablets/hour and up to a maximum of 40 tablets daily (80 mg nicotine), and less dependent smokers used 1 tablet/hour up to a maximum of 20 tablets/day (40 mg nicotine).[74] A significantly higher rate of smoking cessation was achieved with the active treatment than with the placebo at 6 weeks and at 3 and 6 months, and compliance with treatment was high with no serious treatment-related adverse events after 12 months of use. This product was safe and effective and the 2 mg nicotine sublingual tablet (Nicorette® Microtab) was launched in Sweden, with an OTC marketing license on October 6, 1998 by Pharmacia & Upjohn. The pharmacokinetic profile of Nicorette® Microtab was found to be similar to that of the 2 mg Nicorette® nicotine gum.[74] The Nicorette® Microtab is available in the United Kingdom and the course of therapy involves placing one Microtab under the tongue where it dissolves and releases nicotine, which is absorbed through the buccal mucosa.[75] The Nicorette® Microtab is provided in a unique patient-friendly dispenser package where each tablet is individually protected in a blister and dispensed by pressing out at the time of use (Figure 7.6). The course of treatment involves a 12-week program in moderate smokers (i.e., less than 20 cigarettes per day) with 1 sublingual tablet every hour, and in heavy smokers (i.e., greater than 20 cigarettes per day) with 2 tablets every hour; thereafter, a gradual reduction of dosage is recommended with completion after 6 months.[75]

Stoppers®: A nicotine sublingual pellet was originally available as an OTC product in Europe and the United Kingdom in the mid 1990s and distributed by Charwell Pharmaceuticals Ltd. Stoppers contained approximately 0.5 mg of nicotine and were available in chocolate, orange, and peppermint flavors. These were provided in a multiunit dispenser with a plastic top. These pellets were designed to be placed in the oral cavity under the tongue, and were developed to

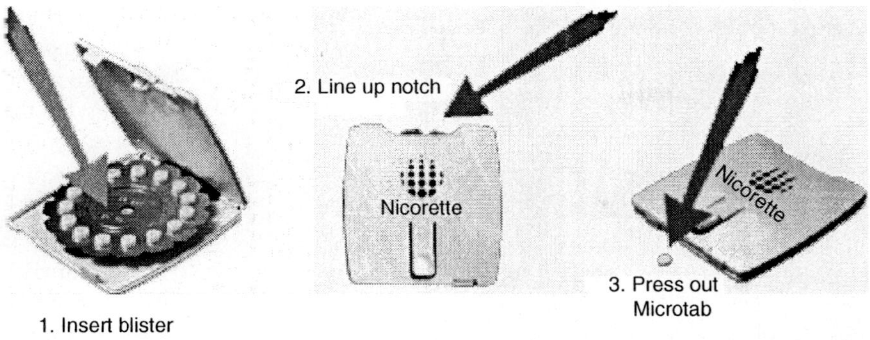

Figure 7.6 Nicorette® Microtab nicotine lozenge and multiunit dispenser.

address the acute craving for nicotine, and were administered on an as-needed basis. However, the content of nicotine in the product decreases over time, perhaps due to the volatility of nicotine, and a fraction is lost every time the container is opened.

Resolution®: The Resolution® pellet-like tablet is manufactured by Phoenix Pharmaceuticals and distributed by Ernest Jackson in the United Kingdom. The tablet contains 0.5 mg of nicotine, together with the antioxidants Vitamins A, C, and E.

Ariva™ Cigalett: Ariva™ is a small pellet that contains tobacco with nicotine levels comparable to a light cigarette developed by Star Scientific. Ariva™ has a sweet taste, and is made from powdered tobacco that is compressed into a pellet, and is sold in child-resistant packs. The tablet dissolves in the mouth like candy.[76]

4. Oral Inhalers

Nicotine oral inhalers provide a method of NRT that most closely mimics the behaviors (i.e., handling and inhalation through the mouth) of cigarette smoking, and provides the smoker with adequate amounts of nicotine to reduce the urge to smoke, and may provide some degree of comfort by simulating the hand-to-mouth ritual similar to smoking. However, clinical evidence suggests that this mode of delivery is less effective compared to nicotine gums or intraoral sprays.[10] Nicotine vapor delivered in oral inhalers is similar to the inhaler technology used to supply bronchial asthma medications. It has been demonstrated that nicotine plasma levels and heart rate increase was dose-related for cigarettes (after 0, 4, 8, and 16 puffs, 0.1 mg nicotine per puff) and nasal spray (0, 1, 2, or 4 sprays at 0.5 mg nicotine per spray) but not for vapor inhaler (0, 30, 60, 120 vapor inhalations, 0.013 mg nicotine per inhalation), indicating limited nicotine delivery from the inhalation device.[77]

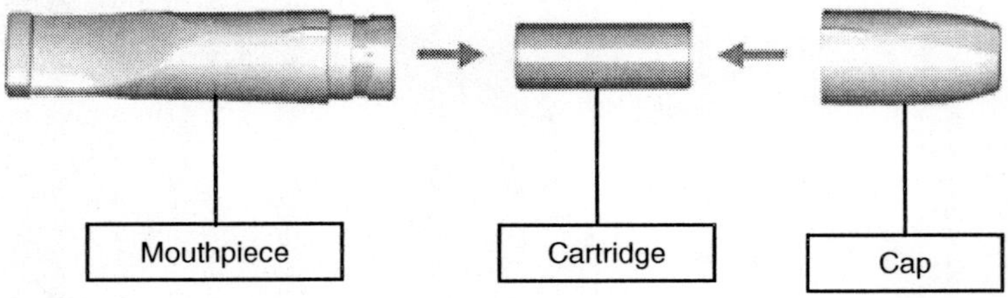

Figure 7.7 Nicorette® Inhalator nicotine delivery device.

Nicotine delivered by inhalation results in unpleasant side effects in 32 to 50 percent of patients involving oropharyngeal irritation and adverse effects including burning throat and nose, watery eyes, runny nose, coughing, and sneezing, and might be expected to limit consumer acceptance.[10] The Nicotrol® Inhalant, with a 4 mg nicotine cartridge, was developed by Pharmacia and Upjohn and approved by the FDA (NDA 20-714) on May 02, 1997, but has since been discontinued in the United States, perhaps due to the lack of patient acceptance due to side effects (i.e., irritation).[78] However, the Nicorette® Inhalator is on the market in Europe; it consists of a mouthpiece, a replaceable nicotine cartridge, and cap (Figure 7.7).[79]

The nicotine oral inhaler is administered as an inhaled vapor from cartridges containing porous plugs impregnated with nicotine. The cartridges contain 10 mg of nicotine and deliver a maximum of about 4 mg of nicotine during use. Nicotine is released from the device when air is drawn through the mouthpiece. Only a small portion of the dose is released with each inhalation, and the amount released depends upon the volume and temperature of air passing through the device. At room temperature the inhaler releases approximately 13 mg of nicotine per 50 ml volume of air inhaled, or about 1 mg of nicotine per 80 inhalations. However, deep inhalation can release up to 4 mg of nicotine per use over about a 20 minute period of active puffing.[10] Oral inhalation over 5 minutes by active rapid shallow sucking/puffing "buccal mode" produces delivery of nicotine comparable to slow deep inhalations "pulmonary mode." The daily dose should not exceed 16 cartridges (about 64 mg of delivered nicotine). The usual course of therapy is a 12-week program that begins with use of 6 to 12 cartridges per day for the first 8 weeks, 3 to 6 cartridges over the next 2 weeks, and finally the dosage is reduced and treatment is terminated over the last 2 weeks.

5. Artificial Cigarettes

Other nicotine replacement products are on the market or have been proposed. For example, nicotine is delivered in aerosol form from the "smokeless cigarette" marketed by Advanced Tobacco Products under the trade name Favor®. However, these modes of nicotine delivery do not result in significant nicotine blood levels in patients after use, and inhalation of these nicotine vapor products may be too irritating to the buccal mucosa or the throat to be tolerated by patients for long-term use as a smoking cessation aid. Other studies have shown that smoking cigarettes that do not provide nicotine may temporarily suppress cigarette withdrawal symptoms, and that the sensory characteristics of cigarettes contribute to the abuse liability of smoking.[80]

6. Lollipops

Nicotine-containing lollipops were introduced into commerce and sold without a prescription as unapproved drug products in the early 2000s by several companies over the Internet as an aid in smoking cessation. Several of these companies received warning letters from the FDA (Ashland Drug, The Compounding Pharmacy, and Bird's Hill Pharmacy) and the products were subsequently removed from the market.[81] The nicotine lollipop is an appealing dosage form as it serves as a surrogate for putting a cigarette in the mouth and allows for a rapid release of nicotine that can be titrated by the patient. The nicotine lollipop marketed by Bird's Hill Pharmacy contained nicotine salicylate combined with a natural sweetener and flavoring agents in a sugar-free candy base, and was available as a 2 and 4 mg dosage. The nicotine lollipop marketed by the Compounding Pharmacy also contained nicotine salicylate combined with a natural sweetener and flavorings in a sugar-free candy base, and was available in a 2 mg dosage strength. A similar nicotine lollipop marketed by Ashland Drug contained a nicotine base combined with a natural sweetener, flavoring agents, and a candy base and was available in 0.5, 1, 2, and 4 mg dosages. The therapeutic regimen for these products was similar in that the high-dosage strength products were for the heavy smoker and lower dosage strength products for the light smoker wishing to quit, and were used as needed to reduce nicotine craving.

These NR products addressed both the craving component of withdrawal by rapid delivery of nicotine, and the ritual of smoking (i.e., the hand-to-mouth fixation) that is a significant component of the smoking process. The lollipops were intended to help smokers quit their tobacco habit by suppressing the symptoms of nicotine withdrawal. Although these products were withdrawn from the market, other companies have evaluated the clinical development of similar nicotine lollipop-type delivery systems.

7. Smokeless Tobacco Products

Chewing tobacco, oral snuff, or tobacco sachets provide another smokeless method for delivery of nicotine to the buccal mucosa. Smokeless tobacco products such as sachets are especially popular in Scandinavia and in the United States and contain ground tobacco in packets that are sucked or held in the mouth. However, use of tobacco sachets results in nicotine blood levels that are more comparable to those resulting from nicotine gum compared to those resulting from smoking a cigarette (see Figure 7.1).[40] Nicotine sachets require approximately 30 minutes of use to attain the C_{max} level of approximately 12 ng/ml, which is less than 50 percent of the C_{max} obtained from smoking a cigarette. The slow absorption of nicotine from sachets may be due to the slow release of nicotine into the mouth and to swallowing a significant portion, which is only partly absorbed by the GIT. The pharmacokinetics and pharmacodynamics of nicotine absorbed from moist snuff placed between the cheek and gum for 30 minutes demonstrate that snuff products are capable of rapidly delivering high doses of nicotine, which can lead to dependence in the same way that smoking does.[82] For these reasons, and the fact that long-term use of snuff can lead to adverse health effects such as oral cancers, cardiovascular disease, and gingival disease, it should not be considered a viable option for smoking cessation. Thus, the smokeless tobacco products provide another alternative to smoking by providing low nicotine plasma levels compared to smoking and maintain lower steady-state nicotine blood levels, however, they do not provide the peak levels needed to satisfy severe craving.

C. Nicotine Nasal Sprays

Nicorette® Nasal Spray is a new form of NRT that rapidly relieves the symptoms of nicotine withdrawal. Nicotrol® Nasal Spray was approved as a prescription drug in the form of a metered unit-dose nicotine nasal spray by the FDA (NDA 20-385) on March 22, 1996 and is marketed by Pharmacia and Upjohn.[83] Previously this was only available by prescription but is now an OTC product available in the United Kingdom.[84] Nicotine nasal spray contains 100 mg of nicotine, a lethal dose, and delivers nicotine faster than an inhaler. The nasal solution has a pH of 7.0 and is provided in a metered-spray pump container. Nicotine is dissolved in a vehicle comprising polysorbate (Tween® 80), phosphate buffers, citric acid, parabens, edentate disodium, sodium chloride, and aroma.[10] The nasal spray has a greater abuse potential because the dose can be increased to the levels producing psychoactive effects.[10] The nicotine nasal spray is well suited to relieve the strong cravings in those who smoke more than 20 cigarettes a day, and seems to reduce withdrawal symptoms and craving, appealing to heavy smokers needing higher doses of nicotine.[85] Nicorette® Nasal Spray was designed to be sprayed into the nostrils where it is rapidly absorbed and some fraction of nicotine may subsequently be absorbed by the postnasal, intraoral mucosa, and GIT. The dosing regimen consists of spraying nicotine into one or both nostrils at each use, maximum of two doses (two sprays into each nostril) per hour for a total of three months; after eight weeks the number of doses should be decreased. The side effects include local irritation, runny nose, sneezing, and watery eyes.[84]

IV. NICOTINE PRODUCTS IN THE R&D PIPELINE

A. Intraoral Products in Development

Several companies have been involved in the research and development of nicotine IODs over the past several decades and the majority of the activity has been in the development of alternatives to transdermal patches, which have a relatively slow onset of action and are not as effective in treating the acute periods of cigarette cravings. Lozenges, sublingual tablets, and gums have been investigated as alternatives to the transdermal dosage forms to address the need to more rapidly deliver nicotine to treat the acute or breakthrough periods of craving. Other faster onset NR products such as nasal sprays and inhalators have been evaluated to address the acute period of craving.

Pharmetrix Corp. was an early innovator in developing the concept of nicotine sublingual tablets and lozenges in the early 1990s and this work resulted in several patents issued to the company and subsequently licensed to Pharmacia AB. The concept of combination therapy to improve smoking cessation therapy using a transdermal patch to provide for maintenance therapy and coadministration of nicotine in the form of a lozenge to address periods of acute craving was introduced by Pharmetrix in the mid 1990s and resulted in the issuance of several patents discussed below.

1. Nicotine Lozenge

Pharmetrix developed a lozenge, which contained 1 mg of nicotine base provided in a unit-dose foil pouch to improve the stability of the product. Typical formulations of the nicotine lozenge are shown in Table 7.7. These lozenges contain 1 mg of nicotine formulated with stabilizing excipients and taste-masking agents.

Table 7.7 Composition of the Pharmetrix Lozenge

	Formulations (mg/lozenge)[a]		
Composition	A	B	C
Nicotine	1.0	1.0	1.0
Mannitol	200	200	0
Xylitol	1309	1316	1400
Mint flavor	20	0	20
Tobacco flavor	0	6	0
Ammonium glycyrrihizinate	15	15	15
Sodium carbonate	5	5	5
Sodium bicarbonate	15	15	15
Hydrogenated vegetable oil	25	30	25
Magnesium stearate	10	10	10

[a] Taken from References 98, 106, and 107.

Taste-masking agents are important to improve the palatability of the lozenge because nicotine has a bitter peppery taste and requires formulation with flavoring agents to improve patient acceptance. The nicotine lozenges were designed to be manufactured by standard powder granulation and tablet compression processes. The lozenge was designed to slowly dissolve when placed in the mouth and sucked on to promote the release of nicotine and its subsequent absorption by the buccal mucosal tissues of the oral cavity. The typical dissolution

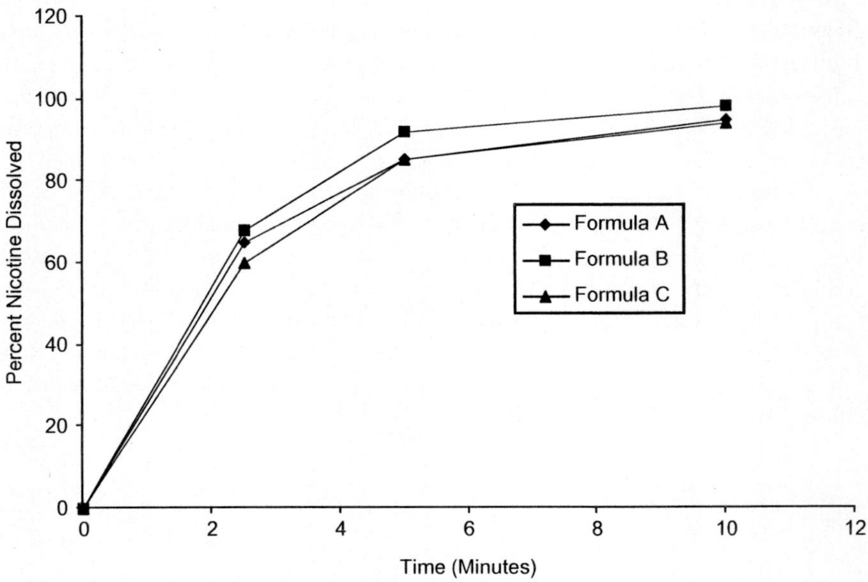

Figure 7.8 In vitro dissolution profile of a nicotine lozenge (compositions from Table 7.7; adapted from Reference 97).

profile of three nicotine lozenge formulations is shown in Figure 7.8. This proprietary technology has been licensed to Pharmacia; however, it has not yet been commercialized.

2. OT-Nicotine

The pharmacokinetic profile and behavioral characteristics of oral transmucosal nicotine are different from the other nicotine replacement products currently on the market. The element of control offered by the OT-nicotine lollipop-type device is different from that offered by the other treatment modalities. The ability to titrate the dose of nicotine and rapidly deliver adequate doses to the individual may more closely address the physiologic dependence and psychological craving effects of cigarettes. These properties of OT-nicotine may be especially effective in the early stages of smoking cessation and in smoking relapse prevention. The OT-nicotine device is comprised of a lollipop-like nicotine candy-based matrix on a stick resembling the length of a cigarette, as illustrated in Figure 7.9.

Anesta has completed several Phase 2 clinical studies on OT-nicotine in more than 400 smokers and nonsmokers. These studies have evaluated the effects of different dosage strengths, the acceptability of different taste formulations, and the ability of OT-nicotine to aid in smoking cessation with limited, short-term use and with eight weeks of therapy. In early studies, OT-nicotine has been shown to: (1) increase blood concentrations in proportion to dose, (2) decrease the desire

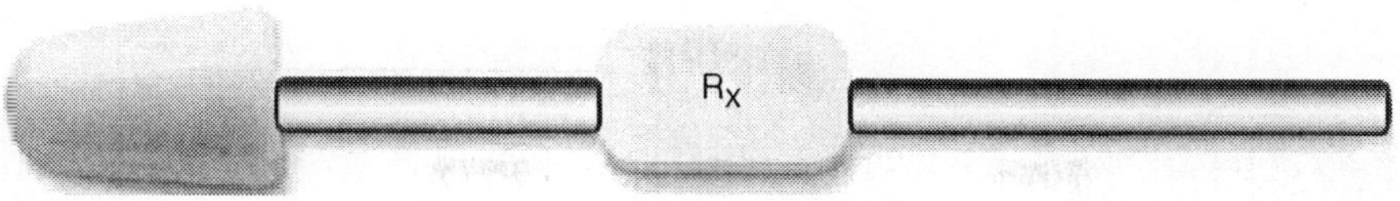

Figure 7.9 Example of OT-nicotine oral transmucosal device.

to smoke at higher doses, and (3) exhibit a high patient acceptance of the product. Side effects observed with OT-nicotine were mild and typical of those seen with nicotine administration. Clinical data from a larger Phase 2 study comparing smoking quit rates over an eight-week period of use showed that smokers using OT-nicotine achieved quit rates of 21 percent, similar to both the nicotine gum (16 percent) and the nicotine patch (21 percent).[86]

3. Nicotine Chewing Gums

Delivery of nicotine has been evaluated in other gums as well. For example, Perfetti, S.p.A. of Milan, Italy has been assigned U.S. patent 5,488,962 on a nicotine chewing gum. The chewing gum contains up to 25 wt percent of a gum base, and nicotine in a dose of 0.3 to 0.4 mg per gum strip, where the gum releases nicotine during chewing over a period of 20 minutes. The gum has the composition outlined in Table 7.8.

Table 7.8 Composition of a Nicotine Chewing Gum

Ingredients[a]	Composition (wt%)
Gum base	25.0
Glucose syrup	18.0
Sucrose	55.5
Glycerol	0.8
Natural flavoring (spearmint oil)	0.7
Nicotine	14.2 mg per 100 grams (0.4 mg per strip of 2.8 g gum)

[a] Adapted from Reference 104.

Delivery of nicotine from the gum was evaluated in a pharmacokinetic study in smokers who abstained from smoking for ten hours. Each subject was provided four pieces of chewing gum containing 0.4 mg of nicotine each (total 1.6 mg nicotine) and chewed for 20 minutes. Samples of saliva and plasma were taken and the pharmacokinetics of nicotine were evaluated and compared to a single piece of Nicorette® 2 mg nicotine gum, shown in Figure 7.10.

Formulation of nicotine polacrilex in a new directly compressible, free-flowing gum base powder excipient (Pharmagum™) and evaluation of its in vitro release has recently been reported.[87] Pharmagum™ (Pharmagum™ M and Pharmagum™ S) is a mixture of polyol(s) and/or sugars with a gum base and is

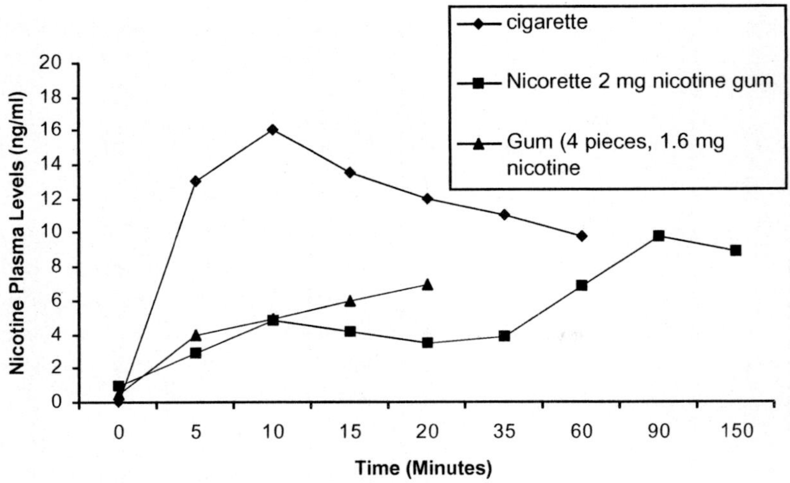

Figure 7.10 Comparison of nicotine plasma levels after smoking and chewing gums (adapted from Reference 32).

available as an excipient ready for formulation. Compositions containing 84.8 percent Pharmagum™ M or S, 2.2 percent nicotine polacrilex, 2.0 percent maganesium stearate, 8 percent sorbitol, and 3.0 percent sodium carbonate were compressed into tablets in a Manesty tablet press. The gums were hard and dissolution of nicotine into artificial saliva was evaluated in a chewing chamber at a temperature of 37°C with a chew rate of 60 chew/minute.[88] Release of nicotine from chewing gums was evaluated compared to Nicorette® 4 mg chewing gum and is shown in Figure 7.11. The release rate of nicotine from the Pharmagum™ base was faster compared to the conventional Nicorette® gum, in part due to the fact that the Pharmagum™ gum crumbles into particles to facilitate drug release, and then tends to allow the wetted gum particles to readhere, compared to the Nicorette®, which remains as a single mass.

4. Nicotine Transmucosal Patches

Nicotine replacement therapy requires a fast release of nicotine followed by a prolonged release of nicotine for maximal efficacy. Formulations of buccal bioadhesive tablets for a more prolonged release of nicotine have been reported.[89] A bilayer buccal adhesive nicotine tablet has been evaluated that provided a drug release pattern combining the fast-release and prolonged-release kinetics and resulted in improved smoking cessation rates.[90] A combination of 20 percent Carbopol 934 and 20 percent hydroxypropylcellulose (HPC) was found to provide suitable adhesion and controlled release of nicotine for up to four hours in human volunteers.

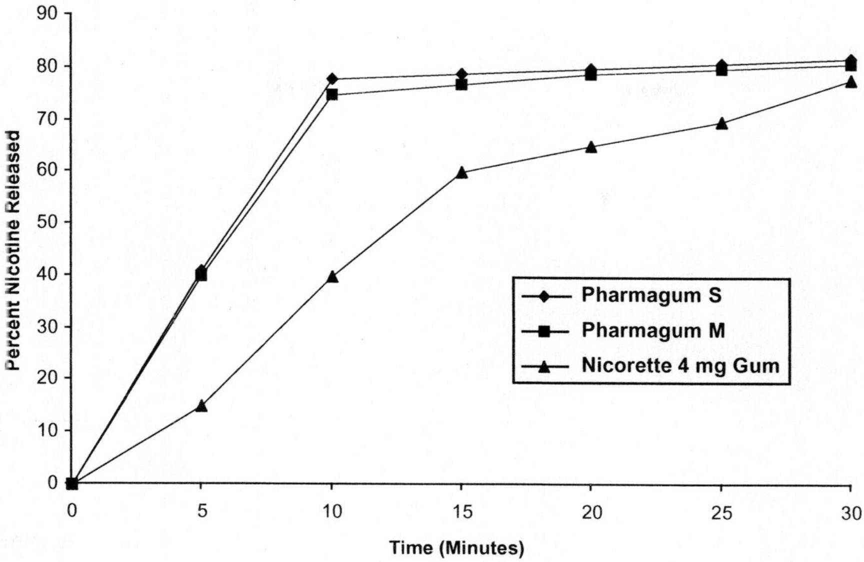

Figure 7.11 Nicotine dissolution from Pharmagum™ formulations and Nicorette® 4 mg gum (adapted from Reference 87).

B. Patent Domain: Nicotine Intraoral Products

The selected key innovator patents either issued or pending in the United States that provide protection for the commercial products and many others in the R&D pipeline on buccal and OT delivery of nicotine from lozenges or other dosage forms (i.e., lozenges, sublingual tablets, buccal patches, quick-dissolving films) are summarized from a recent patent search of issued U.S. patents and published patent applications in Table 7.9.

The important technical aspects of several of the key patents related to the delivery of nicotine from pharmaceutical compositions such as lozenges, sublingual tablets, buccal patches, lollipop-type candy, quick-dissolve films, and other solid IODs are briefly summarized below.

U.S. Patent 3,901,248: Work in the early 1970s by Aktiebolaget Leo in Sweden led to the first patent issued in 1975, claiming a chewable smoking substitute composition comprising 15 to about 80 percent gum base and a nicotine cation exchange resin complex dispersed in the gum. The cation exchange resin complex constitutes up to about 10 percent of the chewing gum composition and affords a nicotine release, when chewed, of approximately that available when smoking a conventional cigarette.[111] A wide variety of strongly acidic (polystyrene type containing sulfonic functional groups), weakly acidic (i.e., methacrylic type containing carboxylic functional groups), and mildly acidic (i.e., polystyrene type

Table 7.9 Key U.S. Patents and Patent Applications Issued and Pending on Buccal Delivery of Nicotine

		U.S. Patents: Delivery of Nicotine from a Lozenge or Buccally [a]			
Number	U.S. Patent No.	Filed (Issued)	Innovator company	Assignee	Title (Reference)
1	6,552,024B1	11-5-1999 (4-22-2003)	Lavipharm Laboratories	Lavipharm Laboratories	Compositions and methods for mucosal delivery (91)
2	6,592,887	6-5-2002 (7-15-2003)	LTS Lohmann Therapie-System AG	LTS Lohmann	Water soluble film for oral administration with instant wettability (92)
3	6,280,761	5-15-1996 (8-28-2001)	Pharmetrix Corp.	Pharmacia AB	Nicotine lozenge (93)
4	6,211,194	4-30-1998 (4-3-2001)	Duke University	Duke University	Solution containing nicotine (94)
5	6,177,096	4-6-1999 (1-23-2001)	LTS Lohmann Therapie-System GmbH	LTS Lohmann Therapie-System GmbH	Water soluble film for oral administration with instant wettability (95)
6	5,948,430	8-1-1997 (9-7-1999)	LTS Lohmann Therapie-System GmbH	LTS Lohmann Therapie-System GmbH	Water soluble film for oral administration with instant wettability (96)
7	5,721,257	6-7-1995 (2-24-1998)	Pharmetrix Corp.	Pharmacia AB	Method and therapeutic system for smoking cessation (97)
8	5,662,920	5-15-1996 (9-2-1997)	Pharmetrix Corp.	Pharmacia AB	Nicotine lozenge and therapeutic method for smoking cessation (98)
9	5,603,947	7-21-1994 (2-18-1997)	Cygnus Therapeutic Systems	Cygnus Therapeutic Systems	Method and device for providing nicotine replacement therapy transdermally/transbuccaly (99)
10	5,599,554	3-19-1996 (2-4-1997)	Procter & Gamble Company	Procter & Gamble Company	Treatment of nicotine craving and/or smoking withdrawal symptoms (100)
11	5,593,684	3-31-1994 (1-14-1997)	Pharmetrix Corp.	Pharmacia AB	Method and therapeutic system for smoking cessation (101)
12	5,549,906	6-26-1993 (8-27-1996)	Pharmetrix Corp.	Pharmacia AB	Nicotine lozenge and therapeutic method for smoking cessation (102)

Table 7.9 Key U.S. Patents and Patent Applications Issued and Pending on Buccal Delivery of Nicotine (continued)

Number	U.S. Patent No.	Filed (Issued)	Innovator company	Assignee	Title (Reference)
	U.S. Patents: Delivery of Nicotine from a Lozenge or Buccally [a]				
13	5,512,306	11-7-1994 (4-30-1996)	Pharmacia AB	Pharmacia AB	Smoking substitute (103)
14	5,488,962	12-2-1994 (2-6-1996)	Perfetti S.p.A.	Perfett, S.p.A.	Chewing gum which is a substitute for tobacco smoke (104)
15	5,453,424	10-19-1995 (8-6-1996)	Pharmacia AB	Pharmacia AB	Smoking substitute (105)
16	5,362,496	8-4-1993 (11-8-1994)	Pharmetrix Corp.	Pharmetrix Corp.	Method and therapeutic system for smoking cessation (106)
17	5,135,753	3-12-1991 (8-4-1992)	Pharmetrix Corp.	Pharmetrix Corp.	Method and therapeutic system for smoking cessation (107)
18	5,110,605	8-21-1990 (5-5-1992	Oramed, Inc.	Oramed, Inc.	Calcium polycarbophil-alginate controlled release composition and method (59)
19	5,048,544	8-10-1990 (9-17-1991)	Mascarelli, R.	None	Cigarette substitute (108)
20	4,967,773	6-26-1987 (11-6-1990	Shaw, A.S.	None	Nicotine containing lozenge (109)
21	4,806,356	4-3-1987 (2-21-1989)	Shaw, A.S.	None	Tobacco product (110)
22	3,901,248	8-15-1974 (8-26-1975	Aktiebolaget Leo	Aktiebolaget Leo	Chewable smoking substitute composition (111)
	U.S. Published Patent Applications				
1	20030119879	10-15-2002 (6-26-2003)	Landh, T., and Nils-Olof, L.	—	Nicotine and chocolate compositions (112)
2	20030087937	10-15-2002 (5-8-2003)	Nils-Olof, L.	—	Nicotine and cocoa powder compositions (113)

[a] Summarized from a patent search using keywords of nicotine and lozenge; and nicotine and buccal from Reference 114.

containing phosphonic functional groups) cation exchange resins were evaluated with nicotine and shown to form a reversibly dissolving drug complex when incorporated into various gum bases and chewing gum compositions. Several compositions were prepared and demonstrated controlled release of nicotine from resin complexes from a chewing gum in chewing studies in humans over a period of 30 minutes, where up to 90 percent of the nicotine was released. Gum compositions were claimed where 0.5 to 4 grams of gum comprise 15 to 80 percent of the formulation, and contain 0.5 to 2 percent of nicotine and 1 to 10 mg of nicotine cationic resin complex or up to 10 percent by weight.

U.S. Patent 5,110,605: This patent issued in 1992 to Oramed, Inc. claims a polymeric composition comprising a reaction complex formed by the interaction of a polycarbophil component with alginic acid or its salt in the presence of a divalent cation in the presence of an active agent. The composition comprises 0.1 to 90 percent calcium polycarbophil and 0.1 to 99 percent of alginic acid or its salt, which forms particles with the active agent passing through a sieve of 100 mesh size. The patent also claims the method of controlled release of an active form of the formulation, but does not specifically mention nicotine in the claims.[59] However, this patent is listed in the *Orange Book,* which provides protection for the Commit™ 2 and 4 mg nicotine polacrilex lozenges marketed by GlaxoSmithKline.[115]

U.S. Patent 5,488,962: This patent is similar to the early work of Aktiebolaget Leo described in U.S. 3,901,248. It was issued as U.S. 5,488,962 to Perfette, S.p.A. in 1996 which claims a chewing gum that is a substitute for tobacco smoke and is formed into three 3 g strips, characterized in that each strip contains 18 to 25 wt percent of a gum base and not more that 0.3 to 0.4 mg of nicotine dispersed in the gum base, and does not contain a cation exchange resin. The invention provides a nicotine-containing gum, which simulates cigarette smoking, without an unpleasant taste, and poor chewing characteristics.[104]

U.S. Patent 5,549,906: This patent issued in 1996 to Pharmetrix Corp. describes a method for smoking cessation therapy utilizing an improved nicotine lozenge to satisfy transient craving. The lozenge contains nicotine, a nonnutritive sweetener, and an absorbent excipient, and is indicated for reducing nicotine craving.[102] The patent claims the buccal delivery of nicotine from a lozenge that dissolves in the mouth within two to ten minutes. The lozenge is formulated with nicotine in an absorbent excipient such as beta-cyclodextrin and a nonnutritive sweetener such as xylitol and ammonium glycyrrhizinate and is buffered to a pH of between 6 and 11.

U.S. Patent 5,593,684: This patent issued in 1997 to Pharmetrix Corp. describes a method for treating conditions responsive to nicotine therapy, and particularly for smoking cessation therapy and for reducing nicotine craving that utilizes transdermal nicotine delivery for obtaining baseline plasma levels coupled with transmucosal administration of nicotine to satisfy transient periods of breakthrough craving.[101] This patent is similar to U.S. Patent 5,362,496; however, the patent claims the concurrent administration of transdermal and transmucosal nicotine, where the transmucosal administration provides transient plasma levels of nicotine about 5 ng/ml above that provided by the transdermal administration. The transmucosal administration of nicotine is provided by a lozenge, tablet, or capsule that dissolves in the mouth within two to ten minutes, and is buffered to a pH of 6.8 to 11.

U.S. Patent 5,599,554: This patent issued in 1997 to Proctor & Gamble describes a cigarette substitute composition comprising an edible lollipop portion containing a smoking substitute such as nicotine and a handle portion attached to the lollipop having the size and shape of a cigarette.[100] The lollipop portion is preferably a hard or semihard candy containing nicotine, sugar, or a sugar-based substitute which is pleasantly flavored with mint.

U.S. Patent 5,721,257: This patent issued in 1998 to Pharmetrix Corp. claims a method for nicotine smoking cessation therapy for reducing nicotine craving that utilizes transdermal nicotine delivery for obtaining baseline nicotine plasma levels coupled with transmucosal administration of nicotine to satisfy craving.[97] The method comprises the steps of first treatment with nicotine by transdermal administration to obtain blood levels of between 5 to 35 ng/ml for at least 12 hours, then a second treatment with nicotine by transmucosal administration through the nasal membranes by administration of an aerosol formulation that provides transient blood levels of nicotine about 5 ng/ml above those provided by transdermal administration.

U.S. Patent 6,177,096: This patent issued in 2001 to LTS Lohman Therapy-System GmbH describes a composition containing therapeutic agents including nicotine as the base or salicylate salt for delivery to the oral cavity in the form of a film, and is an extension of earlier technology patents on rapidly disintegrating sheetlike preparations and thin quick-dissolving (QD) film for intraoral delivery.[96] The film composition comprises at least one water-soluble polymer, at least one polyalcohol, and at least one pharmaceutical ingredient, wherein the composition has mucoadhesive properties and is intended to quickly dissolve and disintegrate upon administration in the oral cavity.[95] A typical film to deliver 1 to 2 mg of nicotine contains the following composition on a dry weight basis: 31.5 percent Kollidon, 42 percent HPMC, 15 percent flavoring agent, 5.5 percent nicotine salicylate, and 6.0 percent caramel liquid.

U.S. Patent 6,592,887: This patent issued in 2003 to LTS Lohman Therapy-Systems GmbH describes a method for release of 1 to 2 mg of nicotine as the base or salicylate salt form from a mucoadhesive film applied into the oral cavity where nicotine is released and delivered to the buccal mucosa.[92] Thin films can be manufactured using a solvent coating and drying process, and mucoadhesive nicotine film compositions claimed in an earlier patent are described.[95]

U.S. Patent 6,552,024B1: This patent issued in 2003 to Lavipharm Laboratories describes a novel flexible QD film containing nicotine that dissolves in the mouth without water.[91] This represents a new class of dosage form for intraoral delivery of nicotine, which is designed to dissolve quickly (within seconds to minutes) when applied to the tongue or sublingually. The dosage form is prepared by either a wet film coating or hot melt coating manufacturing process, where the formulation (i.e., solution for wet film coating or molten semisolid mass for hot melt coating) is cast onto a moving web, dried, and cooled, resulting in a thin flexible film, and then die-cut (into any size or shape) and packaged into unit-dose blister packs or multidose containers. The patent claims a film composition of a water-soluble hydrocolloid, an active agent, and a mucosal adhesion enhancer. The film dosage unit is thin (between 1 to 20 mils), having a short dissolution time (between 10 to 600 seconds), and contains 0.01 to 75 wt percent of an active agent. Hydroxypropylmethyl cellulose is the preferred hydrocolloid, which can be formulated with other excipients such as emulsifying agents, plasticizers, taste-modifying agents, coloring agents, fillers, preservatives, buffering agents, and stabilizers. The composition of a 3 mil thick QD nicotine film on a weight basis comprises: 5.2 percent nicotine, 3.7 percent peppermint, 78 percent HPMC (Methocel E5), 3.7 percent propylene glycol, 2.9 percent aspartame, 2.6 percent citric acid, 3.7 percent Cremphor EL40, and 0.04 percent benzoic acid. The film disintegrates in vitro in 43 seconds and dissolves in 74 seconds.

C. Future Products

Future OT products for NRT are anticipated based on the activity from the patent literature and products in the R&D pipeline. New products to treat the acute periods of breakthrough craving and the acute withdrawal effects of smoking may be addressed with rapid delivery of nicotine in the form of second-generation quick-dissolve lozenges, and QD films. These classes of rapid onset OT dosage forms to deliver 2 to 4 mg of nicotine may be an effective means to better simulate the rapid delivery of nicotine from cigarettes. Other improved and second-generation OT dosage forms address the lower intensity craving over a longer period (hours), and the more severe withdrawal components of nicotine withdrawal may be better treated with other slower dissolving (ten minutes to several hours) nicotine dosage forms such as lozenges, sublingual tablets, gums, and lollipops, as well as OT-nicotine. As the technology continues to evolve, OT dosage forms such as buccal tablets and patches that adhere to the mucosal tissue (1 to 24 hours) may also provide for an alternative to TNS for prolonged and controlled delivery of lower levels of nicotine to maintain low-level craving and withdrawal symptoms in check for periods of from 16 to 24 hours.

V. SUMMARY

Nicotine is a natural alkaloid and one of the few drugs having physicochemical properties that are well suited for delivery and absorption by the lungs (i.e., smoking cigarettes and inhalators), oral mucosal (i.e., smokeless tobacco, sprays, lozenges, sublingual tablets, and gums) and nasal tissues (i.e., sprays), as well as the skin (i.e., transdermal nicotine patches). There are now many nicotine replacement products on the market used as aids to assist in smoking cessation. Transdermal nicotine patches were one of the first classes of NR products on the market for use as a deterrent to smoking and are relatively effective in providing daily maintenance therapy. New oral transmucosal nicotine replacement products have been more recently introduced including gums, lozenges, lollipops, sublingual tablets, inhalators, and nasal sprays, which are very effective in relieving the urge and acute craving component for smoking and these products also effectively address the behavioral component of the ritual of putting tobacco or cigarettes in the mouth. The oral mucosal nicotine products provide for a more rapid onset compared to transdermal nicotine delivery systems and are more effective in addressing the acute craving component of nicotine addiction. Together, transdermal nicotine maintenance therapy along with oral transmucosal products for breakthrough periods of smoking urges may be more effective than either alone in the design of improved and more effective programs and dosing regimens for smoking cessation. The safety and effectiveness of nicotine delivery by the transdermal and intraoral and nasal transmucosal routes are now well established and have gained their place in the armamentarium of products for nicotine replacement therapy. As innovation continues, new and improved NR technologies, dosage forms, and combination formulations in the R&D pipeline and clinical development, such as lozenges, lollipops, buccal patches, QD thin films, and orally disintegrating tablets, may eventually emerge as products for nicotine replacement and will be welcome additions to the therapeutic arsenal directed toward improved smoking cessation programs and nicotine replacement therapy.

ACKNOWLEDGMENTS

The author acknowledges Dr. Tapash Ghosh for his contribution to this chapter and I thank my colleagues at Pharmetrix Corp. and Lavipharm Inc. for their role in publishing and patenting the technologies that we jointly developed and their work cited on the novel nicotine IODs reported in this chapter.

REFERENCES

1. Henningfield, JE, Fant, RV, Gitchell, J, and Shiffman, S. Tobacco dependence. Global public health potential for new medications development and indications. Ann. N.Y. Acad. Sci. 909:247–256, 2000.
2. Kristein, MM. Economic issues in prevention. Prevent. Med. 6:252–264, 1977.
3. Nixon, JC and West, JF. Cost reductions from a smoking policy. Employee Benefits J. 14(1):26–30, 1989.
4. Public Health Service, Office on Smoking and Health, The health consequences of smoking: Cardiovascular disease. DHHS (PHS) Pub No 84-50204. Washington, DC: US Department of Health and Human Services, 1983.
5. Henningfield, JE, Fant, RV, and Tomar, SL. Smokeless tobacco: An addicting drug. Adv. Dent. Res. 11(3):330–335, 1997.
6. The Merck Index, An Encyclopedia of Chemicals, Drugs, and Biologicals, Eleventh ed, Ed. S. Budavari, Merck & Co., Inc., Rahway, NJ, p. 1030, 1989.

7. Roy, SD. Preformulation aspects of transdermal drug delivery systems, transdermal and dermal therapeutic systems: current status. In: Transdermal and Topical Drug Delivery Systems, Eds. TK Ghosh, WR Pfister, and S-I Yum, Interpharm Press, Inc., Buffalo Grove, IL, pp. 139–166, 1997.
8. Nicotine Material Safety Data Sheet, International Chemical Safety Cards, ICSC: 0519, June 7, 2001.
9. Nicotine Material Safety Data Sheet, Acros Organics N.V., Fairlawn, NJ, December 16, 1997.
10. Miscellaneous Autonomic Drugs, Nicotine and Nicotine Polacrilex. In: AHFS Drug Information, Ed. GK McEvoy, American Society of Health-System Pharmacists, Inc. Bethesda, MD, pp. 1367–1384, 2002.
11. Cleary, GW. The first two decades of transdermal drug delivery systems and a peek into the 1990s. In: Dermal and Transdermal Drug Delivery, Eds. R Gurney and A Teuber, Wissenschaftliche Verlagsgessellschaft GmbH, Stuttgart, pp. 13–32, 1993.
12. Lange's Handbook of Chemistry, Fifteenth Edition, Ed. JA Dean, McGraw-Hill, Inc., New York, pp. 1.282 and 8.62, 1999.
13. Drugs: Synonyms and Properties, Second Edition, Ed. GWA Milne, Ashgate Publishing Co., Burlington, VT, p. 563, 2002.
14. Lobenberg, R, and Amidon, GL. Modern bioavailability, bioequivalence and biopharmaceutics classification system. New scientific approaches to international regulatory standards. Eur. J. Pharm. Biopharm. 50(1):3–12, 2000.
15. Horspool, K and Lipinski, CA. Enabling strategies: Advancing new drug delivery concepts to gain the lead. Drug Delivery Tech. 3(7):34–44, October 2003.
16. Gopalakrishnan, M and Donnelly-Roberts, DL. Nicotine therapeutic prospects? Pharmaceutical News. 5(5):16–20, 1998.
17. Volle, RL. Nicotine and ganglion blocking drugs. In: Texbook of Pharmacology, Eds. CM Smith and AM Reynard, WB Saunders, Philadelphia, pp. 116–126, 1992.
18. Matschiner, S, Neubert, R, and Wohlrab, W. The use of ion pairing to facilitate percutaneous absorption of drugs. In: Percutaneous Penetration Enhancers, Eds. EW Smith and HI Maibach, CRC Press, Boca Raton, pp. 407–417, 1995.
19. Hao, J and Heng, WS. Buccal delivery systems. Drug Development and Industrial Pharmacy, 29(8):821–832, 2003.
20. Nielsen, HM and Rassing, MR. Nicotine permeability across the buccal TR146 cell culture model and porcine buccal mucosa in vitro: Effect of pH and concentration. Eur. J. Pharm. Sci. 16(3):151–157, 2002.
21. Oakley, DM and Swarbrick, J. Effects of ionization on the percutaneous absorption of drugs: Partitioning of nicotine into organic liquids and hydrated SC. J. Pharm. Sci. 76:866, 1987.
22. Henningfield, JE, Radzius, A, Cooper, TM, and Clayton, RR. Drinking coffee and carbonated beverages blocks absorption of nicotine from nicotine polacrilex gum. J. Am. Med. Assoc. 264(12):1560–1564, 1990.
23. Tomar, SL and Henningfield, JE, Review of the evidence that pH is a determinant of nicotine dosage from oral use of smokeless tobacco. Tob. Control. 6(3):219–225, 1997.
24. Rathbone, MJ. Oral Mucosal Drug Delivery, Marcel Dekker, New York, 1996.
25. Veillard MM. Buccal and gastrointestinal drug delivery systems. In: Bioadhesion—Possibilities and Future Trends, Eds. R Gurny and HE Junginger, Stuttgart: Wissenschaftliche Verlagsgesellschaft GmbH, 1990.

26. Nair, MK, Chetty, DJ, Ho, H, and Chien, YW. Biomembrane permeation of nicotine: Mechanistic studies with porcine mucosae and skin. J. Pharm. Sci. 86(2):257–262, 1997.
27. Fincham, JE. Smoking Cessation Products, Chapter 36. In: Handbook of Nonprescription Drugs. Ed. TR Covington, American Pharmaceutical Association, Washington, DC, pp. 715–723, 1996.
28. Steuer, JD and Wewers, ME. Cigarette craving and subsequent coping responses among smoking cessation clinic participants, Oncology Nursing Forum. Mar–Apr; 16(2):193–198, 1989.
29. Harrington, N. The craving factor in the treatment of smoking, Br. J. Soc. Clin. Psychol., Nov; 17(4):363–371, 1978.
30. Russell, MA. In Nicotine Replacement: A Critical Evaluation, Eds. Pomerleau, OF and Pomerleau, CS, Alan R. Liss, Inc, New York, pp. 63–94, 1988.
31. Henningfield, JE and Keenar, RM. Nicotine delivery kinetics and abuse liability. J. Consult. Clin. Psychol. 61(5):743–750, 1993.
32. Kumar, R, Cooke EC, Lader, MH, et al. Is nicotine important in tobacco smoking? Clin. Pharm. Therapeut. 21:520–529, 1997.
33. McNeill, A, Foulds, J, and Bates, C. Regulation of nicotine replacement therapies (NRT): A critique of current practices. Addiction 96:1757–1768, 2001.
34. Sweeney, CT, Fant, RV, Fagerstrom, KO, McGovern, JF, and Henningfield, JE. Combination nicotine replacement therapy for smoking cessation: rationale, efficacy and tolerability. CNS Drugs 16(6):453–467, 2001.
35. Anonymous. Use of nicotine to stop smoking. Med. Lett. Drugs Ther. 37:6–8, 1995.
36. Pfister, WR. Transdermal and dermal therapeutic systems: current status. In transdermal and topical drug delivery systems, Eds. TK Ghosh, WR Pfister, and S-I Yum, Interpharm, Buffaio Grove, IL, pp. 33–112 (1997).
37. Nicotine USP and Nicotine Polacrilex USP, Dictionary of USAN and International Drug Names, USP Pharmacopeial Convention Inc., Rockville, MD, p. 610 (2002).
38. Steuer, JD and Wewever, ME. Cigarette craving and subsequent coping responses among smoking cessation clinic participants, Oncol. Nursing Forum. Mar-Apr. 16(2):193–198, 1989.
39. Silagy, C, Mant, D, Fowler, G, and Lancaster T. Nicotine replacement therapy for smoking cessation. Cochrane Database Syst., Rev.3:CD000146, 2000.
40. Benowitz, NL. Pharmacologic aspects of cigarette smoking and nicotine addition, New Eng. J. Med. 319:1318–1330, 1998.
41. Schneider, NG. Nicotine therapy in smoking cessation: pharmacokinetic considerations, Clin. Pharmacokinet. 23(3):169–172, 1992.
42. Patient Package Insert, Nicotrol® NS (nicotine nasal spray), Physicians Desk Reference, Medical Economics Co., Montvale, NJ, 2843–2846, 2002.
43. Henningfield, JE and Keenan, RM. Nicotine delivery kinetics and abuse liability, J. Consult. Clin. Psychol. 61(5):743–750, 1993.
44. Fant, RV, Henningfield, JE, Shiffman, S, Strahs, KR, and Reithberg, DP. A pharmacokinetic crossover study to compare the absorption characteristics of three transdermal nicotine patches. Pharmacol. Biochem. Behav. 67(3):479–482, 2000.
45. Martindale the Complete Drug Reference, Thirty-Second Edition, Ed. K Parfitt, Pharmaceutical Press, London, England, pp. 1481–1482, 1999.
46. Nunn-Thompson, CL and Simon. PA. Pharmacotherapy for smoking cessation. Clin. Pharm. 8:710–720, 1989.

47. Gourlay, SG and McNeil, JJ. Antismoking products. Med. J. Aust. 153:699–707, 1990.
48. Press Release, FDA Announces that No Over-The-Counter Smoking Deterrent Works, Food and Drug Administration, June 2, 1993.
49. Drug Facts and Comparisons. St. Louis, Mo. Facts and Comparisons, 1998.
50. Houtsmuller, EJ, Henningfield, JE, and Stitzer, ML. Subjective effects of the nicotine lozenge: assessment of abuse liability, Psychopharmacology (Berl.) 167(1): 20–27, 2003.
51. Silagy, C, Lancaster, T, Stead, L, Mant, D, and Fowler, G. Nicotine replacement therapy for smoking cessation. Cochrane Database Syst., Rev. 3: CD000146, 2001.
52. Buffery, P. Nicotine replacement therapy at a glance. UKMi, Northwest Medicines Information Centre, Liverpool, UK, UKMi. April 2001.
53. Conway, B. Chewing gum as a drug delivery system. The Drug Delivery Companies Report, Autumn/Winter, pg. 33–35, 2003.
54. Kvist, C, Andersson, SB, Fors, S, Wennergren, B, and Berglund, J. Apparatus for studying in vitro drug release from medicated chewing gums. Int. J. Pharm. 189(1): 57–65, 1999.
55. Houtsmuller, EJ, Fant, RV, Eissenberg, TE, Henningfield, JE, and Stitzer, ML. Flavor® improvement does not increase abuse liability of nicotine chewing gum. Pharmacol. Biochem. Behav. 72(3):559--568, 2002.
56. Nicorette® 2 and 4 mg (nicotine polacrilex) gum. CDER New and Generic Drug Approvals: 1998–2004, January 19, 2004, http://www.fda.gov/cder/approval/index.htm.
57. Nicorette® gum, 4 mg, CDER New and Generic Drug Approvals: 1998–2004, January 19, 2004, http://www.fda.gov/cder/approval/index.htm.
58. Nicorette®, Med. Ad. News, 20(10):5, October 2001.
59. Acharya, RN. A calcium polycarbophil-alginate controlled release composition and methods. US Patent 5,100,605, May 5, 1992.
60. Nicorette® Gum patient information, GlaxoSmithKline, January 19, 2003, http://nicorette.quit.com/nicr_internal/howhelps.asp.
61. Acharya, RN. A calcium polycarbophil controlled release composition and methods. US Patent 5,102,666, April 7, 1992.
62. Nicorette® product and packaging information. Pittsburg, PA, SmithKline Beecham Consumer Healthcare, LP, 1996.
63. FDA CDER New and Generic Drug Approvals: 1998–2004, January 19, 2004, http://www.fda.gov/cder/approval/index.htm.
64. Patient Product Information, Nicotinell® Fruit 2 mg Chewing Gum, April 2001.
65. Kvist, C, Andersson, SB, Fors, S, Wennergen, B, and Berglund, J. Apparatus for studying in vitro drug release from medicated chewing gums. Int. J. Pharm. 189(1): 57–65, 1999.
66. CDER New and Generic Drug Approvals: 1998–2003, http://www.fda.gov/cder/approval/index.htm.
67. Commit™ Lozenge, Material Safety Data Sheet, GlaxoSmithKline, November, 5, 2002.
68. Acharya, RN. A calcium polycarbophil-alginate controlled release composition and methods. US Patent 5,100,605, May 5, 1992.
69. Commit™ nicotine polacrilex lozenge, Electronic Orange Book Approved Drug Products with Therapeutic Equivalence Evaluations, updated December 22, 2003, January 19, 2004, http://www.fda.gov/cder/ob/default.htm.

70. Acharya, RN. A calcium polycarbophil controlled release composition and methods. US Patent 5,102,666, April 7, 1992.
71. Shiffman, S et al. Efficacy of a nicotine lozenge for smoking cessation. Archives of Int. Med. 162(11):1267–1276, 2002.
72. Patient Product Information, Nicotinell® mint 1 mg lozenge, March, 2001.
73. Molander, L and Lunell, E. Pharmacokinetic investigation of a nicotine sublingual tablet. Eur. J. Clin. Pharm. 56:813–819, 2001.
74. Society for research on nicotine and tobacco, nicotine replacement therapy, Pharm. J., 261:7018, October 31, 1998.
75. Nicorette® Microtab patient instructions, GlaxoSmithKline, January 19, 2003, http:// www.nicorette.co.uk/products/microtab.asp.
76. DiPasquale, CB. Virginia Company to market tobacco lozenge, critics charge candy-like nicotine treat will lure children. www.AdAge.com/adnews.cms?newsID=33424, January 26, 2003.
77. Schuh, KJ, Schuh, LM, Henningfield, JE, and Sitzer, ML. Nicotine nasal spray and vapor inhaler: abuse liability assessment. Psychopharmacology. (Berl.) 130(4): 352–361, 1997.
78. Nicotrol® nicotine inhalant, Electronic Orange Book Approved Drug Products with Therapeutic Equivalence Evaluations, updated December 22, 2003, January 19, 2004, http://www.fda.gov/cder/ob/default.htm.
79. Nicorette® Inhalator patient information, GlaxoSmithKline, January 19, 2003, http:// nicorette.quit.com/nicr_internal/aboutquit_fs.asp.
80. Butschky, MF, Bailey, D, Henningfield, JE, and Pickworth, WB. Smoking without nicotine delivery deceases withdrawal in 12-hour abstinent smokers. Pharmacol. Biochem. Behav. 50(1):91–96, 1995.
81. FDA, CDER, Warning Letters and Untitled Letters to Pharmaceutical Companies 2002, http://www.fda.gov/cder/warn/warn2002.htm.
82. Fant, RV, Henningfield, JE, Nelson, RA, and Pickworth, WB. Pharmacokinetics and pharmacodynamics of moist snuff in humans. Tob. Control. 8(4):387–392, 1999.
83. Nicotrol® nasal spray, Electronic Orange Book Approved Drug Products with Therapeutic Equivalence Evaluations, updated December 22, 2003, January 19, 2004, http://www.fda.gov/cder/ob/default.htm.
84. Nicorette® nasal spray patient information, GlaxoSmithKline, January 12, 2003, http://nicorette.quit.com/main.asp.
85. Nicorette® nasal spray patent information, GlaxoSmithKline, January 19, 2003, http://nicorette.quit.com/ncr_internal/aboutquit_fs.asp.
86. OT-Nicotine Product Information, Anestha Corp., January 20, 2004, http://www.xmission.com/anesta/pipeline/nicotine.html.
87. Morjaria, Y, Conway, BR, Irwin, WJ, Barnett, PX, and Chan, RS. Chewing gum as a drug delivery system. Control. Rel. Soc. Annual Meeting Proceed. #651, 2003.
88. European Pharmacopoeia Supplement, 3rd ed., 129–130, 2001.
89. Park, AO, Duguid, CR, and Bain, DF. Formulation of a bioadhesive buccal tablet for nicotine replacement therapy. J. Pharm. Pharmacol. 51 (Suppl Sept):286, 1999.
90. Park, CR and Munday, DL. Development and evaluation of a biphasic buccal adhesive tablet for nicotine replacement therapy. Int. J. Pharm. 237(1–2):215–226, 2002.
91. Chen, L-L, Pfister, WR, Renn, DW, Buranachokpaisan, T, Osborne, J, Tran, HS, and Tao, L. Compositions and methods for mucosal delivery. US Patent 6,552,024B1, April 22, 2003.

92. Zerbe, HG, Guo, J-H, and Serino, A. Water soluble film for oral administration with instant wettability. US Patent 6,592,887, July 15, 2003.
93. Santus, GC. Nicotine Lozenge. US Patent 6,280,761B1, August 28, 2001.
94. Westman, EC, Rose, JF, and Tomlin, KF Solution containing nicotine. US Patent 6,211,194, April 3, 2001.
95. Zerbe, HG, Guo, J-H, and Serino, A. Water soluble film for oral administration with instant wettability. US Patent, 6,177,096, January 23, 2001.
96. Zerbe, HG, Guo, J-H, and Serino, Water soluble film for oral administration with instant wettability. US Patent, 5,948,430, September 7, 1999.
97. Baker, RW, Santus G, and Vintilla-Friedman, S. Method and therapeutic system for smoking cessation. US Patent 5,721,257, February 24, 1998.
98. Santus, GC. Nicotine lozenge and therapeutic method for smoking cessation. US Patent 5,662,920. September 2, 1997.
99. Wong, O, Farinas, K, Cleary, GW, Chiang, C-M, and Xia, J. Method and device for providing nicotine replacement therapy transdermally/transbuccally. US Patent 5,603,947, February 18, 1997.
100. Majeti, S. Treatment of nicotine craving and/or smoking withdrawal symptoms. US Patent 5,599,554, February 4, 1997.
101. Baker, RW, Santus, GC, and Vintilla-Friedman, S. Method and therapeutic system for smoking cessation. US Patent 5,593,684, January 14, 1997.
102. Santus, GC. Nicotine lozenge and therapeutic method for smoking cessation. US patent 5,549,906, August 27, 1996.
103. Carlsson, T and Anderson, SB. Smoking substitute. US Patent 5,512,306, April 30, 1996.
104. Perfetti, G. Chewing gum which is a substitute for tobacco smoke. US Patent 5,488,962, February 6, 1996.
105. Carlsson, T and Anderson, SB. Smoking substitute. US Patent 5,543,424, August 6, 1996.
106. Baker, RW, Santus, GC, and Vintilla-Friedman, S. Method and therapeutic system for smoking cessation. US Patent 5,362,496, November 8, 1994.
107. Baker, RW, Santus, GC, and Vintilla-Friedman, S. Method and therapeutic system for smoking cessation. US patent 5,135,753, August 4, 1992.
108. Mascarelli, R and Mascarelli, G. Cigarette substitute. US Patent 5,048,544, August 10, 1990.
109. Shaw, AS. Nicotine Containing Lozenge. US Patent 4,967,773, June 26, 1987.
110. Shaw, AS. Tobacco product. US Patent 4,806,356, April 3, 1987.
111. Lichtneckert, S, Lundgren, C, and Ferno, O. Chewable smoking substitute composition. US Patent 3,901,248, August 26, 1975.
112. Landh, T and Lindberg, N-O, Nicotine and chocolate compositions. US Patent Application 20030119879A1, June 26, 2003.
113. Lindberg, N-O, Nicotine and cocoa powder compositions. US Patent Application 20030087937A1, May 8, 2003.
114. US Patent and Trademark, USPTO Patent Full-Text and Image Database, 1976 to present, http://patents.uspto.gov/, February 3, 2004.
115. Commit™ nicotine polacrilex lozenge, Electronic Orange Book Approved Drug Products with Therapeutic Equivalence Evaluations, updated December 22, 2003, January 19, 2004, http://www.fda.gov/cder/ob/default.htm.

8

Oral Transmucosal Powder Injection

Sung-Yun Kwon, Terry L. Burkoth, David Longridge, Huw Millward, and Brian J. Bellhouse

CONTENTS

I. INTRODUCTION

A. Medical Rationale for Transmucosal Drug Delivery

Transmucosal drug delivery has the advantages of avoiding presystemic metabolism and the hepatic first-pass effect (1). The oral mucosal tissue is thicker than the epidermis but has relatively high permeability and is rarely involved in the hypersensitivity reaction common in skin. In addition, oral mucosal tissue heals rapidly. For these reasons, transmucosal delivery is a very promising alternative to parenteral delivery in general and relaxes molecular weight and chemical character limitations on the transdermal delivery of macromolecular drugs.

It is clear, however, that the oral mucosal membranes do have some barrier function and the bioavailability of macromolecular drugs applied to the surface can be poor (0.1 to 5 percent) because of limitations of membrane permeability or metabolism at the absorption site (2–8). In order to overcome the diffusion barrier and improve bioavailability, several different approaches have been investigated. These include formulating the drug with absorption enhancers, increasing bioadhesion to prolong contact, and formulating to sustain drug release. More drastic measures include drug derivitization to alter the physicochemical properties and drug modification or inclusions of additives to inhibit enzymatic degradation and to increase transport rate (9–15).

A new drug delivery technology being developed is the injection of solid powder forms of therapeutic compounds into transdermal or transmucosal membranes thus overcoming their barrier function. Minimally invasive injection techniques provide needle-free and pain-free particle injection of traditional drugs, drugs from biotechnology such as proteins, peptides, and oligonucleotides, as well as traditional and genetic vaccines and other therapeutic compounds. Unlike traditional transdermal patch technology, there is no theoretical limit to the molecular weight for delivery and preclinical/clinical studies have shown a very short lag time for the drugs to reach the systemic circulation and thus have the potential to provide therapeutic benefit. Fine drug particles of 20 to 100 μ mass mean aerodynamic diameter are accelerated to supersonic velocities (300 to 900 m sec^{-1}) within a transient helium gas jet and thereby gain sufficient momentum to be ballistically delivered into skin or mucosal sites. Depending on the particle size, density, nozzle design, and helium pressure, penetration depth can be optimized for specific applications. More important, patients feel no pain and there is no tissue damage or adverse reaction. The purpose of this chapter is to introduce the Oral PowderJect® system, the application of needle-free injection technology specifically to deliver powdered drugs into or via oral mucosal tissue and to review the preclinical and clinical feasibility data for this technology.

B. Oral Mucosal Membranes

The oral mucosal membrane is composed of a stratified squamous epithelium quite similar to skin. The total surface area is approximately 100 cm^2. Lipids and glycolipids exuded from the membrane coating granules of the cells pack between flattened cells. These lipids may be organized but are less structured than those found in the stratum corneum in skin. Because of this more loosely packed organization of intercellular lipids, the oral mucosal membranes are generally more permeable than skin (16–18). Internally, cells are filled with proteins and keratin. The main pathway of diffusion across the oral mucosal membrane is intercellular via the lipid route.

There are three structurally distinct oral mucosa as barriers to drug delivery.

1. *Masticatory mucosa* has a keratinized epithelium that covers the gingiva and hard palate. These mucosal tissues are strongly attached to the underlying structure by collagenous connective tissue.
2. *Lining mucosa* is a layer that is not keratinized and covers the sublingual and buccal surfaces and the lining of the lips. Mucous membrane on these surfaces is attached to the underlying structure by a loose, elastic connective tissue. This epithelial layer is significantly different in thickness, turnover time, and barrier properties.
3. *Specialized mucosa* has characteristics of both the keratinized and nonkeratinized epithelium and covers only the dorsum of the tongue.

The oral epithelium is the primary barrier to transmucosal drug delivery and the structure of each of the different types of mucosal tissue has a significant influence on the transport of drugs. The basement membrane between the epithelium and the underlying connective tissue also plays a role as a secondary and minor barrier (17,19–21). This connective tissue does not appear to offer a significant diffusional barrier to most compounds.

For systemic drug delivery, the buccal and sublingual membranes may be useful sites because of their high permeability. The buccal mucosa has a thickness of 580 μm, an area of 50 cm^2, and a turnover time of 13 days, and the sublingual mucosal thickness is 190 μm, the area is 26 cm^2, and the turnover time is 20 days. Generally the sublingual membrane is more permeable than the buccal membrane and has been used for this characteristic as a drug delivery site. The sublingual site has been used historically as a portal for traditional drugs such as the administration of organic nitrates for treatment of angina and ergot alkaloids for treatment of migraine. Commercial formulations for sublingual or buccal delivery include fast-dissolving dosage forms such as Expidet® for oxazepam or lorazepam from Wyeth and Lyocs® for phloroglucinol from Lafon (22).

C. Oral Mucosal Vascularization

Blood is supplied to the oral mucosa primarily from the external carotid artery, which serves several large buccal blood vessels (23). The floor of the mouth, the root of the tongue, and the buccal mucosa of the cheek are considered to be the most highly vascularized areas of the oral cavity. Blood perfusion is high; for example, the buccal blood flow in monkeys was found to be 20 to 25 ml/min per

100 g tissue (24). Moreover, a diversified drainage of venous and lymphatic capillaries allows for rapid exchange of metabolic products and uptake of drugs through the mucosa. The arterial as well as the venous and lymphatic capillaries reach deep into the multilayered epithelium via the infiltrating connective tissue papillae and ensure maximum vascular access to drugs delivered by the oral mucosal route. This high degree of vascularization combined with relative lack of keratinization make these sites the most promising for oral mucosal drug delivery. In contrast to the oral mucosa, the structure, metabolic activity, and blood perfusion of the skin make it a less ideal portal for drug delivery (25).

II. POWDER INJECTION SYSTEM

A. The Oral PowderJect® System

The Oral PowderJect® system is a safe and effective needle-free technique for transmucosal powder delivery. The Oral PowderJect® system, in common with the range of PowderJect delivery systems (26), harnesses the energy from a compressed gas source to accelerate and deliver a range of powdered or particulate compounds to the oral cavity. The ballistic nature of powder injection is a fundamental distinction between the Oral PowderJect® system and diffusion-driven systems such as oral tablets and intraoral patches.

The Oral PowderJect® delivery system (Figure 8.1) is a handheld device for access and drug delivery to the different types of mucosal tissue within the

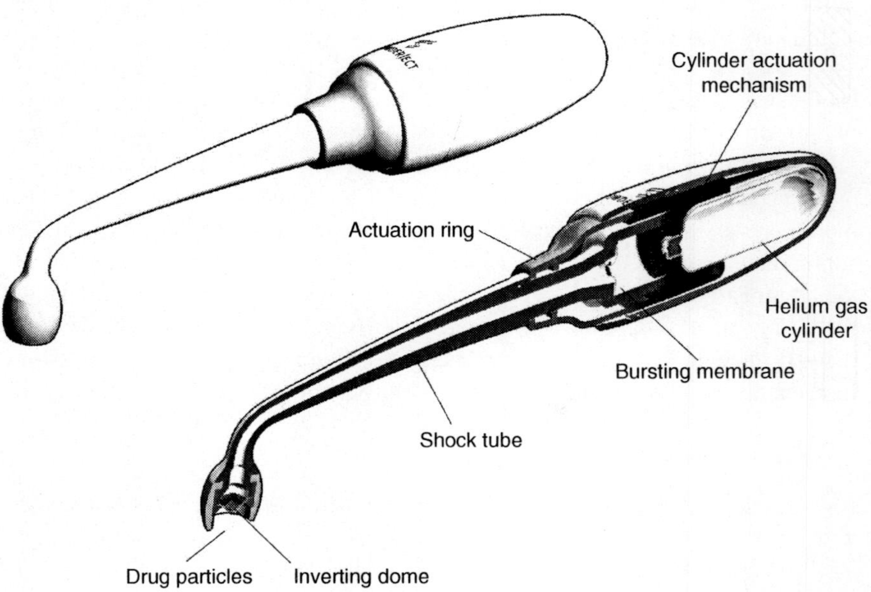

Figure 8.1 Schematic diagram of Oral PowderJect®.

mouth. Prior to actuation, the compliant tip of the device is placed over the relevant area of the oral mucosa. Within the tip a dose of powdered drug is retained on the concave face of an inverting dome, and the dome is held at a predefined distance from the oral mucosa. Following actuation, a gas pressure pulse strikes the flexible dome with sufficient energy to invert the dome. This action accelerates the powdered drug such that the drug leaves the surface of the inverting dome with sufficient velocity to penetrate the oral mucosa. The dome itself does not come into contact with the oral mucosa and the dome stops the high-pressure gas flow from being released into the mouth; hence the Oral PowderJect® system is pain-free and inherently quiet (27,28).

A cross-sectional view of the Oral PowderJect® prototype delivery system employed in the Phase I clinical trials is shown in Figure 8.2, highlighting the four key components.

1. Gas Cylinder

The prototype design incorporates an O-ring sealed piston within a reusable stainless-steel body. The cylinder can be pressurized up to 80 bar, and has a safety factor in excess of 4.0. The compressed gas is released through linear displacement of the piston to provide an annular exit area of approximately 1.0 mm^2. The gas cylinder incorporates a safety catch or clip (not shown) in order to prevent

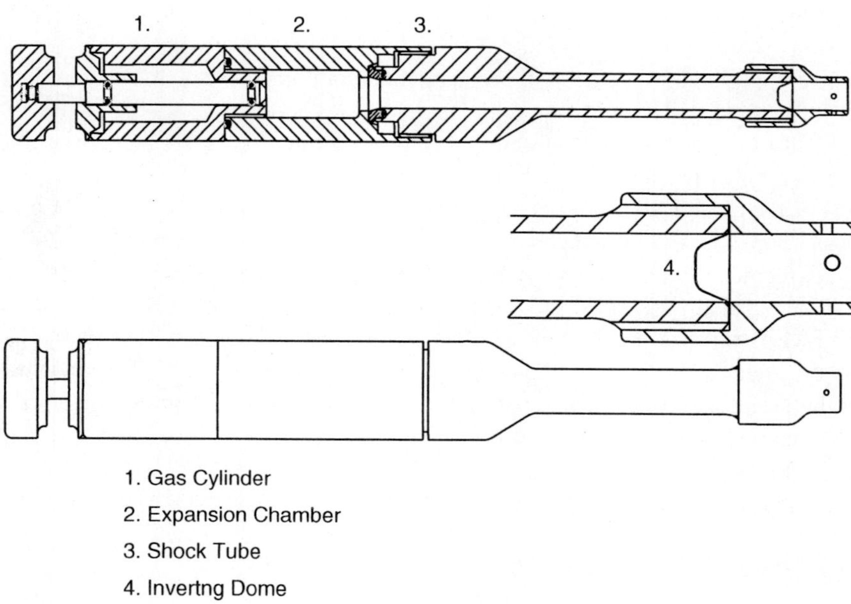

Figure 8.2 Phase I/II clinical prototype design: (1) gas cylinder; (2) expansion chamber; (3) shock tube; (4) inverting dome.

accidental actuation. A disposable aluminum gas cylinder is currently in a late stage of development for high-volume manufacture. This ampoule-type cylinder with a break-off tip will have a similar safety factor to address expected exposure of commercial products to ranges of temperature, external pressure, and impact.

2. Expansion Chamber

The expansion chamber is the transient linking reservoir between the gas cylinder and the shock tube separated by the rupturing membrane. This chamber is designed to collect gas until initiation pressure is reached and contains sufficient volume to drive the gas flow down the shock tube.

3. Shock Tube

The shock tube is a straight or curved tube that allows the tip of the Oral PowderJect® delivery system to be placed at the appropriate point within the mouth. The upstream end holds a bursting membrane, which is typically a thin polymeric film diaphragm of 6 mm diameter; the downstream section of the shock tube holds a soft tip, which in turn holds the inverting dome. The shock tube channels and focuses the transient gas flow from the rapid-opening bursting membrane to the point of application of the energy on the convex face of the inverting dome.

4. Inverting Dome

The inverting dome is the most critical component with the Oral PowderJect® delivery system. It fills a number of key roles: it holds the powdered drug during delivery, accelerates the drug during actuation, and releases the drug at the point of full inversion. The inverting dome is nonporous and nonrupturable so that the tip of the Oral PowderJect® system may be used within the mouth and not release the driver gas. The dome requires properties of high flexibility (i.e., low resistance to inversion) and high impact strength. A number of polymeric materials have been assessed (polyurethane, polyester, etc.) together with various types of fabric reinforcement to increase strength while maintaining flexibility. In addition to the mechanical properties, the inverting dome must hold a uniform distribution of drug particles at the time of injection. To achieve uniform powder distribution on the concave surface, several techniques are being investigated. These include dome surface features such as surface roughness, grooves, or fibers for powder entrapment. Alternatives include attachment via weak adhesive surface coatings or electrostatic attraction. Exemplary shapes, a 6 mm diameter axis-symmetric hemispherical dome and a 16 mm × 6 mm oval dome, are shown in Figure 8.3.

B. Oral PowderJect® Dynamics

The Oral PowderJect® system is a highly transient delivery system in that it relies on the rapid transference of gas energy to particle kinetic energy via the inverting dome. The key gas dynamic feature within the Oral PowderJect® system is the

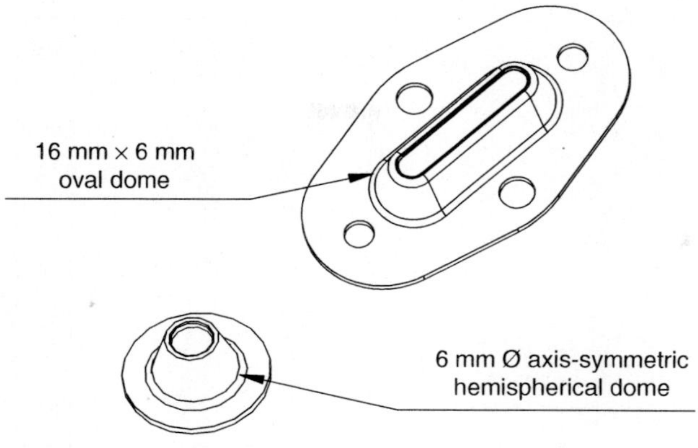

Figure 8.3 Schematic diagram of two different inverting dome designs.

shock wave that is generated by the rapid-opening bursting membrane. The strength and form of the shock wave as it interacts and reflects off the inverting dome control the dome velocity, which in turn dictates the final powder velocity.

The gas dynamic process commences when the gas cylinder is actuated to deliver a fixed amount of driver gas into the expansion chamber. The volume of the gas cylinder is in the range of 2.0 to 5.0 ml and the volume of the expansion chamber is typically 20 to 80 percent of the gas cylinder.

As the pressure rises within the expansion chamber, the downstream section of the expansion chamber is isolated from the shock tube by the bursting diaphragm. The time for the expansion chamber pressure to reach the critical burst pressure of the bursting membrane is of the order of 500 μs. The powder delivery kinetics are independent of this initial rise time. Once the critical burst pressure has been reached, the bursting membrane opens rapidly, resulting in a shock wave traveling at supersonic velocity down the shock tube. A typical shock velocity for the device shown in Figure 8.2 is approximately 900 ms^{-1}.

A range of metallic and polymeric bursting diaphragm materials has been assessed with critical burst pressures in the range 10 to 40 bar (Figure 8.4). The rate of opening of the bursting membrane, the mode of opening, and the effective open area during gas flow control the shock strength (peak pressure). The shock strength is typically in the range 5 to 20 bar. Using a low-molecular-weight driver gas and a high-molecular-weight driven gas downstream increases the shock strength (29). For this reason helium is used as the driver gas within the gas cylinder and air within the shock tube is the driven gas. The shock wave strength at initiation is dictated by the critical burst pressure of the bursting membrane but viscous effects within the shock tube and unsteady expansion waves within the shock tube can reduce the final shock strength (30). For a fixed shock tube geometry, changes in expansion chamber length and diameter can overcome these detrimental effects.

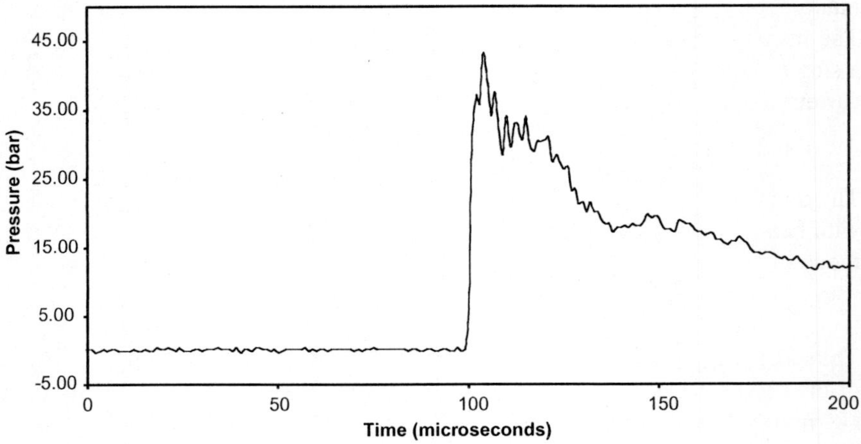

Figure 8.4 Shock wave profile of Oral PowderJect® delivery system.

The interaction of the shock wave with the inverting dome sets up a reflected shock wave pattern within the shock tube and a compression wave downstream of the dome (generated by the rapid motion of the dome). The differential pressure across the dome results in an inversion time for the dome of the order of 100 μs (Figure 8.4). The dome inversion time is very short compared to the time to reach steady-state conditions (≈3 ms). Consequently, the expansion chamber or shock tube can be vented in order to prevent permanent pressurization of the device. In addition to venting upstream of the dome, the tip or air space downstream of the dome can be vented to eliminate compression of the air within the tip, and hence the quality of the tip seal at the oral mucosa is not critical.

C. Design Considerations

There are several factors to be considered in optimization of the system for a specific application.

1. Powder Formulation

The Oral PowderJect® system delivers drug particles in the size range of 20 to 100 μm. The range of particle size is a crucial delivery parameter because both momentum and velocity are dependent on particle size. The particle density and robustness are also important design parameters.

2. Payload

The minimum payload of the current system is currently approximately 1 mg, determined by a current practical limit of pharmaceutical dispensing technology. This methodology must be capable of accurate, high-throughput automated manufacture and 2 to 3 mg per dose is currently the most comfortable range. The maximum payload of current prototype devices is approximately 8 mg for a single shot, but delivery efficiency is reduced at this higher payload limit with current nonoptimized prototypes.

3. Dome Design

The dome material selection is critical as noted above. The trade-off of strength with flexibility and shape is influenced by drug payload, dome geometry, and the mode of filling or drug retention.

4. Target Area

The target area, and hence the payload, can be increased by scaling-up the size of the dome. The current injected area for Oral PowderJect® prototypes is 20 to 100 mm^2. Several different dome shapes have been tested including hemispherical (axis-symmetric) and oval or oblong designs (Figure 8.3).

5. Use Pattern

Two different device designs are available depending on the specific application intended. For acute or single use (and appropriate for any Phase I and II clinical trials and feasibility tests) a single shot, completely disposable design has been developed. Clinical prototypes of this design currently employ reusable stainless steel parts such as gas cylinders for flexibility in testing (Figure 8.2). For a longer course of therapy a multiple-use design with a disposable dome and gas cylinder but with a reusable body, safety features, and a triggering mechanism is being designed (Figure 8.5). Because the performance characteristics of the single-use and reusable device will be identical it is possible to test applications and modifications in the clinic with the single-use disposable prototype.

D. In Vitro Test Methods

A number of in vitro techniques have been developed to characterize device performance for optimization for specific applications and to refine the system. The most appropriate and sensitive test methods include the following.

1. Pressure Characterization

Miniature quartz pressure sensors, with response times of the order of 1 μs, have been employed to measure dynamic gas pressure within the confined spaces of the Oral PowderJect® system (Figure 8.4).

2. High-Speed Imaging

State-of-the-art high-speed imaging equipment, pressure-triggered and incorporating pulsed laser illumination and digital image capture, is used to visualize the modes of dome inversion, dome velocities, and particle delivery profiles.

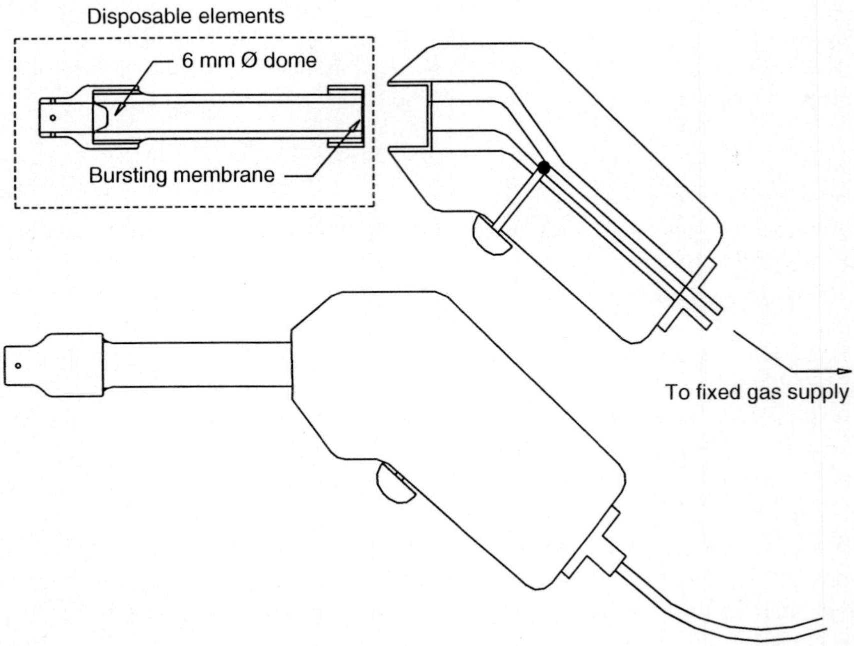

Figure 8.5 Multiple-shot reusable gas supply concept design.

3. Penetration Energy Measurement

A metallized thin-film test method has been developed that is essentially a measure of the damage that high-velocity drug particles can do to a precision thin metal layer deposited on a plastic film substrate. This damage correlates with the particle delivery profile and the characteristic energy of the shock wave. The higher the response from the test, the higher the damage/disruption, indicating that the device as configured imparts more power to the particles. Damage to the film may be measured by electrical resistance change, by light reflectance, or by transmission. The measurement is reproducible and is sensitive to all major device parameters (i.e., shock strength, payload, dome dynamics, powder formulation, etc.). This method can measure the distribution of drug and particle energy across the target area.

4. In Vitro Permeability Measurement

Dissected oral mucosa from a test animal can be used as a diffusion membrane for in vitro study of drug permeation using Franz diffusion cell techniques (31). Various parameters of the powder injection system such as particle formulation, size distribution, payload, and device design variables can be readily evaluated. This can help to single out and optimize the most critical factors influencing drug permeation. Using optical microscopy on the mucosal tissue samples, this technique can be useful to further understand the detailed mechanism of drug delivery by Oral PowderJect® and to optimize the device and formulation.

III. PRECLINICAL STUDIES: DELIVERY OF TESTOSTERONE

The aim of these studies, using a full-size Oral PowderJect® Phase I clinical prototype, was to evaluate the effects of dome loading, shape, and dimensions on the deliverability of a model lipophilic drug, testosterone, to the buccal mucosa of conscious beagle dogs.

A. Experimental Methods

Female beagle dogs of 12.8 to 14.6 kg were acclimated and allowed food and water ad libitum. The dogs (n = 4) underwent a four-week crossover design involving the following treatment regimens:

- Crystalline testosterone (8 mg, Sigma Chemicals) as a fine suspension in 35 percent (w/w) polyvinylpyrrolidone C-30 in water (6 mg/g testosterone concentration) via subcutaneous injection by needle and syringe
- Testosterone (8 mg in 4 × 2.0 mg doses), particle size range 38 to 75 μm, via Oral PowderJect®, 40 bar actuation pressure, 6 mm hemispherical dome
- Testosterone (8 mg in 2 × 4.0 mg doses), particle size range 38 to 75 μm, via Oral PowderJect®, 40 bar actuation pressure, 16 × 6 mm oblong dome
- Testosterone (8 mg in 1 × 8.0 mg dose), particle size range 38 to 75 μm, via Oral PowderJect®, 40 bar actuation pressure, 16 × 6 mm oblong dome

The testosterone was milled and sieved into the required particle size ranges using standard stainless steel mesh sieves. Whole blood (1.0 ml) was drawn at 0, 0.17, 0.5, 1, 2, 4, 6, 12, and 24 hours postadministration, and serum aliquots assayed for testosterone by radioimmunoassay (Diagnostic Products Corporation, CA).

B. Results

Figure 8.6 shows the effect of surface area and payload on the testosterone concentration profile in serum. Bioavailability in the range of 13 to 40 percent was achieved following Oral PowderJect® injection of testosterone when compared to subcutaneous administration. Bioavailability was dependent upon dome loading and geometry. Increasing drug loading above 4.0 mg on the Oral PowderJect® dome resulted in a significant reduction in bioavailability from 30 to 13 percent. This is most probably due to target overload with the current 16 × 6 mm dome. Appearance of testosterone in the systemic circulation was extremely rapid in all cases with T_{max} being approximately ten minutes for each Oral PowderJect® administration. The maximum serum concentration (C_{max}) tended to be higher following Oral PowderJect® administration of testosterone compared to subcutaneous administration. The maximum serum concentration was achieved significantly more

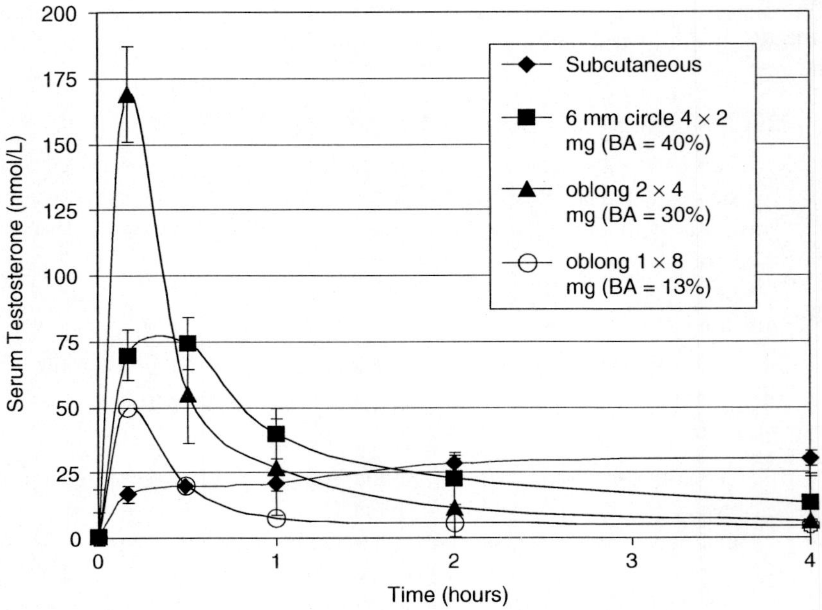

Figure 8.6 Effect of surface area and payload on Oral PowderJect® delivery of testosterone.

quickly following Oral PowderJect® delivery compared to subcutaneous administration via needle and syringe.

C. Discussion and Conclusions

This study demonstrates that the Oral PowderJect® system offers efficient, extremely swift delivery of a hydrophobic drug, testosterone, as measured by the mean bioavailability (mean BA 13 to 40 percent) and time to maximum serum concentration of ten minutes. Bioavailability was dependent upon dome loading and dome geometry; therefore, treatment area and payload should be optimized for maximum bioavailability. Reproducible, rapid onset delivery via this route may offer a significant advantage over other conventional methods of drug delivery, especially when swift, noninvasive therapeutic intervention is desirable.

IV. CLINICAL RESULTS: DELIVERY OF LIDOCAINE HYDROCHLORIDE

Many patients undergoing dental surgery are afraid of needle and syringe injections (32). There are a number of techniques that may be employed to reduce the discomfort of intraoral injections and hence alleviate the problem of needle phobia. The most widely used of these is the application of topical anesthetics. Agents shown to be effective include Xylocaine 5 percent ointment and EMLA® (33,34), although the manufacturer does not recommend the latter for intraoral use. Another method is the use of transcutaneous electronic nerve stimulation (TENS) (35).

A. Oral PowderJect® Clinical Study Objectives

Oral PowderJect®, a novel system for administering powdered drug to the oral mucosa, can be a promising alternative to reduce pain and discomfort from dental surgery. In this clinical study, the following objectives were set to evaluate Oral PowderJect® as an alternative to needle and syringe local anesthetic for dental surgery:

- To observe and assess any damage or irritation to the oral mucosa from Oral PowderJect® delivery of a powdered local anesthetic
- To determine the level of pain or discomfort experienced during an Oral PowderJect® delivery compared to traditional needle and syringe injection
- To evaluate the quality of local anesthesia from a single Oral PowderJect® delivery of powdered local anesthetic compared to a PowderJect® Sham

B. Experimental Methods

Fourteen healthy adult subjects (four males, ten females) aged over 18 years old were enrolled. Clearance from their medical practitioners was sought and a full medical history taken before acceptance. The subjects were advised that they could withdraw from the study at any time for any reason without prejudice to their future medical and dental care. They attended the dental surgery clinic for study assessments on four occasions.

Lidocaine hydrochloride in crystalline form (Martindale, Romford, UK) was sieved to give a range of particle size, 38 to 53 μm mean diameter, for use in the Oral PowderJect®. The drug was analyzed for purity, divided into vials, and sterilized by gamma irradiation. It was then reanalyzed for content. For the conventional needle and syringe administration a plain lidocaine 2 percent solution (USP Graham Chemical Co., New York) was used.

1. Methods: Pain on Administration and Mucosal Damage

This part of the study was a single-stage assessment of pain on delivery and mucosal damage. It compared the pain experienced during a needle and syringe injection with that during delivery via the Oral PowderJect®. Each volunteer received an injection of 0.2 ml (4 mg) of 2 percent plain lidocaine hydrochloride labial to an upper lateral incisor (12 or 22) using a standard infiltration technique with a syringe and 27 gauge needle (Monoject 401®). Within 30 seconds each volunteer received an Oral PowderJect® delivery of powdered lidocaine hydrochloride (payload 3.2

mg ± 20 percent) labial to the opposite upper lateral incisor. The mucosa at the latter site was examined for surface deposit of lidocaine and monitored for one minute postdelivery in order to assess any mucosal damage. The volunteers were asked to score the pain on administration of both injections on a 100 mm visual analogue scale (VAS).

2. Methods: Local Anesthesia Effectiveness

This part of the study was a randomized, double-blind, and placebo-controlled assessment of the quality of local anesthesia. Each volunteer made three separate visits to the dental surgery clinic and was informed that he or she would receive a number of random treatments. The volunteers were in fact to receive one Oral PowderJect® injection per visit to total two active and one sham or one active and two shams over the three visits. The number of active injections and the order of active to sham treatments was randomized among the volunteers in order to blind both them and the investigator, while ensuring each volunteer received at least one active injection.

The investigator administered an Oral PowderJect® injection to the oral mucosa labial to the upper lateral incisor on the left or right (12 or 22). After 50 seconds, the gum labial to the untreated upper lateral incisor was probed without penetrating the mucosa using the back-end of a conventional 30 gauge needle (Monoject 401®) following the method described in Juhlin and Evers (36). The gum at the treated site was probed in exactly the same way 10 seconds later (one minute after the injection). The subject then scored the pain produced by the two probings on standard 100 mm VAS scales.

3. Methods: Statistical Analysis

The Mann–Whitney U-test was used to compare the pain on administration between Oral PowderJect® and conventional needle and syringe. Comparison of the pain upon probing the sites was made using Kruskal–Wallis nonparametric one-way analysis of variance, with post hoc comparisons between sites using the Mann–Whitney U-test. Statistical significance was accepted for $p < 0.05$.

C. Results

1. Pain on Administration

The pain scores for the two different types of administration were markedly different (Figure 8.7). A median pain VAS score of 49.9 was obtained for the needle and syringe, whereas the corresponding value of 2.4 for the Oral PowderJect® was significantly lower ($p < 0.0001$).

2. Tissue Response

All 14 subjects completed the study. The 14 sites subjected to delivery of lidocaine hydrochloride by the Oral PowderJect® were examined after one minute. No damage and no irritation were observed at any of the mucosal tissue sites. There was no

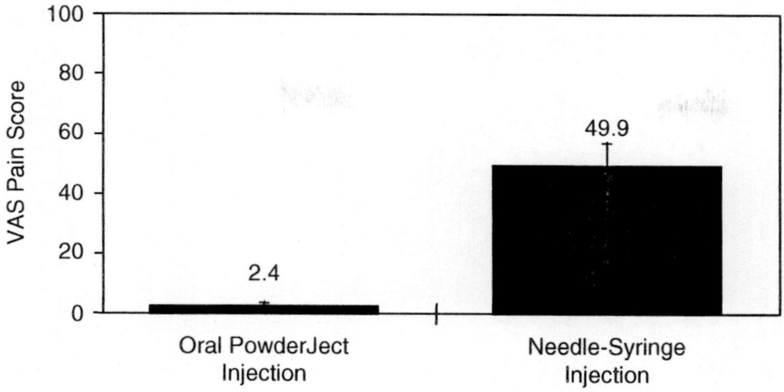

Figure 8.7 Comparison of the pain on administration of lidocaine using a conventional needle and syringe and the Oral PowderJect® (100 = painful injection, 0 = pain-free procedure, $n = 14$).

evidence of any tissue reaction 24 to 48 hours postdelivery. Administrations using the Oral PowderJect® were well tolerated locally and there were no adverse events.

3. Surface Deposit and Residue in the Oral PowderJect® System

No drug residue was observed on the surface of the treated oral mucosa in any of the 14 subjects. Some residual lidocaine hydrochloride was, however, found in the Oral PowderJect® system after each administration. In 11 cases this was classified as minor and in 3 cases as a major proportion of the original payload by visual estimation alone.

4. Quality of Local Anesthesia

The median VAS pain scores upon probing were 55.6 for the control sites, 30.6 for the sites subjected to a sham Oral PowderJect® administration, and 15.0 for those at which lidocaine HCl was administered by Oral PowderJect® (Figure 8.8). There was a highly significant difference among the three treatments ($p < 0.0001$). The reduction in the median pain score from 55.6 for the control site to a value of 30.6 for the sham injected site was statistically significant ($p = 0.0013$), indicating a clear placebo effect. The median pain score of 15 at the lidocaine HCl site was, however, significantly lower than both the sham ($p = 0.0033$) and the control ($p < 0.0001$) sites. The values from the three visits for each treatment site were analyzed and no significant differences were found among visits.

D. Discussion and Conclusions

The first part of this study compared delivery of lidocaine hydrochloride to the oral mucosa in powder form using the Oral PowderJect® system with delivery

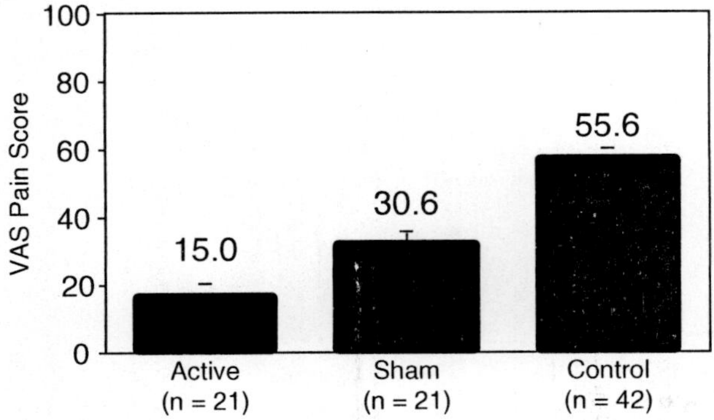

Figure 8.8 Comparison of pain on probing at control, sham, and lidocaine sites (100 = painful injection, 0 = pain-free procedure).

via conventional needle and syringe. The relative levels of pain associated with the two methods of administration clearly show that administration using a needle and syringe is associated with a significant amount of pain (median pain VAS 49.9), and delivery via the Oral PowderJect® system is essentially painless (median pain VAS score 2.4). Furthermore, despite the fact that solid particles of drug were administered at high velocities, no damage was observed upon close examination of the oral mucosa at the site of the Oral PowderJect® injection.

The second part of the study examined the ability of the Oral PowderJect® delivery system to deliver a payload of lidocaine hydrochloride sufficient to provide anesthesia of the oral mucosa to prevent the pain of probing using the back of a dental needle. A double-blind, sham-controlled design was used, as pain perception is both subjective and can vary widely between individuals. Indeed, there was a significant reduction in the pain score at the site of the sham injection when compared with probing at the control (untreated) site, commonly referred to as a placebo effect. This may have been due in part to the fact that the control probing was always performed first, allowing the subjects to prepare themselves for the second probing. It has been suggested that there is a significant correlation between expected pain and that subsequently experienced (37).

Following administration of lidocaine HCl by the Oral PowderJect® delivery system, a significant reduction was observed in the median pain score: from 30.6 at the sham site to 15.0, indicating a local anesthetic effect, despite the significant placebo effect. These observations clearly demonstrate that even though some powdered lidocaine HCl may be observed in the Oral PowderJect® delivery system, a sufficient quantity is delivered to exert its classical effect. The local anesthesia observed following administration of lidocaine hydrochloride by the Oral PowderJect® delivery system had a rapid onset, as evidenced by the reduction in pain scores after one minute. This compares favorably with topical local anesthetic preparations currently used in dental practice. Topical gels, such as lidocaine 5 percent, have been shown to have an effect after two minutes, which increases up to five minutes (38).

Because the quality of local anesthesia assessments were conducted over three separate visits, it was important to demonstrate the lack of a visit effect. The results showed that the pain values obtained for control, sham, and active sites did not vary significantly among visits.

In the present study, pain was assessed by blunt probing with the back of a dental needle and the surface of the oral mucosa was not broken. To further confirm the clinical utility of the Oral PowderJect® delivery system, it will be necessary to investigate its action against a more invasive stimulus, such as a dental injection or other standard painful procedure.

It is concluded that the Oral PowderJect® delivery system can safely deliver sufficient lidocaine hydrochloride to evoke a rapid onset of local anesthesia of the oral mucosa sufficient to prevent the pain of probing with the back of a dental needle. Unlike administration by needle and syringe, this novel method of delivery is associated with a very low level of pain.

V. FUTURE APPLICATIONS

Oral PowderJect® is designed to deliver drugs in powdered form without pain and in enclosed spaces because there is no gas flow. Preliminary preclinical and clinical studies demonstrate that the Oral PowderJect® delivery system can be used for local dental applications as well as systemic drug delivery. Moreover, it is feasible to target specific sites or tissue for drug delivery. For example, the next generation of modified PowderJect® delivery systems will make delivery possible inside the body directly to stomach tissue, lung tissue, and colonic tissue as well as to the inside of blood vessels, to the central nervous system via the olfactory pit, and so on. Such site-specific delivery would be aided by or combined with a suitably modified fiber-optic gastroscope or other minimally invasive surgical instrument.

REFERENCES

1. Davis SS, Illum L, Tomlinson E. Enzymatic barriers to peptide and protein absorption and the use of penetration enhancers to modify absorption. In: Lee VHL, ed. Delivery Systems for Peptide Drugs. New York: Plenum, 1986; 87–104.
2. Nagai T. Adhesive topical drug delivery systems. J Control Rel 1985; 2:121–134.
3. Schurr W, Knoll B, Ziegler R, Anders R, Merkle HP. Comparative study of intravenous, nasal, oral and buccal TRH administration among healthy subjects. J Endocrinol Invest 1985; 8:4–14.
4. Anders R, Markle HP, Schurr W, Ziegler R. Buccal absorption of protirelin: Effective way to stimulate thyrotropin and prolactin. J Pharm Sci 1983; 72:1481–1483.
5. Aungst BJ, Rogers NJ. Comparison of the effects of various transmucosal absorption promoters on buccal insulin delivery. Int J Pharm 1989; 53:227–235.

6. Zang J, Niu S, Ebert C, Stanley TH. An in-vivo dog model for studying recovering kinetics of the buccal mucosa permeation barrier after exposure to permeation enhancers: Apparent evidence of effective enhancement without tissue damage. Int J Pharm 1994; 101:15–22.
7. Oh CK, Ritschel WA. Biopharmaceutical aspects of buccal absorption of insulin. Meth Find Exp Clin Pharmacol 1990; 12:205–212.
8. Steward A, Bayley DL, Howes C. The effect of enhancers on the buccal absorption of hybrid (BDBB) α-interferon. Int J Pharm 1994; 104:145–149.
9. Andreas BS, Andreas SS. Synthesis and *in vitro* evaluation of chitosan-EDTA-protease-inhibitor conjugates which might be useful in oral delivery of peptide and proteins. Pharma Res 1998; 15:2:263–269.
10. Yamamoto A, Taniguchi T, Rikyuu K, Tsuji T, Fujita T, Murakami M, Muranishi S. Effects of various protease inhibitors on the intestinal absorption and degradation of insulin in rats. Pharma Res 1994; 11:10:1496–1500.
11. Langguth P, Merkle HP, Amidon GL. Oral absorption of peptides: The effect of absorption site and enzyme inhibition on the systemic availability of metkephamid. Pharma Res 1994;11:4:528–535.
12. Krishnan BR, Rajagopalan JS, Anderson WL, Simpson M, Ackler S, Davis CM, Ansari AM, Harris TM, Ekwuribe N. Chemical modification of insulin with amphiphilic polymers improve intestinal delivery. Proc Int Symp Control Rel Bioact Mater 1998; 25:124.
13. Caliceti P, Veronese FM. Successful insulin delivery by PEG conjugation. Proc Int Symp Control Rel Bioact Mater 1998; 25:263.
14. Hori R, Komada F, Okumura K. Pharmaceutical approach to subcutaneous dosage forms of insulin. J Pharm Sci 1983; 72:435–439.
15. Choi HK, Flynn GL, Amidon GL. Transdermal delivery of bioactive peptides: The effect of n-decylmethyl sulfoxide, pH and inhibitors on enkephalin metabolism and transport. Pharm Res 1990; 7:1099–1106.
16. Wertz PW, Swartzendruber DC, Squier CA. Regional variation in the structure and permeability of oral mucosa and skin. Adv Drug Del Rev 1993; 12:1–12.
17. Rathbone MJ, Tucker IG. Mechanism, barriers and pathway of oral mucosal drug permeation. Adv Drug Del Rev 1993; 12:41–60.
18. Wertz PW, Squier CA. Cellular and molecular basis of barrier function in oral epithelium. Crit Rev Ther Drug Carr Sys 1991; 8:237–269.
19. Alfano MC, Chasen AE, Masi CWJ. Autoradiographic study of the penetration of radiolabeled dextrans and inulin through non-keratinized oral mucosa *in vitro*. J Periodont Res 1997; 12:368–377.
20. de Vris ME, Bodde HE, Junginger HE. The in-vitro penetration of β-blocking agents through porcine buccal epithelium. Proc of Int Symp Control Rel Bioact Mater 1990; 17:295–296.
21. Kremer MJ, Shade WB, Holbrook NS, Wertz PW, Squier CA. Morphologic location of the buccal permeability barrier. J Dent Res 1992; 71:213
22. Rathbone MJ, Ponchel G, Ghazali FA. Systemic oral mucosal drug delivery and delivery systems. In: Rathbone MJ, ed. Oral Mucosal Drug Delivery New York: Marcel Dekker, 1996.
23. Stablein MJ, Meyer J. The vascular system and blood supply. In: Mayer, J, Squier, CA, and Gerson, SJ, eds. The Structure and Function of Oral Mucosa. Elmsford, NY: Pergamon, 1984; 237–256.

24. Nanny D, Squier, CA. Blood flow in the oral mucosa of normal and atherosclerotic monkeys. J Dent Res 1982; 56 (Suppl) 465.
25. Tauber U. Drug metabolism in the skin: Advantages and disadvantages. In: Hadgraft, J, ed. Transdermal Drug Delivery: Developmental Issues and Research Initiatives. New York: Marcel Dekker, 1989; 99–111.
26. Sarphie DF, Greenford J, Ashcroft SJH, Bellhouse BJ. Transdermal powdered delivery (TPD): *In vivo* determination of particle penetration depth and therapeutic efficacy. Proceedings of Fourth Percutaneous Penetration Conference 1996; 4B.
27. Bell JI, Bellhouse BJ. (1994) EP Patent No. 9426379.5.
28. Bellhouse BJ, Sarphie DF, Greenford JC (1995) EP Patent No. 9502879.1.
29. Anderson JD. Modern Compressible Flow with Historical Perspective. 2nd Edition. New York: McGraw-Hill, 1990.
30. Nabulsi SM, Page NW, Duval AL, Seabrook YA, Scott KJ. A gas-driven gene gun for microprojectile methods of genetic engineering. Measure Sci Technol 1994; 5:267–274.
31. Zhang H, Robinson JR. In vitro methods for measuring permeability of the oral mucosa. In: Rathbone MJ, ed. Oral Mucosal Drug Delivery. New York: Marcel Dekker, 1996; 85–100
32. Kleinknecht RA, Klepac RK, Alexander LD. Origins and characteristics of fear in dentistry. J Am Dent Assoc 1973; 86:842–848.
33. Svensson P, Petersen JK. Anaesthetic effect of EMLA occluded with Orahesive® oral bandages on oral mucosa: A placebo-controlled study. Anesth Prog 1992; 39:79–82.
34. Meechan JG, Donaldson D. The intraoral use of EMLA cream in children: a clinical investigation. J Dent Child 1994; 61:260–262.
35. Quarnstrom F. Electronic dental anesthesia. Anaesth Prog 1992; 39:79–82.
36. Juhlin L, Evers H. Review: EMLA—A new topical anaesthetic. Adv Dermatol 1990; 5:75–92.
37. Burstein A, Burstein M. Injection pain: memory, expectation and experienced pain. NY J Dent 1979; 49:183–185.
38. Holst A, Evers H. Experimental studies of new topical anaesthetics on the oral mucosa. Swed Dent 1985; 9:185–191.

9

Drug Delivery Systems for the Treatment of Periodontal and Dental Diseases

Su Il Yum and Eric Yum

CONTENTS

I. INTRODUCTION

Dentistry is a profession of diagnosing and treating diseased teeth and their surrounding structures. Dentists may treat the teeth or the structures that surround and support them (i.e., gum tissues, bone attachments, and adjacent supporting teeth). The general dentist can treat pathologies in multiple areas, whereas the specialist can remedy the more complicated cases.

Practically all branches of dentistry utilize pharmaceutical products, particularly those of controlled-release dosage forms. We have identified a few specific fields of dentistry where the drug products of our interest are administered. In most instances, the pharmaceutical products, except for a few exceptions, are used as an adjunct to other treatments.

In this chapter, we have reviewed and highlighted the controlled-release pharmaceutical products deployed in periodontics, endodontics, and orthodontics. We begin with a general description of various dental diseases, current treatment methods, and the pharmaceutical products used in each specialty. The drug products are evaluated in terms of in vitro and in vivo functionality, efficacy, and safety. Regulatory and clinical aspects of the pharmaceutical products are briefly reviewed.

II. PERIODONTAL AND DENTAL DISEASES

A. Periodontal Disease

The tooth rests in a bone socket; attachment fibers connect and suspend the tooth to the bone, and gum tissue covers the bone. Figure 9.1 illustrates the basic structure of the human tooth and the areas where various dental diseases may occur.

Periodontics (peri meaning around, dontia meaning tooth) is the specialty that deals with these structures that surround the tooth: gum tissue, bone structures (alveolar bone), and various fibrous attachments (periodontal ligaments). In periodontal disease, the supporting structures become infected, necrotic, and ineffective, therefore losing their ability to anchor teeth.

1. Gingivitis

The mouth possesses numerous cracks and crevices. One popular harbor for pathological bacteria is the space between gum tissue and the tooth surface called the gingival sulcus, or periodontal pockets. Bacteria collect and fester in this moist protected environment. These bacteria, anaerobic in nature, excrete toxins as waste products in the pockets and provoke a response from the body's immune system—hence the gums become inflamed. Inflammation causes the gums to swell; this swelling results in an increase in pocket depth, leading to an increase in bacterial colonies. Gingivitis does not cause loss of attachment fibers or bone supporting structures.

2. Periodontitis

Prolonged infection of gum tissue stresses the periodontal structures. Chronic presence of inflammation may provoke irreversible damage to the periodontal structures: alveolar bone and periodontal ligaments. Periodontitis attacks the supporting structures of teeth, is insidious if untreated, and can result in tooth loss.

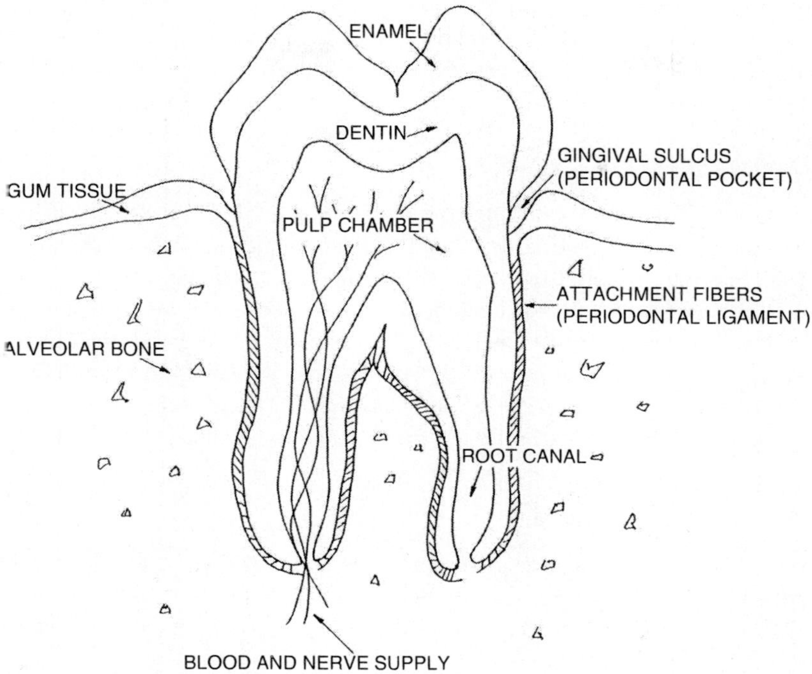

Figure 9.1
Structure of human tooth and the areas where various dental diseases may occur.

3. Scaling and Curettage

Dental treatment for periodontal diseases consists of thoroughly removing all infectious substances. Scaling removes calculus (a calcified material consisting of minerals and bacteria) and plaque (a slimy film of sugars, salts, proteins, and bacteria) from the surfaces of teeth. Scaling is reserved for more routine cleaning and maintenance therapy. Curettage involves more profound cleaning of tooth surfaces usually involving surfaces farther down inside the periodontal pocket. Curettage is needed in many cases where infection is advanced and damage to tissue is substantial. Scaling is reserved for more routine cleaning and maintenance therapy.

4. Periodontal Surgery

If periodontitis progresses, severe destruction can result. Pockets several millimeters deep may result from destruction of bone and attachment fibers. The mere presence of these pockets further promotes the progression of periodontitis unless treatment is initiated. Therefore, the dentist must go in and recontour and remove gingival tissues in order to reduce the pocket depth. Removing loose, pocket-forming gum tissue, a procedure called gingivectomy (common techniques employed), as well as osteoplasty, the recontouring of bone, are used to enhance hygienic cleansing. Indeed, both techniques help eliminate bacterial harbors and create better access for at-home cleaning.

B. Dental Diseases

Bacteria can colonize the supporting structures of the tooth and the tooth surface itself. Although enamel is the hardest calcified material the human body produces, it still falls prey to bacterial infection. Bacteria colonize enamel surfaces much as barnacles do on ship hulls. Once there, bacteria leach acidic toxins that erode enamel. Over time, this process can lead to decay, referred to as caries. If detected early, caries can be removed and filled with materials available to the dentist. However, if left untreated, caries can progress deep into the tooth to more vital structures, such as the tooth pulp. The tooth pulp contains the nervous and vascular systems of the tooth.

1. Operative Dentistry

Operative dentistry deals with the common filling and prosthetic crown. This field encompasses all procedures dealing with the restoration of tooth surfaces destroyed by bacterial decay. The general dentist practices this field of dentistry.

2. Endodontics

Endodontics deals with the internal portion of the tooth or the tooth pulp (endo meaning inside, dontics meaning tooth). The pulp contains the nervous and vascular supply needed to maintain tooth vitality. Often tooth decay erodes the enamel and dentin and invades the pulp. Once there is invasion, the pulp becomes irritated and diseased. The infection produces byproducts, such as pus, which are expressed out of the root tips and into the surrounding bone structures. Over time, the pus levels increase and soon collect to form an abscess. This invasion often leads to necrosis of the vulnerable pulp tissue, leaving the tooth nonvital and, more importantly, the patient in pain.

a. Root Canal Treatment (RCT): Once it is established that the pulp is infected, the endodontist must drill through the tooth bone to access the pulp and the contents of the root canals in order to remove the necrotic tissue. This procedure is referred to as root canal treatment where the removal of infected pulp extends into the roots of the tooth called the root canal. RCT procedures employ no intracanal medicaments and utilize permanent restorative materials. The internal area or infected pulp is bored clean and filled with an inert material to eliminate any voids that may provide for future bacterial invasions.

b. Pulpotomy: This procedure is usually reserved for children who still have their primary teeth or baby teeth. An infected baby tooth provides the dentist with a number of treatment options: the first alternative is to extract the tooth although this may affect intra-arch dynamics. The second alternative is RCT, which is more expensive as well as traumatic to the child. The third option is a pulpotomy. In this procedure, the operator removes only the diseased portion of the pulp, mainly the contents in the pulp chamber. In a pulpotomy, the operator places medicaments and temporary filling materials.

3. Orthodontics

Orthodontics (ortho meaning straighten, dontia meaning teeth) is the specialty where operators deal with the alignment of teeth. Intra-arch and interarch relationships are taken into consideration. The upper jaw, or maxilla, and the lower jaw, or mandible, each carry 16 teeth. The orthodontist is concerned with tooth alignment as well as interjaw relationships. Correcting crooked teeth not only yields aesthetic dividends but functional ones as well. Poorly aligned teeth hinder proper access for daily brushing, which may lead to caries and periodontal diseases, and impede the proper dynamic relationships between jaws during chewing or speaking, and so on. Basic fixed orthodontic treatments include the use of appliances such as brackets, devices cemented to teeth, and wires that fit into brackets. Wires are connected to brackets via rubber-based rings. The operator utilizes a host of physics-based theories to torque, twist, and tilt the teeth needing reorientation. Forces are generated by the use of wires and rubber-based elastics in an infinite menu of combinations.

Patients needing mild or retentive treatment are given removable appliances. Removable appliances utilize the same theories used in fixed treatment; however, the appliance generating tooth movement can be removed much as a mouth guard used in sports. A removable appliance is custom made to fit the patient. A removable appliance contains an acrylic polymer base material with embedded wires protruding from this base. These wires are bent according to treatment goals.

4. Biofilm and Dental Diseases

The development of biofilms and the role they play in corrosion and deposition processes in the industrial water systems share a similar chain of events with the development of periodontal disease. Biofilms consist of microbial cells (algae, fungi, or bacteria) and the extracellular polymer they produce. Algae biofilms provide nutrients (organic carbon) that help support the growth of bacteria and fungi. Algae do not require organic carbon for growth but instead utilize carbon dioxide and the energy provided by the sun to manufacture carbohydrate. Growth and dispersal of algae cells will provide nutrients that can support a larger and more diverse population of microorganisms.

Microorganisms are found in the oral cavity, on skin, and in the gastrointestinal system. Bacteria attach to surfaces by proteinaceous appendages referred to as fimbriae. Once a number of fimbriae have "glued" the cell to the surface of the tooth, detaching the organism is very difficult. Bacteria grow on surfaces to which organic molecules are absorbed thereby providing needed nutrients for bacterial growth. Once attached, organisms begin to produce slime or an extracellular biopolymer film. The biopolymer consists primarily of polysaccharide and water. Gellation of some biopolymers can occur upon the addition of divalent cations such as calcium and magnesium. Biofilms can also lead to the formation of mineral scales. Calcium ions are fixed to the biofilm matrix by carboxylate functional groups present on the polysaccharide. Calcium ions held in place by biofilms on teeth are readily available to react with carbonate or phosphate anions. This process provides crystal growth sites that would not normally be present on a biofilm-free surface. Additionally, biofilms trap precipitated calcium salts and lead to formation of corrosion byproducts that act as crystal growth sites. A typical biofilm-induced mineral deposit is the calcium phosphate scale that forms on the tooth surfaces.

III. PERIODONTAL AND DENTAL PRODUCTS

A. Classification of Drug Delivery Products

1. Chemotherapy for Pulpotomy

In a clinical study conducted by Fuks et al. (1), the effect of ferric sulfate (FS) was compared to that of dilute formocresol (DFC) as pulp dressing agents in pulpotomized primary molars. Ninety-six primary molars in 72 children were treated by a conventional pulpotomy technique. Fifty-eight teeth were treated by an FS solution for 15 seconds, rinsed, and covered by zinc oxide–eugenol paste (ZOE). In another 38 teeth, a cotton pellet moistened with 20 percent DFC was placed for five minutes, removed, and the pulp stumps covered by ZOE paste. The teeth of both groups were sealed by a second layer of intermediate restorative material and restored with a stainless steel crown. Based on the clinical and radiographic examinations 6 to 34 months following the pulpotomy, it was concluded that the success rates of the FS group and the DFC group were not significantly different (93 vs. 84 percent). It may be speculated that the controlled release of ZOE from the paste formulation might have contributed to the reasonably good success rates in both groups.

ZOE and KRI pastes are available for use as temporary cements and restorations, permanent cements, intermediate restoration, and thermal insulation bases. They also serve as sealants following pulpotomy and root canal, and for periodontal dressings. Typical compositions of ZOE and KRI pastes are as follows: ZOE (zinc oxide 2.5 g to 1 ml eugenol: Sultan Chemists, Inc., Englewood, NJ, USA); KRI (Iodoform 80.8 percent, camphor 4.86 percent, P-chlorophenol 2.025 percent, menthol 1.215 percent: Pharmachemie AG, Zurich, Switzerland). To prepare ZOE paste, the two primary ingredients are mixed to a desired consistency just prior to use. As suggested by the lists of ingredients of these cement materials, both ZOE and KRI pastes have anti-inflammatory, antiseptic, and analgesic activity and are resorbable.

When these paste materials are applied to a dentinal cavity, small quantitites of active ingredients (e.g., eugenol, iodoform) diffuse through the dentin to the pulp. The low concentrations of the actives exert anti-inflammatory and local anesthetic effects on the dental pulp, resulting in acceleration of pulpal healing.

2. Local Delivery of NSAIDs in Root Canals

Negm (2) reported clinical efficacy of intracanal delivery of two nonsteroidal anti-inflammatory drugs (NSAIDS): diclofenac and ketoprofen, with and without hyaluronidase for the control of posttreatment pain. In a double-blind study involving 760 subjects, the study was carried out on originally asymptomatic and symptomatic teeth that required endodontic therapy. Endodontic treatment was completed in three visits during which time medications were placed into the canal either at the end of the first visit or the second visit. The formulations evaluated in the study were: diclofenac 75 mg/3cc for IM injection (Voltaren, Ciba-Geigy, Basle, Switzerland), ketoprofen 100 mg/5cc IM injection with 2.5 percent benzyl alcohol (Specia-Groupe, Rhone-Poulenc, Paris, France), and hyaluronidase BP (ovine): Hyalase powder 1500 i.u. per ampule (Laboratories Fisons SA, Le Trait, Rouen, France). The above medications were directly injected into the canals or mixed with hyaluronidase before injection. This was done by dissolving the powder content of each ampule of Hyalase in 0.1 cc of the medication. With a 16 gauge needle, about 0.1 cc of solutions was then delivered to each canal.

Statistical analysis of the pain score data revealed that both diclofenac and ketoprofen significantly reduced the mean pain score in all cases and were significantly superior to the placebo until the end of the study. Postendodontic pain occurred with less frequency when hyaluronidase was added to the NSAIDS but the differences between the diclofenac and ketoprofen were insignificant.

3. Formulations Used in Orthodontia

Treatment of the oral microflora, caries, and formation of white spot lesions or decalcified areas are concerns for patients receiving treatment with fixed orthodontic appliances. Several treatment methods using controlled delivery systems are available for some or all of these risk factors related to orthodontic treatment. Ogaard et al. (3) demonstrated the efficacy and safety of a fluoride varnish (Fluor Protector®) and an antimicrobial varnish (Cervitec®, 1 percent Chlorhexidine/1 percent thymol). The test group of subjects received both fluoride and antimicrobial varnishes and the control group received the fluoride varnish only. The treatment started at prebonding every week for three weeks, at bonding, and six weeks following the bonding of the orthodontic brackets to the teeth. During the three-week prebonding period, the mean visible plaque index, gingival bleeding index, and Streptococci mutans in plaque decreased in both groups. At bonding and after twelve weeks following bonding, the mean level of Streptococci mutans in plaque was significantly lower in the Cervitec group than in the control group. Similarly, favorable clinical results were observed by Twetman et al. (4) with

Cervitec for the control of Streptococci mutans in plaque adjacent to bonded orthodontic brackets. The fluoride-releasing products also include the elastomeric ligature ties (Fluor-I-Ties®) and adhesives that are used for bonding as well. The fluoride delivery systems seem to be effective in the prevention of demineralization and enhance remineralization of enamel through calcium fluoride and fluorapatide formation, and hence reduce white spots around orthodontic brackets (5,6).

The chewing gums containing polyol (xylitol) were determined to be effective for the reduction of Streptococci mutans and dental caries in children wearing fixed orthodontic appliances (7). The gums were given to the patients for daily use after meals and snacks. The gums contained an unspecified amount of xylitol. A 2.7 gm of gum was given to the children at a time with a daily maximum amount of 10.5 gm of gum. After a 28 day trial, the plaque and saliva levels of Streptococci mutans were significantly reduced in most cases by 13 and 33 percent in groups that received xylitol gums.

4. Chemotherapeutic Agents for Periodontitis

Periodontal disease appears to arise from the interaction of pathogenic bacteria with a susceptible host. The main aims of disease management have been to establish a high standard of oral hygiene and professionally and thoroughly to debride the root surface. Chemical agents have been considered and administered primarily as an adjunctive therapy. The therapeutic agents and commercial products that have been available for the local chemotherapy targeted at the periodontal pockets include tetracycline hydrochloride (Actisite®), chlorhexidine gluconate (PerioChip®), minocycline hydrochloride (Dentomycin® and Periocline®), and metronidazole benzoate (Elyzol®). Table 9.1 illustrates the physicochemical properties of these therapeutic agents and Table 9.2 describes attributes of the commercial products containing these agents.

Chemotherapy may be directed at subgingival plaque using these antimicrobial agents possibly in conjunction with anti-inflammatory agents. Although limited, at the present time, to clinical or animal testing, several nonsteroidal anti-inflammatory agents and other immune-response modifiers have been shown to be effective in reducing breakdown and in promoting healing, including bone regeneration. The NSAIDs that have been evaluated include oral flurbiprofen (8), oral ibuprofen (400 to 800 mg tablets) (9), and topical ketoprofen (1 percent cream for local treatment) (10). Other chemotherapeutic agents that have been evaluated as an adjunctive treatment include oral antibacterial agent augmentin (11) and triclosan gel (12).

5. Fluoride and Antimicrobial Products

It has been shown clinically that long-term fluoride treatment of teeth can prevent demineralization (white spots), enhance remineralization of enamel through calcium fluoride and fluor apatite formation, and prevent caries. The sources of fluoride are fluoride-releasing elastomers (e.g., Fluor-I-Ties®, Ortho Arch Co.,

Table 9.1 Properties of Chemotherapeutic Agents for Treatment of Periodontitis

Chlorhexidine Gluconate or Digluconate
MW 505 (base); 701 (gluconate salt); 897 (digluconate salt)
Solubility in water @20°C: 0.08% (w/v) for base; >50% for digluconate salt
LD50 in mice (mg/kg): 22 iv; 800 oral
Incompatible with soaps and anionic materials and anionic surfactant
Active against gram-positive and gram-negative bacteria; more effective against gram-positive bacteria; most active at neutral or slightly acidic pH
Chlorhexidine gluconate: dental gel—1%; mouth wash—0.2%; indications are prevention and treatment of gingivitis and plaque
5% v/v solution has pH 5.5–7
MP 134°C for base
Minocycline Hydrochloride
MW 457 (Base); 494 (HCl Salt)
Yellow crystalline powder
Soluble in water
1% solution has pH 3.5–4.5
Oral dose 100 mg every 12 hrs yields 2–4 μg/ml plasma concentration in humans
Tetracycline derivative with clinical uses similar to those of tetracycline HCl
Tetracycline Hydrochloride
MW 444 (Base); 481 (HCl Salt)
Yellow, hydroscopic, crystalline powder, darkens in moist and when exposed to strong sunlight
Freely soluble in water (1 in 10 H_2O)
pH 2.1–2.3 in 2% water solutions
LD50 oral in rats; 6443 mg/kg
Stable in neutral and in alkaline solutions
DEC 170°C (base); DEC 214°C (HCl salt)
Metronidazole Benzoate
MW 171 (base); 275 (benzoate salt)
Yellowish white powder
Practically insoluble in H_2O
Slightly soluble in alcohol
Anticipated to be a carcinogen
MP 160°C (base)

Inc., Hoffman Estate, IL) and polymer-based varnish (e.g., Fluor Protector®, Ivoclar-Vivadent®, Schaan, Liechtenstein). The fluoride compounds are either silane fluoride (Fluor Protector) or sodium fluoride (Duraphat, DFL, Spain). The inactive vehicles include a natural resin base (Duraphat) and a polyurethane polymer (Fluor Protector). The drug loading ranges from 0.7 percent (Fluor Protector) to 2.5 percent (Duraphat). The fluoride varnish products have become the most widely used preparations for topical fluoride application in Europe (13).

Table 9.2 Commercial Products Indicated for Periodontal Diseases

Product	Manufacturer	Delivery	Contents	Method of Use	Indication	How It Works
Periocline®	Sunstar Inc., Osaka, Japan	Syringe	Minocycline Hydrochloride (2%); Hydroxyethyl cellulose; Aminoalkyl Methacrylate Copolymer RS; Triacetin; Magnesium Chloride; Glycerine	Three to four applications at two-week intervals	For adults with moderate to severe chronic periodontitis as an adjunct to scaling and root planing where pockets are 5 mm or more in depth	Drug diffusion from gel that becomes dispersed
Elyzol®	Dumex, Ltd., Copenhagen, Denmark	Syringe	Metromidazole benzoate 25%; Glycerol monooleate; Triglyceride; Sesame seed oil	Two separate applications, one week apart	Adjunct to conventional therapy	The preparation comes as a semisolid suspension which first liquifies at body temperature; then when the mixture comes into contact with gingival fluid (water), it sets in a liquid crystalline state and transforms into a sticky gel that becomes dispersed

Table 9.2 Commercial Products Indicated for Periodontal Diseases (continued)

Product	Manufacturer	Delivery	Contents	Method of Use	Indication	How It Works
PerioChip®	Perio Products, Ltd., Jerusalem, Israel	Manually place the pieces of film by dental professional	Chlorhexidine gluconate 2.5 mg; Formaldehyde crosslinked Byco protein matrix (5 mm × 4 mm × 0.3 mm)	Application every three to six weeks	Adjunct to scaling and root planing for periodontal pocket of at least 5 mm	Drug diffusion from matrix that erodes
Actisite®	Alza Corporation, Palo Alto, USA	Fiber is inserted into the periodontal pockets and a cyanoacrylate adhesive should be used to help secure the fiber in the pocket	Tetracycline hydrochloride 25%; Ethylene/vinylacetate copolymer (23 cm fiber with 0.5 mm diameter contains 12.7 mg of drug)	At the end of ten days of treatment, all fibers must be removed; fibers lost before seven days should be replaced	An adjunct to scaling and root planing; repeat fiber applications have not been studied	Drug diffusion from fiber matrix

Cervitec (Ivoclar-Vivdent, Schaan, Liechtenstein) is an antimicrobial varnish indicated for the prevention of periodontal diseases. The varnish contains 1 percent chlorhexidine and 1 percent thymol in a polymer base. The Cervitec varnish has been evaluated also for the treatment of oral microflora, caries, and gingival conditions in patients receiving treatment with fixed orthodontic appliances (14).

B. In Vitro Mechanisms and Functionality

1. Periodontal Products

There are two different types of vehicle or matrix delivery systems for products used to treat periodontal disease:

- Nonerodible matrix (e.g., ethylene vinyl acetate (EVA) copolymer)
- Disintegrating or erodible matrix (e.g., hydroxyethyl cellulose (HEC), glycerol monooleate (GMO), and protein gels)

Drugs used in the treatment of periodontal disease have a range of solubility characteristics including those that are:

- Highly soluble or miscible with water, such as chlorhexidine gluconate, tetracycline HCl, and minocycline HCl
- Practically insoluble, such as metronidazole benzoate

The loading dose of drug in the matrix or gel may be:

- In large excess above solubility
- At or just above solubility

Therefore, different combinations of vehicle characteristics, drug–water solubilities, and drug loading vis-a-vis the drug's solubility in the vehicles throughout the functioning lifetime of the dosage form would give rise to different drug release mechanisms, particularly during the pseudo steady-state or true steady-state periods. If the drug loading is in excess above the drug's solubility in the vehicle, during the initial periods following administration of the dosage form, the release would be dominated by dissolution of excess drug particles on the surface and release of dissolved drug molecules on the surface layers of the dosage form. After the initial burst period, the subsequent drug release kinetics is determined by the properties of the vehicle, drug water solubility, and drug-loading-to-solubility ratio.

As excess drug dissolves away from the surface layers, which become free of excess drug, the properties of vehicle and drug in the deep layers will dictate the mechanism of drug release.

Actisite® is a noneroding matrix that is in the form of a fiber containing tetracycline HCl, which has moderate water solubility (about 100 mg per ml or greater). Following application of Actisite, there is an initial burst of the drug representing dissolution of drug particles on the surface. The subsequent drug release is controlled by both osmosis and diffusion. As water migrates into the fiber, diffusion and osmotic pumping increase, and the release rate then declines as a function of an inverse square root of time. Elyzol® and PerioChip® have similar pseudo steady-state drug release characteristics. Following an initial burst of drug release due to dissolution of drug particles on the surface, both products become a viscous gel due to hydration (PerioChip®) or softening and melting (Elyzol®). The dissolution or erosion of vehicles in both products plays a key role in determining the drug release kinetics along with other factors such as dissolution and diffusion (or osmosis) of drug in the vehicles.

Disintegration of these dosage forms is certainly possible, and will affect drug release characteristics due to the increasing surface area for dissolution and diffusion. Perocline, due to its relatively low viscosity and low drug concentration, is anticipated to release the drug rather quickly. The diffusivity of the drug in this vehicle is high, and due to the low drug loading (below solubility) the release rate can precipitously decay over a short period of time.

Table 9.3 highlights the qualitative in vitro functionality of selected commercial periodontal products.

2. Products for Pulpotomy and Root Canals

The release rate of the actives from ZOE and KRI pastes are high immediately following their application. The release rate will slow down as the paste goes through a setting process in situ, and will further decrease as the concentration of the actives in the cement decays with time. In one study, maximal in vitro release of eugenol into phosphate buffer from ZOE cement was attained within five hours and the total amount released was 5 percent of the total eugenol loading. The release rate precipitously decreased after the first five hours (15).

Another in vitro study compared the antimicrobial and cytotoxic effects of KRI and ZOE used in primary tooth pulpectomies (16). L929 Mouse fibroblast cells were used for the cytotoxicity tests. The authors determined that ZOE produced better antimicrobial activity than KRI paste. It was interpreted that the result might represent a variation in the diffusion rate of the actives through the in vitro microbial-culture media. The authors used *S. faecalis* as the representative microorganism in their in vitro testing, which is thought to be most prevalent in pulpal infection. It was also concluded that both ZOE and KRI pastes had high cytotoxicity regardless of the setting time in the 24 hour direct cell–medicament contact test. ZOE cytotoxicity decreased to control levels after only one day of setting in the direct contact experiments, compared with greater than seven days for KRI paste.

3. Products for Fluoride and Antimicrobial Therapies

In vitro fluoride release from 200 fluoride-containing elastometric ligature ties (Fluor-I-Ties) was performed in distilled water up to 180 days (17). Fluor-I-Ties released significant amounts of fluoride initially (fraction undefined). Fluoride release was characterized by an initial burst of fluoride during the first and second

Table 9.3 Qualitative Estimation of In Vitro Drug Release Mechanisms of Commercial Periodontal Product

Commercial Products	Vehicle Property	Drug Solubility in Water	Drug Loading	Probable Drug Release Mechanisms
Actisite® (Tetracycline-HCl) (fiber)	Ethylene/Vinyl Acetate Nonerodible	Moderate	Large excess above solubility	Dissolution, osmosis, and diffusion
Elyzo® (Metronidazole Benzoate) (gel)	Plasticized Glycerol Mono-Oleate Erodible and disintegrating	Low	Large excess above solubility	Dissolution, disintegration, erosion, and diffusion
PerioChip® (chlorhexidine gluconate) (film)	Crosslinked protein matrix Erodible and disintegrating	High	Slightly above or at solubility	Dissolution, erosion, disintegration, and osmosis and diffusion
Periocline® (minocyline-HCl) (gel)	Hydroxyethyl cellulose gel Erodible and soluble	High	Below solubility	Diffusion

days, followed by an exponential decrease. By the end of the second week 88 percent of the initial fluoride loading was released from the elastomer. It was concluded that for optimum clinical benefit, Fluor-I-Ties should be replaced monthly.

Cervitec® varnish was investigated in vitro for its antimicrobial activities against different gram-positive and gram-negative bacterial strains as well as yeast using the agar diffusion inhibitory test (18). The results of the study supported the functionality of Cervitec® and demonstrated that chlorhexidine and thymol diffuse out of the varnish and are active against various oral pathogens such as *Streptococcus mutans, Actinobacillus actinomycetem,* and *Candida albicans.*

C. In Vivo Pharmacokinetics and Pharmacodynamics

1. Periodontal Products

a. Actisite®: In the periodontal pockets, the Actisite tetracycline fiber provides a mean gingival fluid concentration of 1590 μg/ml tetracycline throughout the ten day treatment period. Concentration in saliva immediately after fiber placement (around the nine selected teeth) was 50.7 μg/ml and declined to 7.6 μg/ml at the end of ten days (19).

In a controlled 60 day clinical trial, 113 adult patients with periodontitis (age 25 to 88) with a mean pocket depth of 7.2 mm received supragingival cleaning followed by one of four treatments, randomized to a single tooth per quadrant. These treatments were

1. Actisite fiber for 10 ± 2 days
2. Control fiber for 10 ± 2 days
3. Scaling and root planing
4. No treatment

Teeth treated with Actisite fiber were found to have significantly reduced probing depth and bleeding on controlled force probing. Probing depth reductions were greater in deep (≥7 mm) than in moderate (≤ 6 mm) sites (19). In this study, immediately following therapy both Actisite fiber and scaling and root planing produced significant reduction in the number of sites infected with probable periodontal pathogens (compared to untreated controls; *Fusobacteriums nucleatum, Porphyrsomonas gingivalis, Prevotella intermedia, Eikenella corrodens, Campylobacter rectus,* and *Actinobacillus actinomycetemcomitans*).

In another similar study, the effects of scaling and root planing alone, and scaling and root planing followed by Actisite fiber treatment, were compared (19). Subjects (age 32 to 75) had a baseline pocket mean depth of 6.4 mm. Two nonadjacent sites with pockets with bleeding on probing were selected for treatment and followup at one, three, and six months. Analysis of results showed that adjunctive fiber therapy with scaling and root planing provided significantly greater reductions in probing depth and bleeding on probing than scaling and root planing alone at followup visits.

b. Elyzol®: Concentrations ranging from 103 to 1297 μg/ml of metronidazole were recorded in inflamed pockets treated with Elyzol (metronidazole benzoate) (20,21). Effective concentrations were maintained for 24 to 36 hours. Systemic levels of metronidazole between 0.2 and 1.3 μg/ml were measured after the administration of 29 to 103 mg of the gel.

Two studies involving a total of 230 patients compared the metronidazole gel without scaling with a single root planing alone (20,21). The metronidazole and root planing reduced pocket depth similarly to approximately 1.0 to 1.5 mm with benefit lasting throughout six months of followup. Both treatments reduced the proportion of pathogens in the gingival flora; the metronidazole gel was more effective against *A. actinomycetemcomitans*. The susceptibility of organisms to metronidazole remained unchanged.

Another study determined the time that the gel matrix material persists in periodontal pockets after one application of Elyzol® Dentalgel (22). Gingival crevicular fluid was absorbed with filter paper and was assayed for the amount of glycerine monooleate (GMO) and oleic acid (a degradation product of GMO). In only 1 out of 12 patients, GMO was detectable within the pocket 24 hours following the application. GMO was found no longer than 12 hours in the remaining patients.

c. PerioChip®: PerioChip® (23,24) is a sustained-release device composed of a crosslinked protein containing chlorhexidine gluconate as the active agent. The release profile of chlorhexidine from the degradable films was altered by the amount of the active agent incorporated into the film and by the crosslink density of the polymer.

Perio Products Ltd. has completed multicenter Phase 3 trials in Europe and the United States. The U.S. trials consisted of double-blind studies that compared the PerioChip® as an adjunct to scaling and root planing with scaling and root planing alone and scaling and root planing in conjunction with a placebo. Results indicated that the patients who were treated with the product had a 40 to 50 percent greater reduction in the periodontal pocket depth when compared with the control groups. Approximately 10 percent more patients reported mild to moderate pain during the first 24 to 48 hours following PerioChip® insertion compared with the control group.

d. Perocline®: Distribution of several periodontopathic bacteria in adult periodontitis and the microoganisms' in vitro susceptibility to the active agent in Perocline were evaluated in conjunction with its efficacy (25,26). The patients with periodontal pockets of equal to or greater than 4 mm depth were treated with the minocycline hydrochloride gel once a week for four weeks. The lesions were clinically examined after one and four weeks of Periocline gel administration into the pockets. The distribution of the subgingival micoorganisms included *Capnocytophaga sputigena* (37.1 percent), *Pevotella intermedia* (22.6 percent), *Porphyromonas gingivalis* (22.6 percent), *Fusobacterium nucleatum* (20.1 percent), *Actinobacillus actinomycetemcomitans* (9.7 percent), and *Eikenella corrodens*

(4.8 percent). One to three bacteria species were contained in 76.8 percent of the sites. The concentration of the drug applied to deep periodontal pockets inhibited the growth of most of these microorganisms investigated. There were no significant differences among sites with strains exhibiting low or normal susceptibility to minocycline-HCl.

In a separate clinical and microbiological effect study, the local administration of Periocline gel as an adjunct to scaling and root planing was evaluated for its efficacy. The sites had probing depths greater than 5 mm and loss of attachment greater than 2 mm before the treatments. After scaling and root planing, Periocline was injected into the pockets. The control sites were irrigated with saline following scaling and root planing. The treatments were repeated once a week for 4 consecutive weeks. At the end of 12 weeks, clinical conditions (pocket depth, attachment levels, bleeding on probing) improved in both groups; significantly greater improvements were obtained in the test group. Microbiological study revealed that Periocline effectively eliminated periodontopathic Gram-negative bacteria (25,26).

2. Products for Pulpotomy and Root Canals

a. ZOE and KRI Pastes: The ZOE cement is one of the most inert materials used in dental practices. Unlike other cement materials, such as phosphate-bonded cements, its effect on the pulp is mild and it causes only a slight reduction in odontoblasts. Such a mild inflammation decreases after a few days and the odentoblast layer recovers after a number of weeks. Thus, ZOE is well suited to protect pulp from the more irritating phosphate-bonded cements.

The free eugenol released from the ZOE matrix as a result of hydrolysis of ZOE gives rise to persistent efficacy. It is believed that zinc oxide and eugenol, in the presence of water, form zinc eugenolate salt, which hydrolyzes by the in situ water to zinc hydroxide and eugenol. The free eugenol content of the set cement is probably very low.

In a clinical efficacy study, relative rate of success of endodontic treatment of nonvital primary molars using ZOE and KRI paste was compared (27). Out of more than 70 necrotic primary molars filled either with ZOE or with KRI paste, after 12 to 48 months posttreatment, the overall success rate for KRI paste was 84 versus 65 percent after ZOE, which was statistically significant. However, the authors concluded that the overall success rates of first and second primary molars were similar for both materials. For the primary molars, KRI paste was more efficacious. The higher viscosity of ZOE and its lower flow properties were potential reasons for the lower success rate with the ZOE paste in the first molars.

3. Products for Fluoride and Antimicrobial Therapies

In a study conducted by Wilson and Gregory (28), patients with fixed orthodontic appliances were fit with fluoride-releasing elastomers. Their saliva samples were collected at 1 week intervals for 13 weeks. During the control phase when the subjects were fitted with conventional elastomers, the subjects demonstrated no significant changes in the percentage of *S. mutans*. However, after the fluoride-releasing elastomers were placed, the percentage of salivary *S. mutans* decreased significantly. The effect of fluoride on the oral pathogen lasted for about 2 weeks. Wilson and Love (29) also showed that after one month of treatment with the fluoride-releasing elastomers, the enamel of the test teeth of orthodontic patients was significantly harder compared with the control group.

In a two year clinical trial, Seppa (30) applied either Duraphat® or Fluor Protector® two to four times a year to the teeth of children in a low-fluoride area and determined that the fluoride content of the enamel remained elevated for at least two years after discontinuation of treatment with either one of the varnishes. However, the author determined that the caries-preventive effect did not continue after the applications were stopped, indicating that the varnish applications need to be continued as long as caries are a problem.

The effect of Cervitec® varnish on the levels of *Streptococci mutans* in plaque adjacent to bonded orthodontic brackets was evaluated in children using a split-mouth technique with a placebo varnish control (31). Both varnishes were applied four times during a three week period and plaque was collected after the onset of treatment. The results indicated that the proportion of *Streptococci mutans* within the plaque microflora was significantly lower on the test sides than on the opposite sides at the 1 week and 1 month examinations, suggesting that the pathogen in plaque from orthodontic patients can be suppressed effectively by topical administration of the chlorhexidine–thymol varnish. In a similar study, it was determined that microbial vitality, as assessed by a vital fluorescence technique, was significantly reduced 48 and 72 hours following the chlorhexidine varnish treatment, but reported that the inhibitory effect by the varnish could not be detected 12 weeks after varnish application (32).

D. Regulatory and Clinical Aspects

Products intended for the treatment of dental and periodontal disease must meet all regulatory requirements pertaining to safety and efficacy under the conditions of use. A basic set of pharmacological, toxicological, and biopharmaceutical testing should be conducted and documented on raw materials, active agents, and finished product formulations. Format and content of chemistry, manufacturing, and controls (CMC) for the dental and periodontal products are expected to be the same as those applicable to the other pharmaceutical dosage forms filed as a New Drug Application (NDA) or Abbreviated New Drug Application (ANDA).

There are regulatory guidelines available for different aspects of the products. For example, the U.S. Food and Drug Administration (FDA) has issued guidelines for clinical evaluation of drugs to prevent dental caries (33) and clinical evaluation of drugs to prevent, control, and/or treat periodontal disease (34). These guidelines suggest basic concepts that should be considered when developing suitable clinical protocols.

The active agents in the formulations administered for periodontal and dental diseases in most cases are old drugs. This will have an advantage over new chemical entities in that some of the preclinical testing such as carcinogenicity, reproduction, mutagenicity, and ADME studies are not required. Relatively simple pharmacology and safety studies such as adverse reactions and effect studies pertaining to the therapeutic indications are a part of the basic requirements. Most of the formulations administered for dental and periodontal diseases except for injections are topical products. It is expected that some subchronic topical toxicological testing will be required by the regulatory agencies.

The clinical pharmacology studies may have to document the concentrations of the active ingredient in the suitable tissues or sites being treated. The drug concentration in saliva and/or systemic circulation for the duration of treatment is relevant also. To the extent that controlled drug release and effectiveness of therapy are claimed, it would be required to demonstrate in vitro and in vivo the bioavailability of the active drug for the duration of product use. For products indicated for the treatment of periodontal diseases, investigation into the effects of the actives on the microorganisms is also required. It should be pointed out, however, that the clinical significance of the microbiological consequences due to the active agents has not been determined and is not known.

In the design of clinical protocols, there are two key aspects that need to be defined clearly. The first is the relevant information for the active, excipients, packaging, dose, stability, oral clearance rate, and the bioavailability of the active agent. The second is the variables selected for measurement and rationale for their choice, which are discussed in detail in the FDA guidelines for clinical evaluation of drugs to prevent dental caries (35).

IV. SUMMARY

A number of pharmaceutical products are being used and new products are still in the development pipeline. For the treatment of periodontal and dental disease, for example, a new periodontal gel may soon be introduced to the worldwide market (Atridox®, Atrix Laboratories, Fort Collins, Colorado). Pharmetrix, Division of TCPI (Menlo Park, California) has conducted preclinical evaluations of an erodible periodontal gel containing microparticles that release tetracycline HCl into the periodontal pocket similar to Atridox. As can be seen in this chapter, periodontal diseases appear to be an important target area for controlled-release pharmaceutical products. Matrix or gel products used in endodontic and orthodontic treatment procedures can be improved by increasing the substantivity of the dosage forms and the drug release kinetics to achieve improved efficacy and safety.

Chemotherapy of periodontal and other dental diseases faces several formidable challenges and obstacles. Because of the unique anatomical and physiological features of the oral cavity such as large fluid production (saliva and gingival fluids) and relative motions or friction between the dosage form and tissues and food, the retention of dosage form is poor and clearance of active agents is rather large, resulting in the need for frequent dosing. Currently marketed pharmaceutical products for the treatment of various dental diseases require procedures that are time consuming or inconvenient, limiting the widespread usage of the products. These problems also compromise physician/patient compliance. Therefore, for the new products to be successful in the market it is essential that the products embody significant improvement in one or more of these attributes associated with the current dosage forms.

ACKNOWLEDGMENTS

The authors wish to thank Drs. William Pfister and Tapash Ghosh for encouraging us to prepare this important chapter for this book. We also sincerely thank and are deeply grateful to Ms. Joan Schuman for typing and Ms. Diana Southard for making all editorial changes on this manuscript.

REFERENCES

1. Fuks AB, Holan G, Davis JM, Eidelman E. Ferric sulfate versus dilute formocresol in pulpotomized primary molars: Long-term follow-up. Pediatr Dent 1997; 19(5):327–330.
2. Negm MM. Effect of intracanal use of nonsteroidal anti-inflammatory agents on post-treatment endodontic pain. Oral Surg Oral Med Oral Pathol 1994; 77(5):507–513.
3. Ogaard B, Larsson E, Glans R, Henriksson T, Birkhed D. Antibacterial effect of a chlorhexidine–thymol varnish (Cervitec) in orthodontic patients: A prospective, randomized clinical trial. J Orofac Orthop 1997; 58(4):206–213.
4. Twetman S, Hallgren A, Petersson LG. Effect of an antibacterial varnish on mutans Streptococci in plaque from enamel adjacent to orthodontic appliances. Caries Res 1995; 29(3):188–191.
5. Wiltshire WA. Determination of fluoride from fluoride-releasing elastomeric ligature ties. Am J Orthod Dentofacial Orthop 1996; 110(4):383–387.
6. Trimpeneers LM, Dermaut LR. A clinical evaluation of the effectiveness of a fluoride-releasing visible light-activated bonding system to reduce demineralization around orthodontic brackets. Am J Orthod Dentofacial Orthop 1996; 110(2):218–222.
7. Isotupa KP, Gunn S, Chen CY, Lopatin D, Makinen KK. Effects of polyol gums on dental plaque in orthodontic patients. Am J Orthod Dentofacial Orthop 1995; 107(5):497–504.
8. Bragger U, Muhle T, Fourmonsis I, Lang NP, Mombelli A. Effect of the NSAID flurbiprofen on remodeling after periodontal surgery. J Periodontal Res 32(7):575–582.
9. O'Brian TP, Ruszkowski MT, Wolff LF, Hinrichs JE, Hargreeves KM. Effect of a non-steroidal anti-inflammatory drug on tissue levels of immunoreactive prostaglandin E2, immunoreactive leukotriene, and pain after periodontal surgery. J Periodontal 1996; 67(12):1307–1316.

10. Li KL, Vogel R, Jeffcoat MK, Alfano MC, Smith MA, Collins JG, Offenbacher S. The effect of ketoprofen creams on periodontal disease in rhesus monkeys. J Periodontal Res 1996; 31(8):525–532.
11. Haffajec AD, Socransky SS, Dibart S, Kent RL, Jr. Response to periodontal therapy in patients with high or low levels of P. gingivalis, P. intermedia, P. nigrescans and B. forsythus. J Clin Periodontal 1996; 23(4):336–345.
12. Furuichi Y, Ramberg P, Krok L, Lindhe J. Short-term effects of triclosan on healing following subgingival scaling. J Clin Periodontal 1997; 24(10):777–782.
13. Seppa L. A Review. Proc Finn Dent Soc1991; 87(4):541–547.
14. Ogaard B, Larsson E, Glans R, Henriksson T, Birhed D. Antimicrobial effect of chlorhexidine–thymol varnish Cervitec) in orthodontic patients: A prospective, randomized clinical trial. J Orofac Orthop 1997; 58(4):206–213.
15. Lindquist L, Otteskog P. Engenol: Liberation from dental materials and effect on human diploid fibroblast cells. Scand J Dent Res 1980; 88(6):552–556.
16. Wright K, Barbosa S, Araki K, Spangberg L. In vitro antimicrobial and cytotoxic effects of Kri 1 paste and zinc oxide-eugenol used in primary tooth pulpectomies. Pediatric Dentistry 1994; 16(2):102–106.
17. Wiltshire WA. Determination of fluoride from fluoride-releasing elastomeric ligature ties. Am J Orthod Dentofacial Orthop 1996; 110(4):383–387.
18. Petersson LG, Edwardsson S, Arends J. Antimicrobial effect of a dental varnish, *in vitro*. Swed Den J 1992; 16(5):183–189.
19. Actisite® (tetracycline hydrochloride) periodontal Fiber Product Information. ALZA Corporation (1998).
20. Author unknown. Antibiotic gels for periodontal disease. Drug Ther Bull 1994; 32(6):43–44.
21. Soskolne A, Freidman M. Intra-periodontal pocket drug delivery systems. In: Rathbone MJ, ed. Oral Mucosal Drug Delivery. New York: Marcel Dekker, 1992: Chap. 14.
22. Stoltze K. Elimination of Elyzol 25% dentalgel matrix from periodontal pockets. J Clin Periodontal 1995; 22 (3):185–187.
23. Steinberg D, Friedman M, Soskolne A, Sela MN. A new degradable controlling release device for treatment of periodontal disease: *In vitro* release study. J Periodontal 1990; 61 (7):393–398.
24. Perio Products Ltd, Jerusalem, Israel 1998 Perio Chips Platform.
25. Hagiwara, S, Takamatsu N, Tominaga Y, Umeda M. Subgingival distribution of periodontopathic bacteria in adult periodontitis and their susceptibility to minocycline-HCl. J Periodontal 1998; 69 (1):92–99.
26. Nakagawa T, Yamada S, Oosuka Y, Saito A, Hosaka Y, Ishikawa T, Okuda K. Clinical and microbiological study of local minocycline delivery (Periocline®) following scaling and root planing in recurrent periodontal pockets. Bull Tokyo Dent Coll 1991; 32 (2):63–70.
27. Holan G, Fuks AB. A comparison of pulpectomies using ZOE and KRI paste in primary molars: A retrospective study. Pediatr Dent 1993; 15(6):403–407.
28. Wilson TG, Gregory RL. Clinical effectiveness of fluoride releasing elastomers I: Salivary streptococcus mutans numbers. Am J Orthod Dentofacial Orthop 1995; 107(3):293–297.
29. Wilson TG, Love B. Clinical effectiveness of fluoride-releasing elastomers II: Enamel microhardness levels. Am J Orthod Dentofacial Orthop 1995; 107(4):379–381.

30. Seppa L. Studies of fluoride varnishes in Finland. Proc Finn Dent Soc 1991; 87(4):541–547.
31. Twetman S, Hallgren A, Petersson LG. Effect of an antibacterial varnish on mutans Streptococci in plaque from enamel adjacent to orthodontic appliances. Caries Res 1995; 29(3):199–191.
32. Weiger R, Friedrich C, Netachil L, Schlagenhauf U. Effect of chlorhexidine-containing varnish (Cervitec) on microbial vitality and accumulation of supragingivel dental plaque in humans. Caries Res 1994; 28(4):267–271.
33. U.S. Department of Health, Education, and Welfare. Guidelines for the clinical evaluation of drugs to prevent dental caries. Public Health Service, Food and Drug Administration, issued November 1978.
34. U.S. Department of Health, Education, and Welfare. Guidelines for the clinical evaluation of drugs to prevent, control, and/or treat periodontal diseases. Public Health Service, Food and Drug Administration issued November 1978.
35. U.S. Department of Health, Education and Welfare. Guidelines for the clinical evaluation of drugs to prevent dental caries. Public Health Service, Food and Drug Administration issued November 1978.

10

Oral Transmucosal Systems for Local Anesthetics: Dental and Oral Surgical Premedication

Juan A. Mantelle, David Lucking, David P. Kanios, and Gina Fernandez

CONTENTS

I. INTRODUCTION

This chapter describes the typical local anesthetics used in dental practice along with their advantages and limitations. Additional points to consider in the selection of mucobioadhesives and transoral mucosal patch vehicles are described. Finally, a summary of the clinical development program and highlights of selected clinical studies leading to FDA approval are described.

Anesthetics applied to the oral mucosa have been investigated for over half a decade to provide relief of toothaches and related pain. Patents dating back as early as 1939 describe anesthetic therapy to the oral mucosa in such vehicles as ointments, gels, pastes, injectables, and more recently transoral patches. A particular embodiment dating back to 1939 described in a U.S. patent (1) claims anesthetic ointments containing aminobenzoates applied to mucosal membranes without causing irritation. In the 1940s, Curtis (2) made different anesthetic compounds described in a U.S. patent related to using local and surface anesthesia that included a mixture of two or more anesthetics whereby the noted improvement came from pH adjustment of the preparation for mucosal application. In 1950, a U.S. patent (3) described a therapeutic device for application to the area of the gums and the oral cavity.

As Shrontz (3) opened the door to new anesthetic drug delivery techniques, researchers began to patent various compositions, matrices, and dressings that would assist in the delivery of the anesthetic drugs. Many studies were performed in evaluating adherence to moist tissues as this is essential to proper delivery of the anesthetic agent. In 1966, Davis (4) of Astra Pharmaceuticals was issued a Swedish patent describing the delivery of lidocaine from a film capable of providing effective local anesthesia. This film, in which the anesthetic was saturated at the time of application, could not deliver the drug as effectively as can be done from today's systems, where the drugs are fully solubilized. However, this invention was seen to provide better dosing control than the ointments and/or gels available at the time.

In 1972, a U.S. patent (5) issued to Alza Corporation described an adhesive patch that delivered oxytocin through the oral mucosa. This particular invention was comprised of a backing membrane, a specific amount of drug to be delivered, and a pressure-sensitive adhesive coating capable of adhering to the oral mucosa for extended periods of time. With this form of delivery, it was found that controlled amounts of drug could be delivered in a way that constituted a significant improvement over state-of-the-art buccal tablets and lozenges because lagtimes and the predictability of dosage could be better controlled. Following this advance, in 1976, a U.S. patent described a "reservoir" system that delivered anesthetics.

In 1976, another U.S. patent (7) disclosed direct dosage forms for buccal administration to the interior surfaces of the mouth. Tsuk (7) developed a device similar to that described in the 1972 patent issued to Zaffaroni (5) that describes the use of bioadhesives with various medicaments for oral mucosal application.

In 1985, a U.S. patent assigned to Astra (8) disclosed the use of prilocaine, tetracaine, etidocaine, lidocaine, or bupivacaine bases dissolved by means of eutectic mixture. In working with such anesthetic compositions, Broberg (8) found that higher concentrations of local anesthetics and thus enhanced delivery could be achieved from these low-melting eutectic mixtures. Later that year, a European patent publication (9) disclosed the "soft patch" concept. The soft patch is a delivery system consisting of a polyhydric alcohol, a tackifier, and an oleaginous substance that, as claimed, would permit a more accurate control of drug dosage. Delivery systems/patches, as described by Kigasawa (9), consisted of embedded anesthetic agents in ethylene/vinyl acetate copolymer, synthetically woven fabrics, bonded polyester, or polyamide fabric.

In 1987, a U.S. patent assigned to Johnson and Johnson (10) introduced yet another version of a delivery system that consisted of an extruded single or multilayered bioadhesive thin film useful in intraoral controlled drug delivery. Anesthetics, analgesics, anticaries agents, anti-inflammatory agents, antihistamines, antibiotics, antibacterials, and fungicides are classes of drugs that were contemplated for incorporation in these sustained-release films. Although these compositions may be effective in routine short-term treatment, full benefit was anticipated to be derived for long-term periodontal disease treatment. In 1993, U.S. Patent 5,234,957, assigned to Noven Pharmaceuticals, Inc. (11), described the incorporation of anesthetic bases alone or in combination with their corresponding acid salts into polysaccharide adhesive compositions capable of adhering to the oral mucosa for prolonged periods of time for alleviation of pain. This patent also described an anesthetic mixture with a flexible polysaccharide bioadhesive carrier that provided a suitable platform for prolonged usage. Mantelle (11) described the use of both amides and/or esters for local anesthetics alone or in combination with each other for dental pain relief.

The oral mucosa has been recognized as an excellent site for local and systemic delivery of various drugs for many years. But to date, DentiPatch®, formulated using Noven's patented technology (11), is the only transoral anesthetic delivery patch on the market. Further research is presently ongoing which will no doubt yield new commercial systems in the area of transoral mucosal delivery.

II. LOCAL ANESTHETIC AGENT SELECTION

Local anesthetic agents are pharmacologically active compounds that block nerve conduction when applied in therapeutically effective amounts. Local anesthetics are generally classified as esters, amides, or benzoic acid derivatives and can be administered to the buccal mucosa as either the free base or their corresponding salt form. Anesthetic free bases tend to be lipid soluble and as such can effect a quicker onset of anesthesia because they permeate the lipophilic mucosal tissue faster compared to the corresponding salt form of the drug. Once absorbed, local anesthetics act to inhibit depolarization and ion conduction of the nerve fiber thereby blocking pain perception. Conversely, the salt form of the drug, which is hydrophilic, can permeate the lipoprotein nerve membrane only after the buffering capacity of the mucosa converts the salt to the free base, the end result being a delayed onset of anesthesia. Selection of the proper anesthetic agent for application to the oral mucosa is the first critical step towards the formulation of an effective dosage form for dental anesthesia. Factors that the formulator must consider in developing an effective formulation for treatment for tooth pain include:

1. Onset time
2. Duration of anesthesia
3. Toxicity/degradation profiles
4. Relative potency

Efficacy of a topically applied oral mucosal local anesthetic preparation will ultimately depend on the anesthetic concentration as well as the relative potency of the anesthetic selected. To be effective, the anesthetic preparation should contain a sufficiently high concentration of the anesthetic, preferably in the free base form, so as to effect a rapid onset. A given dose of the formulation should also contain enough local anesthetic agent to allow for a prolonged delivery to optimize the duration of the anesthetic effect. As with other therapeutic agents, local anesthetics vary significantly in their relative potencies. Hence, selection of an agent with the correct potency for the targeted clinical indication is critical to the efficacy of the dosage form.

A. Onset of Local Anesthesia

The term "onset of local anesthesia" as used herein refers to the time required to achieve the maximum effect on the individual sensory nerves. As described above, onset of local anesthesia principally depends on the lipid solubility, molecular weight, quantity, and relative potency of the anesthetic agent. Therefore, anesthetics with higher lipid solubility and lower molecular weights will typically have a more rapid onset of local anesthesia compared to water-soluble, higher-molecular weight compounds. Table 10.1 summarizes the physicochemical properties of various local anesthetic agents along with their mucosal permeation properties such as onset and duration of local anesthesia.

Local anesthetic preparations as used in the oral mucosa for dental procedures are designed to effect a very localized (as opposed to systemic) blocking of stimulus-induced nerve conduction. Local anesthetics, as a class of compounds,

Table 10.1 Local Anesthetic Agents

	Local Anesthetic [a]	Onset of Anesthesia	Duration of Effect	pK_a (at 25°C)	Molecular Weight	Chemical Classification
1.	Benzocaine [a]	1	Short	2.5	165.19	Ester
2.	Lidocaine [b]	2–5	Medium	7.9	234.33	Amide
3.	Mepivacaine [b]	1.5–4	Short	7.6	246.34	Amide
4.	Bupivacaine [b]	4–8	Long	8.1	288.43	Amide
5.	Etidocaine [b]	2–5	Long	7.7	276.42	Amide
6.	Dibucaine	3–10	Long	N/A	343.92	Amide
7.	Tetracaine [b]	3–8	Long	8.5	264.83	Ester
8.	Prilocaine [b]	2–5	Medium	7.9	220.31	Amide
9.	Procaine	5–10	Short	8.9	236.30	Ester
10.	Dyclonine [b]	<10	Medium	N/A	289.43	Ketone
11.	Cocaine	2–5	Long	5.6	303.35	Ester
12.	Pramoxine	3–5	Short	N/A	293.39	Ester

[a] Commonly used topical local anesthetics for oral mucosal applications.
[b] Used as injectable agents in specific dental procedures.
Source: Adapted from References 12 to 14.

have a rapid onset of action (typically less than ten minutes) (Table 10.1). To shorten the onset time by means of permeation-enhancing agents generally results in minimal, if any, improvement. Formulations of mucosal local anesthetic preparations are best served by adjustment of the pH so as to ensure that the anesthetic is predominantly in the base form and ensuring that the base is properly solubilized at the target concentration in the preparation. Whenever possible, occlusion of the preparation at the time of application will ensure that the onset time is shortened and prevent the preparation from being washed away by the salivary flow, compared to nonoccluded formulations.

B. Duration

The term "duration of local anesthesia" as used herein is meant to signify the period of time during which the local anesthetic agent measurably blocks nerve conduction. Commonly, prolongation of local anesthesia in the dental practice has been achieved by the addition of vasoconstrictors such as catacholamines (i.e., epinephrine) to the injections so as to cause constriction of blood vessels thus preventing the anesthetic from being washed away from the site of injection by the circulatory system. Because catacholamines are not particularly effective when applied topically, such a prolongation is minimal in topical mucosal applications. Prolongation of local anesthesia upon topical mucosal application can be achieved by one or more of the following strategies:

1. Increased loading of the anesthetic in the formulation (i.e., higher concentration)

2. Increased contact time and larger dose per application site (i.e., mg/cm^2)
3. Selection of right anesthetic agent (i.e., higher mucosal tissue partition coefficient)

The first approach was discussed earlier and can be achieved by increasing the amount of anesthetic agent per dose to an amount greater than that which can be readily absorbed during the required period of local anesthesia. This higher concentration leads to a thermodynamic driving force that can be sustained for a longer period of time. This can be achieved by either increasing the drug concentration or administering a larger volume or dose of local anesthetic product.

Increased contact time as an approach to prolong the local anesthetic effect provides the formulator the most versatility because it allows for the use of short-, medium-, or long-acting anesthetic agents. The common denominator for prolongation of anesthesia via increased contact time is the use of formulations capable of mucoadhesion because these formulations enable the dosage to stay in place thus achieving the desired duration. Mucosal adhesion can be achieved either by the use of in situ bioadhesive film formers or by the use of transdermal/transmucosal patch technology.

The use of in situ bioadhesive film formers has been made commercially available by Zila, Inc. (39) with benzocaine among other active agents. These systems work by application to the mucosal tissue of a solution containing the bioadhesive, the active agent, and other excipients in the presence of a volatile polar solvent such as ethanol. Upon application to the mucosa, the solvent evaporates leaving behind the drug containing bioadhesive. In the case of this product, the prolongation of anesthesia is directly proportional to the solubility of the bioadhesive in salivary fluid as well as the amount of solution being administered. These factors make the dosing and duration variable from person to person due to variable salivary secretion (e.g., volume and flow) among individuals. The use of patch technology in the administration of anesthetic agents, namely, lidocaine, to the buccal mucosa was first introduced commercially by Noven Pharmaceuticals, Inc. in 1996 under the tradename DentiPatch® (40).

Patch technology, by virtue of the unit dose concept, provides dental practitioners with a fixed amount of anesthetic agent for application to a finite area (2 cm^2 containing 46.1 mg of lidocaine in the case of DentiPatch®). Because polar volatile solvents are not required with this type of dosage form, the number of bioadhesives and anesthetic agents to select from is significantly reduced. Hence, bioadhesives that are more resistant to solubilization by salivary fluids can be used, resulting in the ability to extend application time, and thus effective anesthesia, beyond the 15 to 45 minutes attainable with the in situ film formers. In addition, patch technology enables the use of occlusive backings that significantly enhance the rate and extent of anesthetic permeation.

The third option for prolongation of local anesthetic effect is the selection of an anesthetic agent that by virtue of its chemical nature provides extended duration of anesthesia. For example, bupivacaine, etidocaine, dibucaine, and tetracaine all produce a longer duration of local anesthesia than mepivacaine, lidocaine, or prilocaine. In addition, and as proposed by Broberg et al. (8), Chang et al. (15), and Mantelle (11), effective prolongation of the local anesthetic effect can be achieved with specific mixtures of anesthetic agents. Broberg et al. (8) have accomplished prolongation of local anesthesia via the formation of a eutectic mixture of local anesthetics (EMLA®) whereby the mixture of two anesthetic bases, for example, lidocaine and prilocaine, results in a lower melting point than either compound by itself. As a result of this lowered melting point, higher concentrations of solubilized drug can be attained, thus enhancing the driving force for permeation. Chang et al. describe the use of mixtures of the free base and salt of the same drug (i.e., lidocaine and lidocaine HCl) in order to enhance solubility, onset, and duration of local anesthesia over the use of either the salt or the base by themselves. Mantelle (11) has shown that by mixing the base with the corresponding salt, both the fast onset of the shorter-acting anesthetic base and subsequent absorption of the longer-acting salt can be achieved from a single patch composition. The result of this combination is a preparation that has fast onset and prolonged delivery of the anesthetic combination. Which modality or formulation approach is used to extend the duration of anesthesia will ultimately depend on the intended clinical use and the regulatory pathway envisioned. Effective anesthesia for as little as five minutes or greater than 24 hours is all within the scope of the aforementioned formulation approaches.

C. Toxicity/Degradation

Local anesthetics, by their very nature, are all toxic drugs with clinical safety margins typically lower than drugs in other therapeutic categories when it comes to systemic toxicity potential (14). Among the most common systemic toxicity occurrences are CNS toxicity leading to depression, often without the cortical stimulation commonly associated with cocaine; and cardiovascular events primarily associated with antirhythmic effects. These systemic toxicity events can be traced, in most instances, back to either accidental venipuncture and subsequent intravenous injection or to patients with compromised metabolism. Hence, mucosal application provides the formulator with a safer route of administration for a more varied selection of local anesthetic agents.

Reactions associated with oral mucosal applications are usually associated with repeated exposure whereby local contact dermatitis can result with any of the anesthetics used in dental practices. The reason for the excellent safety record of anesthetics as used in dental practices, either topically or by injection, is a result of careful dose monitoring and placement by the practitioners. This is a factor that must be considered when formulating novel preparations for either mucosal or injectable applications so that adequate safety margins can be assured.

1. Esters

Ester-type local anesthetic agents are cleared from the systemic circulation by both hepatic and plasma metabolism, hence the half-lives for the majority of these drugs are very short relative to amide-type agents. In the plasma, hydrolysis by the plasma pseudocholinesterase enzyme is extremely rapid. Hence, in patients with normal plasma pseudocholinesterase enzyme levels, toxic reactions will be shorter-lived compared to amide-type local anesthetics, whereas in metabolically compromised individuals these reactions can be prolonged significantly. Of the ester-type local anesthetics used in dentistry, tetracaine is the most toxic followed by procaine and benzocaine, which have very little toxicity potential. In rare, yet significant, occasions all ester-type local anesthetics can elicit allergic reactions. Thus, as a general rule, patients with compromised plasma cholinesterase levels or known allergic dispositions should not be given ester-type local anesthetics.

2. Amides

Amide-type local anesthetics are much more widely used in dentistry than their ester counterparts at least in part because of the absence of allergic events. Contrary to the ester-type local anesthetics, the amides are cleared almost exclusively by the hepatic route, which in turn leads to longer half-life in the systemic circulation. Systemic side effects associated with the amide-type local anesthetics are very similar to those seen with the ester-type in that they are typically associated with CNS depression and/or antirhythmic cardiovascular effects. Of the amide local anesthetics commonly used in dentistry, prilocaine differs from lidocaine, mepivacaine, etidocaine, and bupivacaine in that it is a toluidine derivative whereas the others are xylidine derivatives. These chemical differences have been associated with the formation of methemoglobin as a result of prilocaine hydrolysis to ortho-toluidine. Although this methemoglobin formation is unlikely in the doses used in dentistry, it is still a factor that must be considered by the formulator.

In summary, selection based on toxicity criteria would point the formulator in the direction of the amide-type local anesthetics because onset and duration are both manipulable for use in dentistry and this selection can avoid allergic reactions in predisposed patients.

3. Ketones

Dyclonine is the only ketone-derived local anesthetic agent routinely used in topical applications (i.e., Sucrets® lozenges) and it is considered the safest of the topical anesthetics from the standpoint of the absence of systemic toxicity and oral mucosal injury after topical applications of low concentration solution. Dyclonine, given by oral mucosal application where low drug concentrations can be utilized, presents the formulator with an excellent choice for quick onset and medium duration when amides cannot be used.

D. Relative Potency

Relative potencies of topical anesthetics used in mucosal applications were evaluated by Tucker and Mather (14) and the resulting ranking established as follows:

tetracaine, bupivacaine, and etidocaine were ranked as approximately equipotent and four times as potent as lidocaine, mepivacaine, and prilocaine. Lidocaine, in turn, is twice as potent as procaine. Dyclonine, in other studies, has been listed as being approximately equipotent to lidocaine. Benzocaine is one tenth as potent as lidocaine. Given the above relative potencies, the decision for the formulator is whether to use the more potent (i.e., bupivacaine or etidocaine) or the lesser potent lidocaine, mepivacaine, or dyclonine. Here again, the decision should be based on the clinical indication sought and the duration of action being targeted; see Table 10.2.

III. BIOADHESIVE POLYMERS

Bioadhesives for transmucosal delivery systems (TMDS) are categorically selected from three distinct groups including: (a) natural products, (b) modified natural products, and (c) synthetic products. Natural products consist of materials such as natural gums, gelatin, and starches. Modified natural products consist of water-soluble materials such as alkyl and hydroxyalkyl ethers of cellulose and starch, carboxylates–carboxymethyl cellulose, and mixed ethers of starch and cellulose. Finally, bioadhesives may be chosen from the synthetic products that include materials such as polyvinyl alcohols, polyvinyl pyrrolidones, polyvinylmethyl ethers, polyacrylic acid and its salts, polysiloxanes, ethylene oxide polymers, and various copolymer systems (16).

Selection of a single group or combinations of bioadhesives are at the discretion of the formulator in order to achieve the desired physical and chemical properties sought during development of the TMDS. Examples of various TMDS and formulation approaches are discussed for each bioadhesive product group.

A. Natural Bioadhesive Products

Gallopo and Dills (17) describe the use of natural products, preferably xanthan gum, for use in the construction of a rigid matrix for bioadhesion to mucous membranes. Xanthan gum in combination with a solid polyol, which is used as an adhesion promoter, is combined with other excipients and active ingredients to form a tablet that can be applied to moist tissues in the mouth. This bioadhesive matrix tablet adheres to the mucosal tissue in the mouth and releases the active ingredient over time as the tablet hydrates and dissolves. Mantelle (11), on the other hand, incorporates natural products, preferably karaya gum, along with polyhydric glycol(s), excipient(s), and drug(s) to form a flexible solid-state matrix for application onto the mucosal tissue. The objective is to create a TMDS that would adhere to the oral mucosal tissue, conform to the topography of the applied area, and deliver the active ingredient directly under the area of application. A water-swellable natural product was selected as the bioadhesive and further incorporation of matrix modifiers determined the end-product physical state and drug delivery profile required to achieve the desired biopharmaceutical properties.

Table 10.2 Chemical Structures of Local Anesthetic Agents

Benzocaine	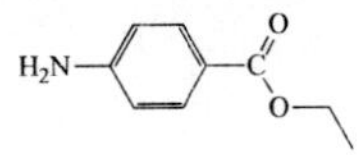
Procaine	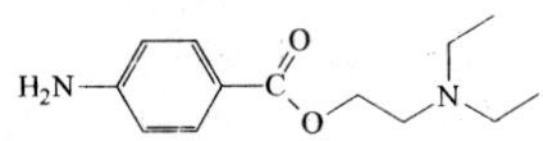
Cocaine	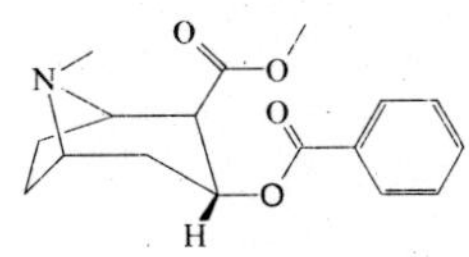
Tetracaine	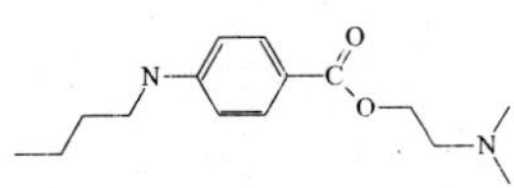
Prilocaine	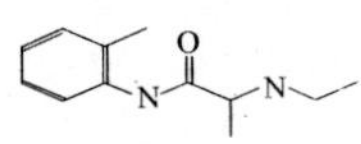
Lidocaine	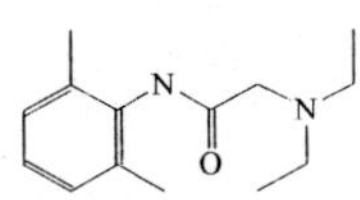
Mepivacaine	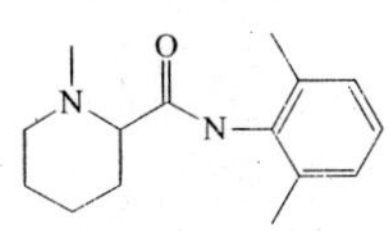
Dibucaine	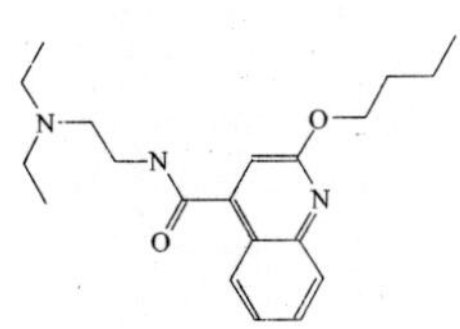
Bupivacaine	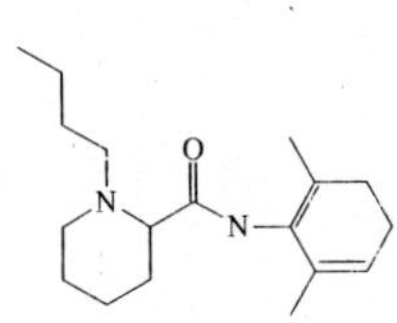
Etidocaine	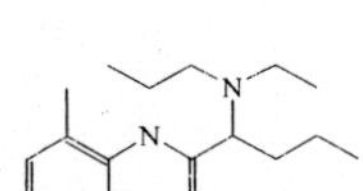

Table 10.2 Chemical Structures of Local Anesthetic Agents (continued)

Dyclonine	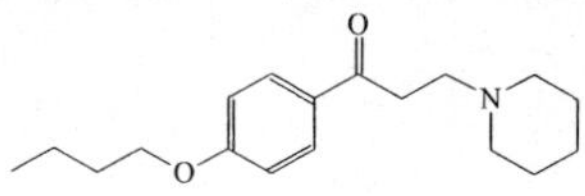
Pramoxine	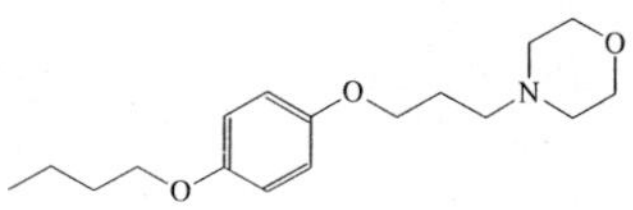

B. Modified Natural Bioadhesives

The use of a modified natural product in the development of a bioadhesive is described by Estes (18). The water-soluble composition is essentially carboxymethyl cellulose or alginate gum that is crosslinked via a polyol mixture, preferably propylene glycol and glycerin. After application, as the bioadhesive hydrates in the presence of moisture, the amount of crosslinking determines the release rate of the active ingredient incorporated into the system. Furthermore, the inclusion of natural gums or starches into the described flexible solid-state matrix increases the internal cohesivity of the platform to provide variations to its internal strength for processability and durability during application.

C. Synthetic Bioadhesives

The use of synthetic products as bioadhesives is by far the most varied and investigated. Synthetic excipients offer the formulator water-soluble and swellable polymers of variable molecular weight, functionality, and electrical charge (i.e., anionic, neutral, and cationic). Although the literature is too numerous for us to cover all facets of synthetic bioadhesives, the examples selected are based on their novelty or range of applications.

Jenkins and Leslie (19) described the use of cellulose materials that are used in the construction of a rigid structure matrix for bioadhesion to the mucosa of the oral or nasal cavity. Here the drug, an aliphatic alcohol, and a water-soluble hydroxyethyl cellulose are blended, coated with the cellulose derivative (hydroxypropyl cellulose), and then compressed into a finite shape. The solid matrix, when applied to the mucosa, allows sustained release of the drug over an extended period while the cellulous polymers maintain adhesion and structural stability.

Schiraldi et al. (10) described the use of cellulose polymers in combination with ethylene oxide, a plasticizer, and drug(s) to form a bioadhesive that is extruded in a single- or multilayer film for application to the wet mucosa. This flexible solid-state matrix utilizes water-soluble hydroxypropyl cellulose and water-insoluble polymers such as ethyl or propyl cellulose to form the carrier matrix for the bioadhesive. By varying the ratio of the polyethylene oxide, water-soluble cellulose, and water-insoluble cellulose it was shown that bioadhesion and release rate could be altered over a wide range for selected drugs.

A novel approach for the use of synthetic products as bioadhesives is described by Kigasawa (19) who developed a pliable soft buccal matrix. The matrix comprises a combination of drug(s), water-soluble protein(s), and a carboxyvinyl polymer and/or a fatty acid ester that results in a controlled-release bioadhesive that can be applied to various intraoral topographies. The soft buccal matrix can be deformed upon application to ensure total surface contact to the intended area. Furthermore, Robinson (20) described the use of a carboxyvinyl polymer as a bioadhesive that when combined with a drug can be in a dry form, a semisolid, or a liquid suspension, and can be applied to the mucosal membrane, internally or externally.

A final example of synthetic bioadhesives is described by Takayanagi and Sawai (21), who employ various combinations of synthetic polymers to produce bioadhesives. These include mixtures of polyvinyl pyrrolidones, cellulose derivatives, polyacrylate salts, and natural bioadhesives along with drug and softening agents that form flexible monolithic or multilayered solid-state matrices that are applied to the oral mucosa.

Although no one example cited above is preferential, the broad range of synthetic polymers in mucosal adhesion applications provides the formulator with myriad possible formulation alternatives.

D. Biocompatability

Biocompatability must also be addressed during the selection of appropriate bioadhesive polymers for pharmaceutical applications. Whether a single bioadhesive or combinations of bioadhesive polymers are chosen for the proposed dosage form, the applied matrix must be nontoxic and nonirritating. For the most part, the materials described in the above examples are well characterized in toxicological studies and deemed nontoxic. Review of manufacturer and published literature (22,23) will usually support the safety and feasibility of pursuing the particular bioadhesive excipients that the formulator has selected. The concern for toxicity is more important in the selection of drugs that are incorporated into the matrix and the dose of the drug that is being delivered.

To a larger extent, the formulator has more control over the choice of bioadhesives and excipients that have low irritation. Here, particular attention must be paid to the area and duration of application as well as the drug and TMDS. Irritation may be a result of functionality or electrochemical charge of the bioadhesive or a mechanical irritation due to the adhesion of the bioadhesive on the mucosal membrane. In the spirit of pursuing physical testing of the bioadhesive (20), the formulator must also be aware that an overly aggressive bioadhesive may cause stripping of mucosal tissue upon removal of the matrix.

E. Matrix Compatibility

The final aspect that must be explored during development of the TMDS is the mutual compatibility of the materials comprising the composition. As is the case for the described matrices, the combination of the bioadhesive(s), drug(s), and excipient(s) must be both thermally and physically stable over the period of the intended shelf life of the product and application period. Therefore, the formulator must pay particular attention to the chemical and physical properties of all materials selected. These include melting and boiling points, molecular weight, viscosity, chemical functionality, and electrochemical charge. For example, a TMDS may appear physically stable, yet on the molecular scale, the drug may be degraded, or as in the case of local anesthetics, the drug may crystallize out by the bioadhesive selected, which in turn results in reduced product stability and may alter the local anesthetic action. Another example of stability failure of the TMDS is a case where, due to the solubility characteristics of the bioadhesive, the drug diffuses out over time along with the excipient into the packaging material. Those involved with the development of TMDS should always be aware of these potential problems that can lead to short- and long-term matrix incompatibility.

IV. VEHICLE SELECTION

As described earlier, local anesthetic agents need not rely on permeation enhancement for oral mucosal delivery. With onset times for these agents being on the order of less than five minutes and efficacy being evaluated in terms of their localized action, permeation enhancement via chemical agents is generally unnecessary.

Vehicle selection for mucosal anesthetic preparations should therefore be based on: (a) anesthetic agent solubilization, (b) effect on the bioadhesive selected, and (c) lack of irritation/sensitization potentials. The solvent or cosolvent system should be designed so as to ensure the full solubilization of the anesthetic base in quantities large enough to provide the onset and duration of local anesthesia required by the therapeutic indication. Polyhydric alcohols such as propylene glycol, dipropylene glycol, and polyethylene glycols have been considered particularly useful for this purpose.

Polyhydric alcohols are also particularly useful for purposes of plasticizing many different bioadhesive agents. In the case of the DentiPatch® system, there are three different polyhydric alcohols used, each having a different function. Dipropylene glycol is used as a drug solvent whereas propylene glycol has the dual function of drug solubilization and bioadhesive plasticization. The third polyhydric alcohol, glycerin, is used as the primary plasticizer for the karaya gum bioadhesive. A plasticizer serves to provide the final composition with the pliability and softness required to effectively adhere at the site of application for the intended wear period.

Other vehicles that have been evaluated for use in the manufacture of mucosal anesthetic preparations include fatty acids, esters, and alcohols as well as other surface active agents. Typically, these other vehicles are used for specific agents in order to further enhance their solubility and thus their rate and extent of permeation.

Uses of generally recognized as safe (GRAS) and approved pharmaceutical excipient ingredients will assist the formulator immensely in navigating the regulatory pathway to FDA approval, because questions regarding toxicity and or irritation are less likely to become an issue. The formulator should, whenever possible, select GRAS-listed ingredients in preparation of the final formulation while always keeping in mind the indication being sought. For example, short-term or sporadic use of the formulation will typically not result in irritation whereas a product intended for chronic use requires assurance that the solvent/co-solvent system selected will not cause irritation or induce a sensitization reaction with prolonged use.

V. DENTIPATCH® DEVELOPMENT

Introduced in the U.S. market in late 1996, DentiPatch® was the first commercial bioadhesive patch (TMDS) designed specifically for application to the buccal mucosa. DentiPatch® is a lidocaine transoral delivery system designed and clinically proven to reduce the perception of pain associated with needle insertion.

A painless alternative to needle insertion in the dental practice was sought for a long time as large numbers of patients are anxious about injection of local anesthetics. In order to accomplish this goal, it was felt that local anesthesia should be attained quickly (within 3 to 5 minutes) and last for up to 60 minutes while utilizing patch technology to ensure the localized action.

Local anesthetic agent selection, as described earlier, led to the use of lidocaine inasmuch as it has a fast onset, medium duration of action, and is a commonly used amide for topical applications within the dental practice. Other local anesthetic agents that could have been incorporated for this type of indication would have been prilocaine, mepivacaine, and/or dyclonine.

Bioadhesion with DentiPatch® was achieved via the use of a natural, plasticized, and water-swellable polysaccharide (e.g., karaya gum). Pliability and complete solubilization of 20 percent (w/w) lidocaine base in the finished product was achieved by incorporation of the various polyhydric alcohols into the dosage form, which in turn provided the necessary flexibility and drug loading to the bioadhesive portion of the product to ensure the practitioner's ability to position the system at the site requiring anesthesia and subsequently achieve the intended therapeutic effect. A custom-designed backing layer comprising a nonwoven fabric with an occlusive polyester film provides the DentiPatch® product with the occlusion and shape-retaining properties required for prolonged application to the oral mucosal tissues.

DentiPatch® comprises a blend of lidocaine, lecithin, propylene glycol, dipropylene glycol, glycerin, karaya gum, aspartame, and spearmint flavor. These components are mixed in order to achieve a viscous blend which is then coated at a predetermined thickness onto the backing which consists of a polyester/EVA film laminated to a polyester/rayon nonwoven fabric. The coated composition is then passed through drying ovens in order to accelerate the gelling process. Once

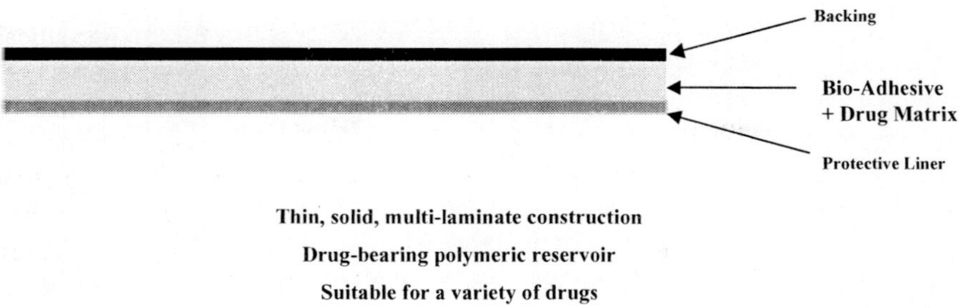

Figure 10.1 DentiPatch® configuration.

fully gelled, the composition is laminated to the release liner prior to die-cutting and packaging for commercial distribution (Figure 10.1).

Early studies were conducted with supplies generated in a laboratory scale with batch sizes ranging from 500 to 5000 grams. Subsequent scale-ups went to 45 and 150 kg batches for the commercialization stages of development. The main issues encountered in scale-up were associated with viscosity regulation and the subsequent gelling of the product wherein the need for tighter controls in the area of temperature regulation were addressed.

VI. CLINICAL STUDIES

A. Overview

It is estimated that 120 million people in the United States do not seek dental treatment on a regular basis due in part to fear and discomfort associated with the injection of local anesthetics into the oral mucosa. In addition, the American Dental Association estimates that there are more than 10 million dental phobics who avoid dental visits at all costs due to their fear of pain.

The use of topical local anesthetics prior to intraoral injection of local anesthetics is a widely accepted practice, because injection is often accompanied by discomfort or pain. Topical agents, including gels, ointments, and sprays, have been used for many years to anesthetize the oral mucosa surface. Each of these dosage forms has distinct disadvantages. Sprays make it difficult to control the amount of material expelled and may result in systemic effects. Ointments and gels may not be confined to the desired site of application, or become diluted in the mouth, resulting in abbreviated or ineffective anesthesia.

B. Background Information

Noven Pharmaceuticals, Inc. has developed a lidocaine transoral delivery system (DentiPatch®) for lidocaine administration to the oral mucosa at a specified site to maximize the drug's therapeutic potential and thus enable some patients to experience needle-free dentistry.

A topically applied patch for dental anesthesia has several potential advantages over conventional delivery techniques. These include improved patient compliance and acceptance, reduction of cross-infection risks, and elimination of bleeding problems associated with injections in hematologically compromised patients. Transmucosal administration of anesthetics allows a rapid decline in anesthesia after the patch is removed. In addition, the patch may be used in place of infiltration anesthesia for periodontal curettage. This may permit the use of the patch by dental hygienists who are not authorized to perform infiltration anesthetic injections.

Any reliable method of evaluating pain should meet three criteria: (1) the stimulus should be measured in physical units to permit quantitation, (2) the sensation must be clearly detectable, and (3) tissue damage should be absent or minimized.

Electrical testing methods, such as those used in Adriani et al. (24,25), employed a pulsatile direct current for testing the effectiveness of topical local anesthetics applied to the tip of the tongue. The endpoint was loss of tingling upon electrical stimulus. However, this method has been found to yield variable thresholds because of possible polarization of the anesthetic substance and the confounding of the simple sensory thresholds with sensations of taste and muscle stimulation (26). In addition, the tongue is neither anatomically nor physiologically representative of the usual areas where topical anesthesia is employed clinically.

The teeth have also been used for experimental determination of the effectiveness of local anesthetics by use of an electrical stimulus. Some researchers have used this method, but others feel that electrical stimulation is not an acceptable modality because it does not occur except under artificial experimental conditions. In addition, the application of the lidocaine transoral delivery system will be for a variety of soft tissue procedures and hence a procedure that evaluates the extent of soft-tissue anesthesia would have more clinical relevance.

Pressure algometry (i.e., quantitative measurement of pressure that induces pain or discomfort) has proved useful in evaluating pain sensitivity in normal tissues. Using a plastic probe, Schmid (27) found that forces of 45 pounds per square inch elicited a pain response when periodontal pockets of various depths were probed. The use of pressure to induce pain has direct clinical relevance to periodontal curettage and root planing procedures.

Among devices capable of delivering measurable mechanical stimuli such as pressure is the commercially available Florida Periodontal Probe (41). The Florida Periodontal Probe permits determination of periodontal pocket depth using a constant force applied to a defined surface of the probe tip. The Florida Periodontal Probe has been modified so that the constant force component has been replaced with a pressure-sensitive device. During usage the tip of the probe is handheld and its tip applied to either the surface of the gingiva or into the gingival sulcus. The amount of pressure (i.e., force) applied to induce pain is translated through a digital encoder and transmitted to a CRT screen and a printer.

In an early evaluation of the patch using pressure algometry, Heins et al. (42) demonstrated that a single application of the patch containing lidocaine in concentrations ranging from 5 to 25 percent (w/w) resulted in an anesthetic effect that permitted application of the maximum force prescribed in the protocol (approximately 184 grams/cm^2). There was a dose-response increase in the pain threshold among placebo, 5, 10, 20, and 25 percent (w/w) lidocaine systems. Based on these results, the 10 and 20 percent lidocaine patches were chosen for evaluation in reducing the pain associated with needle insertion into the mucosa.

C. Evaluation of DentiPatch® in Reducing Needle-Insertion Pain

1. Study Design

The pain associated with dental needles has been reported to be the most fear-provoking procedure in dentistry (28). To reduce the pain associated with needle insertion, dentists often apply topical local anesthetics to the oral mucosa prior to giving injections of local anesthetics. A number of placebo-controlled clinical trials have attempted to study the effectiveness of topical local anesthetic agents with mixed results. Of the eight published placebo-controlled trials, only three have demonstrated topical local anesthetic efficacy (29–31) with the other five concluding that topical local anesthetic agents were no more effective than the placebo. The studies where it was concluded that topical local anesthetic agents were no more effective than placebo had a number of design flaws including the use of a 25 gauge needle, injection of local anesthetic after needle penetration, contact with periosteum, short duration times (15 to 45 seconds) of contact between the mucosa and topical anesthetic agent, and the use of phenol or benzocaine as the active topical local anesthetic agent. By contrast, in the studies claiming efficacy, needles no larger than 27 gauge were used, contact with the underlying periosteum was avoided, and 5 percent lidocaine was used as one of the topical anesthetics.

In 1996, the U.S. Food and Drug Administration approved a mucoadhesive patch containing lidocaine (DentiPatch®, lidocaine transoral delivery system). Although units containing 10 and 20 wt percent lidocaine were approved for clinical use, the currently marketed product contains 46.1 mg of lidocaine (20 wt percent concentration) contained in a bioadhesive matrix. It is a 3 × 1 cm patch which is 2 mm thick and is designed to be applied directly to the mucosa where anesthesia is desired. Lidocaine diffuses from the patch into the oral mucosa while the patch is affixed. It provides site-specific delivery of lidocaine, maximizing the therapeutic effect of the drug, while limiting the dilution of the medication. Onset of local anesthetic effect can occur as early as 2.5 minutes and the patch can be left in place for up to 15 minutes. The anesthetic effect lasts for about 40 minutes after the patch is removed following a 15-minute period of wear. The current FDA-approved indication of the lidocaine transoral delivery system is the production of mild topical local anesthesia of the accessible mucous membranes of the mouth before superficial dental procedures.

Table 10.3 VAS Scale

100 mm

|..|

No Pain   Pain as bad as it could be

VERBAL PAIN SCALE

0 = None
1 = Mild
2 = Moderate
3 = Severe
4 = Very Severe

The patch can also be used to reduce the pain associated with injections of local anesthetics in the oral tissues. Four clinical investigations were conducted to evaluate the efficacy and safety of the two dosage formulations, 23 mg and 46.1 mg, of the lidocaine transoral delivery system in reducing the pain associated with insertion of 25 gauge needles to the level of the bone. Plasma concentrations of lidocaine were also measured because it is desirable that the patch not appreciably add to the systemic local anesthetic concentrations achieved by subsequent injections. All four clinical studies were basically of similar design. The first two studies involved a small number of study participants (less than 50), therefore they are not described here.

The pivotal studies consisted of two randomized, placebo-controlled double-blind studies in adults. The population consisted of normal healthy volunteers of either sex between the ages of 18 and 65 years, in good general health, and with no known contraindications to lidocaine or other local anesthetics. Female participants had to be of nonchildbearing potential (i.e., surgically incapable of conception or postmenopausal), and be willing to abstain from sexual intercourse or use an acceptable double-barrier method of contraception. Prior to entry into the study, subjects were instructed in the use of pain scales, a visual analogue scale (VAS), and a five-point verbal pain rating scale as shown in Table 10.3.

Following baseline physical and dental examinations, there was an initial screening visit where the subjects received a single blind placebo patch applied to the buccal mucosa for 10 minutes at a site 2 mm above the mucogingival junction over the maxillary and mandibular premolars. To assess pain intensity, a 25 gauge needle was inserted at an angle of 45° to a depth of 2 mm apical to the mucogingival junction where the point of the needle contacted the bone and the needle was then immediately removed. The subjects were then required to immediately record their degree of pain using the previously described VAS and verbal pain scales. A minimum score of 2 on the five-point verbal pain scale was usually required for entry into the double-blind phase of the study. This procedure reduced the placebo responders enrolled in the study and is an acceptable form of enriching the study population. Investigators also assessed the occurrence and severity of local irritation at the patch placement site by visual inspection and by comparing intraoral photographs taken immediately before and at 30 minutes and 48 hours after patch removal.

For safety purposes, subjects who entered the double-blind portion of the study underwent a series of medical and laboratory tests, including a complete physical examination, a 12-lead electrocardiogram, a serum chemistry evaluation, a complete blood count with differential and platelet counts, a complete urinalysis, screens for drug abuse, and a serum pregnancy test for women. All tests had to be within normal limits or not clinically significant if not within the normal limits. Participants had to be negative for drugs of abuse and pregnancy.

Within 14 days of completing the screening qualification visit and the medical evaluations, 122 subjects were randomized into the double-blind phase of the study. This phase involved two visits during which placebo patches or active treatment patches that contained 10 or 20 wt percent lidocaine were placed for 15 minutes on the buccal mucosa of the maxillary premolar region, 2 mm apical to the mucogingival junction. During this phase of the study neither the investigator nor the study participants were aware of which treatment was being used.

Baseline pain scores were established by having participants rate the pain they experienced from a 25 gauge needle stick to the level of the bone using both the 100 mm VAS and the five-point verbal pain score immediately before the patch was placed. After placement of the double-blind patch, additional needle sticks were performed at 5, 10, and 15 minutes while the patch was in place. Each successive insertion was made slightly distal or mesial to the preceding one to minimize tissue damage. Immediately after each needle stick, participants rated the pain using the visual analogue scale and the verbal pain scale.

To measure plasma lidocaine concentration, venous blood samples (10 ml) were obtained through an indwelling catheter in the antecubital fosa immediately before patch placement and at 15 and 45 minute (30 minutes after patch removal) time intervals. These samples were spun at 3000 rpm in a refrigerated centrifuge and stored for later gas liquid chromatographic analysis.

Intraoral photographs were taken to monitor mucosal irritation at the patch application site before patch placement, and at 30 and 48 hours after patch removal.

Local irritation was rated using a four-point scale as follows: 0 = no irritation; 1 = minimal (blood vessels raised above normal levels); 2 = moderate (beet redness of mucosa; individual blood vessels not discernable); and 3 = severe (blister formation and necrosis evident).

2. Statistical Analyses

The primary efficacy variable was the change from baseline pain scores (both VAS and verbal) at each time point (5, 10, and 15 minutes). A mean reduction of 16 mm in the VAS pain score was considered to be clinically significant.

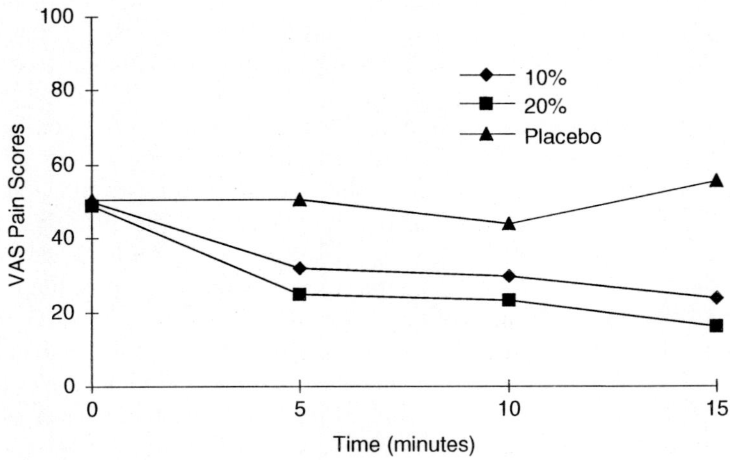

Figure 10.2 Mean change in VAS scores.

3. Results

The data from 116 participants were evaluated for efficacy. There were no significant differences among treatment groups in baseline needle stick pain scores. The mean changes from baseline VAS pain scores are shown in Figure 10.2.

Both the 10 and 20 wt percent lidocaine patches significantly reduced needle stick pain ($p < 0.001$) compared with placebo at all evaluation time points. The onset of anesthesia was within 5 minutes for both active treatment groups. Peak anesthetic effect occurred at the 15-minute time point for both active treatment groups. There was a mean reduction of 49 and 65 percent in VAS pain scores for the 10 and 20 wt percent patches, respectively.

Figure 10.3 illustrates the mean lidocaine plasma concentrations achieved with the 20 percent active patch at 15 and 45 minutes. For comparison, the plasma lidocaine concentrations reported by Goebel et al. (32) after the injection of a single cartridge of (1.8 ml) of 2 percent lidocaine plus 1:100,000 epinephrine is presented. The peak blood levels achieved after an injection of 2 percent lidocaine plus epinephrine are approximately nine times higher than that achieved with the 20 percent patch.

Side effects were evaluated in all 122 participants who entered the double-blind phase of the study. Sixty-six participants reported a total of 96 adverse events. Not all adverse events were related to the study drug. Adverse events were distributed equally between the 3 treatment groups; 21 in the placebo group ($n = 40$); 24 with the 10 percent lidocaine group ($n = 42$); and 21 in the 20 percent lidocaine group ($n = 40$). Many side effects, such as hematoma and stomatitis and pain at the injection site, were procedure-related side effects caused by the multiple needle insertions. If procedure-related events are excluded, the most

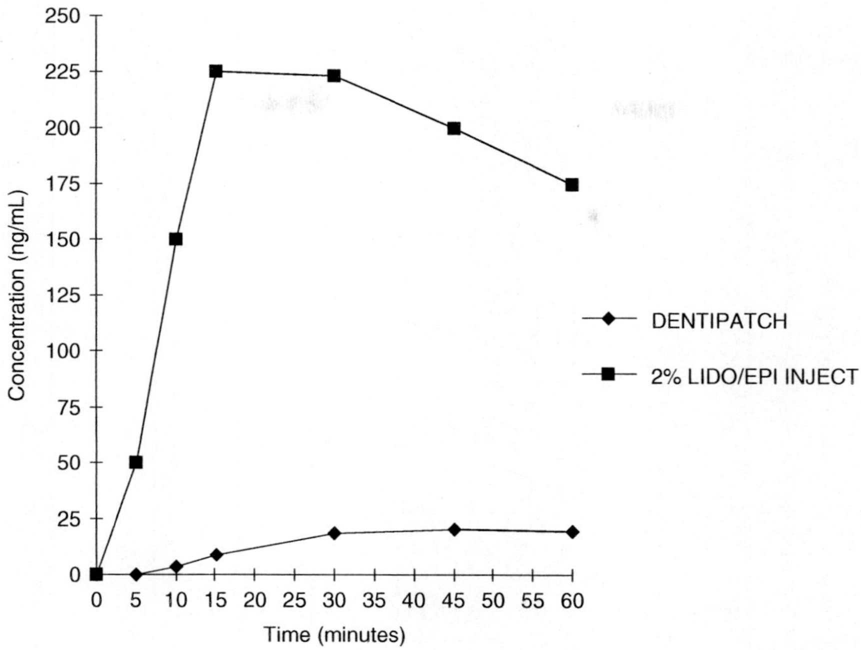

Figure 10.3 Comparison of lidocaine blood plasma levels: DentiPatch[(R)] vs. injection.

common adverse event was unpleasant taste. Taste perversion was equally distributed among all treatment groups. There were no significant differences between active and placebo groups in the occurrence or severity of side effects. Laboratory tests were conducted at the end of the study and were similar to values obtained during the screening phase of the study.

Irritation assessments conducted 30 minutes after patch removal were similar among treatment groups. Minimal irritation was observed in 15 percent of the placebo group, 10 percent of the 10 wt percent lidocaine group, and 8 percent of the 20 wt percent lidocaine group. No participant had irritation scores worse than minimal irritation.

In the second study, the data from 100 participants was evaluated for efficacy. The study was similar to the previous study except the patch was tested at both maxillary and mandibular sites in a crossover fashion and needle sticks were performed at 2.5, 5, 10, and 15 minutes after patch placement. An additional needle stick was conducted at 45 minutes to test the duration of the anesthetic effect. Results for the mandibular and maxillary arches were similar and are shown in Figures 10.4 and 10.5, respectively.

For the mandibular sites the 10 and 20 wt percent transoral systems were statistically significantly better than placebo in the reduction of pain as recorded

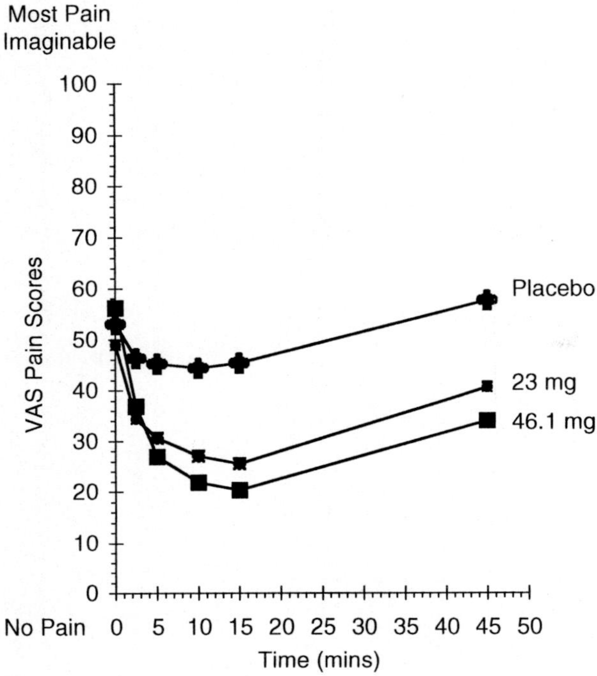

Figure 10.4 Mean VAS pain scores: mandibular.

at 5, 10, 15, and 45 minutes postapplication ($p < 0.0001$). The change from baseline at 2.5 minutes was statistically significant for the 20 wt percent group and approached significance for the 10 wt percent group ($p = 0.0514$).

The decreases in pain scores for both 10 and 20 wt percent groups in the maxillary sites were statistically significantly greater than that seen with placebo at 5, 10, 15 minutes and, at 45 minutes, for the 20 wt percent group. There were no statistically significant differences from placebo at 2.5 minutes. The change in VPS scores followed the same general pattern as the VAS pain data. Participants treated with active patches displayed significantly greater decreases ($p = 0.02$) in verbal pain scores at all evaluation time points.

The development of mucoadhesive patches containing lidocaine represents a major development in the intraoral delivery of topical local anesthesia. The piercing of mucosa by dental injections is accompanied by significant trepidation in many patients. The lidocaine transoral delivery system represents the first FDA-approved topical anesthetic that is highly effective in reducing the pain associated with the insertion of 25 gauge needles through the mucosa to the level of the bone.

Although the total content of lidocaine base in the 20 wt percent patch is 46.1 mg, the total amount of drug absorbed after a 15-minute application is far less; the injection of a cartridge of 2 percent lidocaine (36 mg) plus 1:100,000

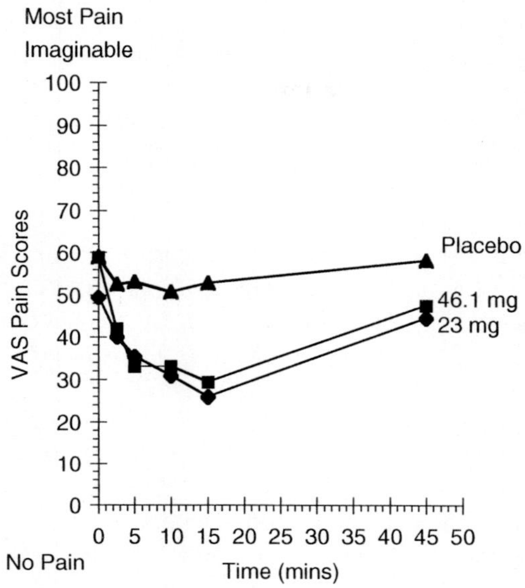

Figure 10.5 Mean VAS pain scores: maxillary.

epinephrine produces lidocaine blood levels that are nine times greater than those achieved with the patch. In addition, local anesthetic blood levels achieved after the application of a 5 percent lidocaine ointment are at least double those achieved after 15 minutes of mucosal contact with the 20 wt percent transoral patch (DentiPatch package insert, Noven Pharmaceuticals, Inc., Miami, FL). With local anesthetics, the risk of adverse systemic reactions is directly correlated with how much drug is absorbed into the systemic circulation (34–36).

D. DentiPatch® Anesthetic Effects during Scaling and Root Planing

1. Study Design

Some patients require local anesthesia for scaling and root planing. A recent study evaluated the anesthetic effect of DentiPatch®, a lidocaine transoral delivery system, during thorough scaling and root planing (37). Data from 20 males and females between the ages of 23 and 60 were evaluated in a double-blind, two-way crossover study. Participants were enrolled in the study if their pocket-probing depths were between 4 to 6 mm, and their gingival index was ≤2 (38). DentiPatch® or a placebo patch was placed 2 mm above the mucogingival junction of the first and second maxillary premolars for 15 minutes immediately prior to the initiation of scaling and root planing (one treatment per visit). Treatment was randomly assigned and participants received their alternate treatment patch at the subsequent visit held following a seven-day washout period. Pain was evaluated using both a VAS and a VPS with the following descriptors: 0 = no pain, 1 = mild, 2 = moderate, 3 = severe, and 4 = very severe. For the VAS, "no pain" was scored on the far left at 0 mm and "pain as bad as it could be" on the far right at 100 mm.

2. Results

Pain scores were recorded after the first scaling and root planing (SRP) in 20 patients and at the completion of scaling and root planing in 15 patients. After the initial SRP stroke, the mean VAS score for DentiPatch® was 20 and for placebo was 31.5. The mean VPS scores were 1.2 for DentiPatch® and 1.5 for the placebo. Following completion of scaling, DentiPatch® was statistically significantly superior to the placebo with a VAS score of 18.3 compared to 35.1 for placebo ($p < 0.01$). Using the VPS scale, it was also found to be statistically significantly superior with a score of 1.2 compared to 1.7 for the placebo ($p < 0.05$).

The results of this study show that application of the DentiPatch® system to the oral mucosa prior to scaling and root planing can produce an anesthetic effect that reduces the pain associated with this procedure. Because dental hygienists, as well as dentists, perform SRP, the use of the DentiPatch® system may negate the need for the dentist to administer local anesthetic by injection. This represents a significant saving of time and inconvenience for all concerned: the patient, the dentist, and the dental hygienist. The DentiPatch® system can be easily applied by a hygienist as soon as the patient is seated and will become effective while the hygienist updates the patient's history, taking 5 to 15 minutes.

VII. SUMMARY

The descriptions and examples of bioadhesives provided in this chapter were intended to bring about awareness of the choices that can be made during the development of TMDS. Bioadhesives can be selected from several distinct product groups: natural, modified natural, and synthetic. Each of these types of bioadhesives, when formulated with drug(s) and excipient(s), can produce very different physical properties and drug-related profiles of the oral transmucosal dosage form. Furthermore, the selection of bioadhesive(s), drug(s), and excipient(s) must be evaluated to ensure that the TMDS is biocompatable with and adheres to the intended area of application. Finally, studies should be conducted to ensure matrix compatibility during the early stages of development of TMDS in order to achieve chemical and mechanical stability.

In order to achieve the degree of transmucosal local anesthesia required for various dental procedures, the selection of the local anesthetic agent, bioadhesive carrier, vehicle and permeation enhancer system, and clinical indication are all co-dependent variables that must be addressed simultaneously by the formulator. In order to succeed, the finished product must provide the patient with the required onset, duration, and depth of anesthesia while minimizing or eliminating any systemic or local side effects. The class of local anesthetics, as incorporated into TMDs has been shown to be the first-line approach in providing the benefits of local anesthesia for the widest array of indications in dental and oral surgical procedures.

In addition to reducing the pain associated with buccal infiltration injections, transoral lidocaine patches have other potential clinical uses in dentistry. Their use before palatal injections and mandibular block injections should reduce the pain caused by these injections in many patients. In a number of soft-tissue procedures, the soft-tissue and periosteal anesthesia produced by the lidocaine patch may be sufficient to substitute for conventional injections. These procedures include gingival scaling and curettage, localized gingivectomies, selected biopsy procedures, and suture removal.

Minor restorative and orthodontic procedures that do not involve tooth reduction but which still cause soft-tissue pain, such as cervical bonding procedures, crown try-ons and cementations, orthodontic band adaptions, and rubber dam clamp placements, are also ideal situations for the placement of transoral lidocaine patches.

In conclusion, mucoadhesive patches containing lidocaine are effective and safe in reducing the pain associated with needle insertions through the oral soft tissues. Their use is also recommended for reducing the soft-tissue pain associated with several minor dental procedures.

REFERENCES

1. Tisza, Edmond T., US Patent 2,142,537 assigned to Rare Chemicals, Inc., Patent Issue Date January 3, 1939, Anesthetic Ointment.
2. Curtis, David, US Patent 2,277,038, Patent Issue Date March 24, 1942, Anesthetic Preparation.
3. Shrontz, Don C., US Patent 2,501,544 assigned to Shellmar Products Corporation, Patent Issue Date March 21, 1950, Therapeutic Product.
4. Davis, S.G., Mass., Swedish Publication 17747/70 assigned to Astra Pharmaceuticals, Inc., Publication Issue Date December 27, 1972, Manufactured Films Containing Anesthetics.
5. Zaffaroni, Alejandro, Calif., US Patent 3,699,963 assigned to Alza Corporation, Patent Issue Date October 24, 1972, Therapeutic Adhesive Patch.
6. Zaffaroni, Alejandro, Calif., US Patent 3,948,262 assigned to Alza Corporation, Patent Issue Date April 6, 1976, Novel Drug Delivery Device.
7. Tsuk, Andrew G., Martin, Frederick H., both from NY., US Patent 3,972,995 assigned American Home Products Corporation, Patent Issue Date August 3, 1976, Dosage Form.
8. Broberg, Berndt F.J., Hans, Evers C.A., Both from Sweden, US Patent 4,562,060 assigned to Astra Lakemedel Aktiebolag, Patent Issue Date December 31, 1985, Local Anesthetic Mixture for Topical Application, Process for Its Preparation, as well as, Method for Obtaining Local Anesthesia.

9. Kigasawa, Kazuo, Ohtani, Hideski, Tanska, Makoto, all from Tokyo, European Patent Application 0,159,168 assigned to Takeda Chemical Industries, Application Date October 23, 1985, Soft Patch Drug Preparation.
10. Schiraldi, Michael T., Rubin, Howard, both from NJ., Perl, Martin M., NY., US Patent 4,713,243 assigned to Johnson and Johnson Products, Inc., Patent Issue Date December 15, 1987, Bioadhesive Extruded Film for Intra-Oral Drug Delivery and Process.
11. Mantelle, Juan A., FL., US Patent 5,234,957 assigned to Noven Pharmaceuticals, Inc., Patent Issue Date August 10, 1993, Compositions and Methods for Topical Administration of Pharmaceutically Active Agents.
12. Drug Facts and Comparisons, 1990 Edition, J.B. Lippincort Company, St. Louis, MO., Page 601.
13. Merck Index, 1989-11th Edition, Merck & Co., Rahway, NJ.
14. Tucker, G.T., Mather, L.E., Clinical Pharmacokinetics of Local Anesthetics, ADIS Press Australasia Pty Ltd., 1979 Edition, Pages 242–278.
15. Chang, Yunick, Patel, N.J., Dinesh C., UT., US Patent 4,888,354 assigned to Theratech, Inc., Patent Issue Date December 19, 1989, Skin Penetration Enhancement Using Free Base and Acid Addition Salt Combinations of Active Agents.
16. Anderson, A.T., Water Soluble Resins (R.L. Davidson, M.Sitting, Ed.).
17. Gallopo, A.R., Dills S.S., US Patent 4,915,948 (1990) assigned to Warner-Lambert Company.
18. Estes, D.K.,US Patent 3,640.741 (1972) (assigned to Hollister Incorporated).
19. Kigasawa, K., others, US Patent 4,572,832 (1986) (assigned to Grelan Pharmaceuticals Co., Ltd.).
20. Robinson, J., US Patent 4,615,697 (1986) (assigned to Bio-Minetics, Inc.).
21. Takayanagi, H., Sawai, Y., US Patent 4,765,983 (1988) (assigned to Yamanouchi Pharmaceutical Co.).
22. International Cosmetic Ingredient Dictionary, Fifth Edition.
23. Handbook of Food, Drug, and Cosmetic Excipients (Smolinske, S.C., author)
24. Adriani, J., Zepernick, R., Clinical effectiveness of drugs used for topical anesthesia. J.A.M.A. 188:711, 964.
25. Adriani, J., Zepernick, R. Arens, J., Authement, E., Comparative potency and effectiveness of topical anesthetics in man. Clinical Pharmacol. and Therap. 5: 49, 1964.
26. Giddon, Donald B., Quadland, M. et al., Development of a method for comparing topical anesthetics in different application and dosage forms. J Oral Therapeutics and Pharmacology, 4 (4), 1968.
27. Schmid, M., Eine neue Parodontalsonde. Med. Diss. Zurich University, 1967.
28. Kleinknecht, R.A., Klepac, R.K., Alexander, L.D., Origins and characteristics of fear of dentistry. JADA 1973;86(4):842–848.
29. Yacoob, H.B., Noor, G.M., Malek, S.N., The pharmacological effect of Xylocaine topical anesthetic—A comparison with a placebo, Singapore Dent J 6(2):55–57, 1981.
30. Holst, A., Evers, H., Experimental studies of new topical anesthetics on the oral mucosa. Swed Dent J 9:185–191, 1985.
31. Rosivack, R.G., Koensgsberg, S.R., Maxwell, K.C., An analysis of the effectiveness of two topical anesthetics. Anes Prog 37:290–292, 1990.

32. Goebel, W.M. et al., Pooled injection data adapted from J Oral Med 35(4):91–96, 1980.
33. Hersh, E.V., Houpt, M.I., Cooper, S.A. et al., Analgesic efficacy and safety of an intraoral lidocaine patch. J Am Dent Assoc 127(11):1626–1634, 1996.
34. Malamed, S.P., Handbook of Local Anesthesia, ed 4, St. Louis, CV Mosby Co., 1996.
35. Jastak, J.T., Yagiella, J.A., Donaldson, D., Local Anesthesia of the Oral Cavity. Philadelphia, WB Saunders Co., 1995.
36. Hersh, E.V., Helpin, M.I., Evans, O.B., Local anesthetic mortality: Report of case. ASDC J Dent Child 58(6):489–491, 1991.
37. Stubbs, P., Doerfler, C. PPD Pharmaco, Inc. Austin, TX: Noven Pharmaceuticals, Inc., Miami, FL. Presented at the Annual Meeting of the American Society for Clinical Pharmacology and Therapeutics, San Diego, CA. March 1997.
38. Loe H., Silness J., Periodontal disease in pregnancy, Acta Odont Scand 1963:21:533–551.
39. ZilaBrace® and ZilaDent®; Distributed by Zila Pharmaceuticals, Inc.; Phoenix, Arizona 85014.
40. DentiPatch®; Distributed by Noven Pharmaceuticals, Inc.; Miami, Florida 33186.
41. Florida Probe Corporation; 1820 NE 23rd Avenue; Gainesville, Florida 32609.
42. Lidocaine Patch for Periodontal Anesthesia; Heins, P.J., D.D.S.; Maruniak, J.W., B.S., R.D.H.; Hassell, T., D.D.S. Presented at the International Association of Dental Research, Glasgow, Scotland, July 1–4, 1992.

11

Quick-Dispersing Oral Drug Delivery Systems*

Kuldeepak Sharma, William R. Pfister, and Tapash K. Ghosh

CONTENTS

*No official support or endorsement of this chapter by the Food and Drug Administration is intended or should be inferred.

I. INTRODUCTION

The oral route of drug administration is the most common and convenient for patient use. Tablets and capsules have emerged as the most popular solid oral dosage forms used today that include conventional and controlled-release tablets as well as hard and soft gelatin capsules. However, many patients have dysphagia or difficulty in swallowing tablets and hard gelatin capsules and therefore do not take medication as prescribed by physicians. It is estimated that 35 percent of the general population, 30 to 40 percent of elderly nursing home patients, and 25 to 50 percent of patients hospitalized for acute neuromuscular disorders and head injuries have dysphagia (1). The main causes of dysphagia include esophageal disorder such as achalsia, Gastro Esophageal Reflux Disease (GERD), cardiovascular conditions such as aneurysm, autoimmune diseases such as Sjogren's syndrome, Auto Immune Deficiency Syndrome (AIDS), thyroid surgery, radiation therapy to head and neck or oral cavity, and other neurological diseases such as cerebral palsy (1).

A detailed survey was conducted to determine the proportion of patients having difficulty in swallowing tablets and to identify the reasons for the difficulty (2). More than 26 percent of patients mentioned problems in swallowing tablets. A prominent complaint was the size of the tablet, followed by the surface, form, and taste of the tablets. Twice as many women as men experienced swallowing problems. Elderly patients (>70 years) had less difficulty than younger patients in swallowing tablets (2). Pediatric and geriatric patients in particular experienced the greatest difficulty in swallowing tablets as well as people who are ill and supine in bed and those patients who are busy traveling without having access to water.

New and novel oral drug delivery systems that dissolve or disperse quickly in a few seconds after placement in the mouth without water can alleviate the problem of swallowing tablets. They offer substantial advantages over ordinary tablets, are more convenient to administer inasmuch as drinking water is not required, and enhance the potential for improved compliance in patients who have difficulty in taking tablets (2). Quick-dispersing oral drug delivery systems (QD) are defined as oral drug delivery systems that dissolve or disintegrate within seconds to a few minutes after placement in the mouth and do not require water to aid swallowing. The QD systems include tablets, caplets, wafers, films, granules, and powders. When QD are placed in the mouth, the dosage form disintegrates instantaneously or within a few minutes releasing the drug, which dissolves or disperses in the saliva. The saliva containing the medicament is then swallowed and the drug is absorbed in the normal way. Some fraction of the drug may be absorbed from pregastric sites such as the mouth, pharynx, and esophagus as the saliva passes down into the stomach. In these cases, the bioavailability of drugs from QD may be greater compared to the standard oral dosage forms. The many benefits of QD are summarized in Table 11.1. Many drug delivery companies are developing QD products based on different formulation and manufacturing technologies (Table 11.2). Only the first four QD technologies are reviewed in this chapter.

This chapter provides a comprehensive review of four different quick-dispersing oral delivery systems (i.e., WOWTAB®, Zydis®, Orasolv®, and Shearform™) and is divided into six sections. The introduction includes definitions, benefits, and description of quick-dissolving/dispersing products. The various technologies used to achieve quick-dissolving/dispersion in the oral cavity are reviewed. The formulations and excipients used in these technologies are discussed. The process and packaging issues pertaining to laboratory scale and commercial manufacturing and packaging considerations are described. The in vitro and in vivo preclinical and clinical testing of these products are included. Finally, new products, challenges, and future trends for these technologies are discussed.

Table 11.1 Benefits of Quick-Dispersing Oral Drug Delivery Products

Clinical
Improved oral absorption
Faster onset of action
Minimized first-pass effect
Improved bioavailability
Medical
No tablet or capsule to swallow or chew
Better taste, no water needed
Improved safety and efficacy
Improved compliance
Technical
Accurate dosing compared to liquid products
Contain sugars and other GRAS excipients
Improved stability due to better packaging
Use common process and conventional equipment
Business
Unique product differentiation
Value-added product line extension
Provide exclusive marketing
Extend patent protection

Table 11.2 The Major Quick-Dissolving and Dispersing Oral Drug Delivery Systems

Technology	Company	Patented	Commercial Products	Comments
WOWTAB®	Yamanouchi	Yes	Yes	Compressed tablet Good dose tablets Disintegrate in 15 seconds
Zydis®	Scherer DDS	Yes	Yes	Freeze-dried tablet Blister packed Disintegrate in 10 seconds
OraSolv®	CIMA Labs.	Yes	Yes	Effervescent tablet Loosely compressed Disintegrate in 60 seconds
Shearform™	Fuisz Technologies	Yes	In development	Thin fiber matrix Loosely compressed tablet Disintegrate in 10 seconds
Lyoc™	Farmalyoc	Yes	In development	Freeze-dried wafer Blister packed Disintegrate in 10 seconds
Quicksolv®	Janssen	Yes	In development	Freeze-dried tablet Blister packed Disintegrate in 10 seconds
EFVDAS®	Elan	Yes	In development	Effervescent tablet Loosely compressed Disintegrate in 60 seconds

II. WOWTAB® TABLETS

A. Technology

Yamanouchi Pharmaceutical Co. Ltd., Japan has developed and commercialized a quick-disintegrating "Without Water Tablet" (WOWTAB®) technology. WOWTAB® is a tablet that has sufficient hardness to maintain physical and mechanical integrity of the dosage form prior to contact with saliva (3). WOWTAB® consists of commonly used tablet excipients, which are Generally Recognized As Safe (GRAS) materials. WOWTAB® when placed in the mouth rapidly becomes soft by absorption of saliva and disintegrates or dissolves within 15 to 20 seconds. WOWTAB® disintegrates or dissolves more quickly when pressure between the upper jaw and tongue or a licking movement is applied to the tablets. WOWTAB® is manufactured using conventional granulators, tablet machines, and packaging equipment, which ensure excellent drug content uniformity and batch-to-batch reproducibility (4). WOWTAB® tablets are produced by a standard tablet compression molding process.

The combination of poor compressible saccharides having a high dissolution rate with good compressible saccharides having a slow dissolution rate is employed to achieve quick dissolution and adequate hardness characteristics of WOWTAB®. The poor compressible saccharides have a hardness of less than 2 kg when compressed under pressures of 10 to 50 kg/cm^2 per punch. Examples of poor compressible saccharides include lactose, mannitol, glucose, sucrose, and xylitol (4). The good compressible saccharides have a hardness of 2 kg or more when compressed under pressures of 10 to 50 kg/cm^2 per punch. Examples of good compressible saccharides include maltose, maltitol, sorbitol, and oligosaccharides such as lactosucrose (4).

Apart from combination of saccharides with poor and good moldability properties, additional additives such as disintegrants, binders, sweeteners, flavors, lubricants, and coloring agents may be formulated in WOWTAB® products. There are no particular limitations on the type of drugs that can be formulated in WOWTAB® products, because both water-soluble and water-insoluble drugs can be included.

The process of preparation of WOWTAB® tablets is simple and involves the addition of an active ingredient to poor compressible saccharide. The mixture is granulated with a good compressible saccharide by spraying an aqueous solution of a good compressible saccharide on the mixture of active ingredient and a poor compressible saccharide. The resulting granules are mixed with excipients and compressed into tablets. The particle size distribution of granules in WOWTAB® tablets can vary from 10 to 1000 microns. These tablets are humidified and dried to further improve their hardness and mechanical strength. These WOWTAB® tablets have sufficient strength and hardness (more than 3 kg) for handling and transporting in bottle packages (4). WOWTAB® products also have advantages for delivery of peptides such as desmopressin or calcitonin to the buccal mucosa for absorption because these drugs, formulated in conventional oral dosage forms, have poor bioavailability due to degradation in the digestive tract. WOWTAB® tablets can be applied to nonpharmaceutical products such as diagnostic agents, functional foods, vitamins, and dental plaque-removing agents (4).

B. Processing and Packaging

A typical process (4,5) for WOWTAB® tablets requires conventional granulator and tableting machines as shown in Figure 11.1. The mixing of active ingredients with saccharides and other additives is carried out in a vertical mixer or blender. The granulation is carried out in a fluidized bed granulator to obtain granules having a desired particle size using common operating conditions such as spray pressure of 0.3 to 2 kg/cm^2 and at a temperature of 30°C. The hardness of the tablets is further improved while maintaining the fast disintegration properties by conditioning with humidity and drying. Typical humidity and drying conditions are shown in Figure 11.1. They have sufficient hardness (more than 3 kp) and strength for handling and packaging in blister packs or bottles. Manufacturing of WOWTAB® tablets may follow any of the following processes.

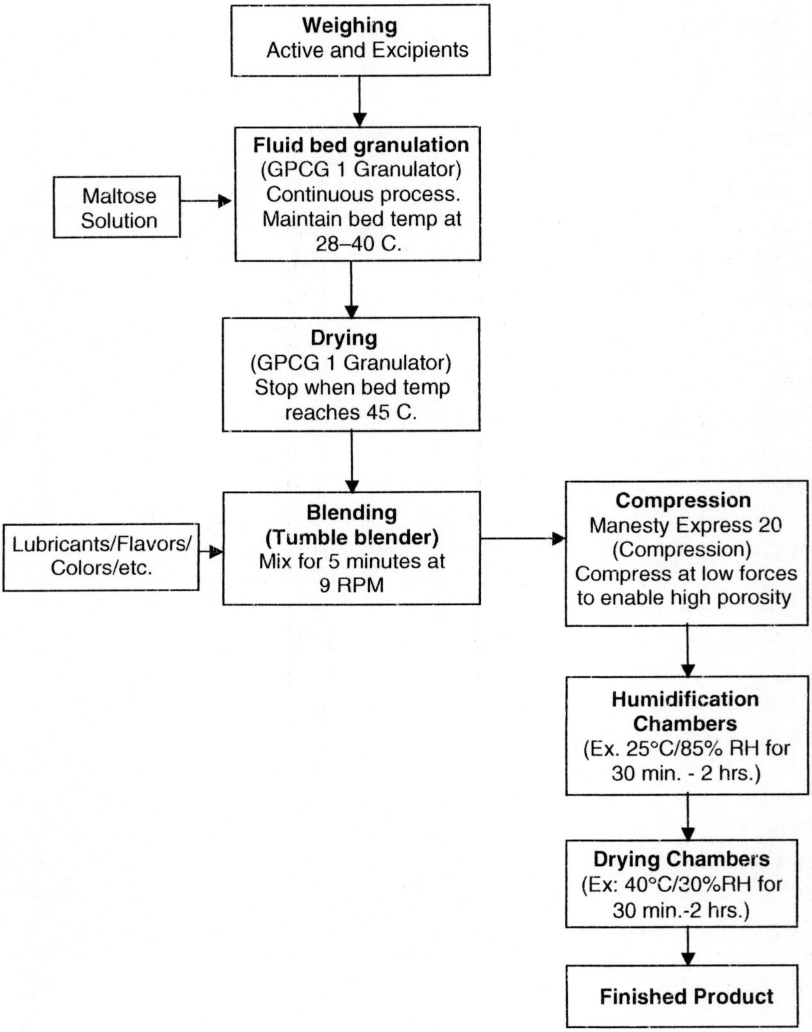

Figure 11.1 WOWTAB® typical tablet manufacturing process.

Process 1

An active ingredient is added to a poor compressible saccharide and the resulting mixture is granulated with a good compressible saccharide. The resulting granulation is subjected to compression molding to obtain QD tablets.

Process 2

A poor compressible saccharide is granulated with a good compressible saccharide to form granules. The granules are mixed with an active ingredient and the mixture is compression-molded to obtain QD tablets.

Process 3

A poor compressible saccharide is granulated with a good compressible saccharide. Separately, an active ingredient is granulated with a good compressible saccharide to obtain granules. The granules are mixed and compressed to obtain QD tablets.

Process 4

A poor compressible saccharide (central core) is coated with a good compressible saccharide (first layer), coated with an active ingredient (second layer), and the resulting product granulated with a good compressible saccharide (third layer). The resulting granules are compressed to obtain QD tablets.

Process 5

A poor compressible saccharide is coated with an active ingredient. The resulting granules are granulated with a good compressible saccharide and then compressed to obtain QD tablets.

C. Formulations

The following examples are provided to illustrate the application of WOWTAB® formulations and are summarized from the patent literature (4,5).

Example 1

A formulation of 20 g of famotidine, 270 g of lactose, 40 g of mannitol, 8 g of aspartame, and 2 g of sodium citrate was mixed and granulated in a fluidized bed granulator using 16 g of maltose dissolved in 144 g of water. The inclusion complex was prepared using 0.34 g of menthol and 2.46 g of beta-cyclodextrin in hot water and the suspension was sprayed onto the granules. After drying, 0.5 wt percent of magnesium stearate was blended and the resulting granules were compressed into tablets using a rotary tableting machine with a 10 mm diameter punch and under a pressure of 133 kg/punch. The resulting tablets weighed 150 mg and disintegrated in 15 seconds in the buccal cavity in three healthy volunteers.

Example 2

A formulation of 3.5 g of gilbenclaimde and 396.9 g of mannitol was mixed and subjected to granulation with 21 g of maltose in 189 g of water using a fluidized bed granulator. After drying, 0.5 wt percent magnesium stearate was blended and the resulting granules (mean particle diameter of 127 microns) were compressed into tablets using a rotary tableting machine with a 10 mm diameter punch under a pressure of 319 kg/punch. The resulting tablet weighed 300 mg and had a disintegration time of 15 seconds and a hardness of 3.0 kp.

Example 3

A formulation of 500 g of acetaminophen was subjected to granulation using 25 g of maltose as a 10 wt percent maltose aqueous solution in a fluidized bed granulator. Sixty-three grams of the resulting acetaminophen granules (mean particle diameter of 127 microns) were mixed with 235.5 g of the previously prepared mannitol granules (mean particle diameter 134 microns) and 1.5 g of magnesium stearate. This acetaminophen mixture was compressed into tablets using a rotary tableting machine with a 10 mm diameter punch under a pressure of 319 kg/punch. The resulting tablets weighed 300 mg, had a hardness of 3.0 kp, and a disintegration of 15 seconds in the buccal cavity in three healthy volunteers.

Example 4

A mixture consisting of 487.5 g of mannitol and 162.5 of lactose was subjected to granulation using a fluidized bed granulator. The fine particles were coated under spray pressure of 3 kg/cm^2 using 139 g of 10 percent maltose aqueous solution. Coating was also carried out under the same conditions using a solution prepared by dissolving 138 mg of YM934 (2-(3,4-dihydro-2, 2-dimethyl-6-nitro-2H-1, 4 benzoxazin-4-yl) pyridine N-oxide) in 50 ml of methanol. Thereafter, granulation was carried out under a spray pressure of 1.3 kg/cm^2 using 98 g of a 20 percent maltose aqueous solution. After drying, 0.5 wt percent magnesium stearate was blended and the resulting granules (mean particle diameter of 161 microns) were compressed into tablets using a rotary tableting machine with a 10 mm diameter punch under a pressure of 319 kg/punch. The resulting tablet weighed 294 mg and had a hardness of 4.5 kp.

D. Performance and Clinical Testing

The following methods are used to study the performance and properties of WOWTAB® tablets (4).

1. Hardness Test

The hardness of the WOWTAB® tablet is measured using a harness meter (manufactured by Scheleuniger). Tablet hardness measured in the lengthwise direction is used as an index of tablet strength. Tablet hardness varies depending on the size and shape of the tablets. Tablet hardness should be more than 2 kg for blister packing and more than 3 kg for packaging in bottles. The result of a typical hardness test is shown in Table 11.3 (4). Each hardness value is the average of three to ten tablets.

2. In Vitro Disintegration Test

The in vitro disintegration of WOWTAB® tablets is measured in accordance with the disintegration test described in the Japanese Pharmacopoeia (6). The complete disintegration of the tablet is defined as that state in which any residue of the tablet remaining on the screen of the test apparatus is a soft mass having no palpable firm core. Each test is carried out using six WOWTAB® tablets.

Table 11.3 Performance Testing of WOWTAB® Tablets

WOWTAB® Placebo Formulation[a]	Tableting Pressure (kg/punch)	Tablet Hardness (kg)	In Vivo Disintegration Time (sec)
Mannitol:Maltose (20:1)	303	5.9	15
Lactose:Maltose (20:1)	334	5.3	15
Mannitol:Lactose:Maltose (10:10:1)	338	3.7	16
Mannitol:Oliosaccharide (20:1)	441	3.6	20

Source: Reference 4.

3. In Vivo Disintegration Test

The disintegration and dissolution of WOWTAB® tablets in healthy volunteers was conducted by placing a tablet in the buccal cavity. The time required for the complete disintegration or dissolution of the tablet without water in the buccal cavity was recorded. The tablet was allowed to move gently from side to side or up or down in the buccal cavity without biting. Other subjective performance-related parameters of WOWTAB® tablets such as smoothness, taste, and flavor were also evaluated in these studies. The typical WOWTAB® tablet has an average disintegration time of 10 to 15 seconds in human volunteers as shown in Table 11.3 (6).

E. Commercial and Future Products

Yamanouchi Pharmaceutical Co. has developed and is marketing Gaster OD™ containing famotidine, an H_2 antagonist for the treatment of peptic ulcer, in Japan. The company also has launched Nasea OD™ containing ramosetron, a 5-HT_3 receptor antagonist for the treatment of nausea and vomiting. The company is also developing WOWTAB®-based QD products using other drugs such as Tamsulosin for the treatment of Benign Prostatic Hyperplasia (BPH) for the Japanese market (7).

Yamanouchi Shaklee Pharma is working with several companies in the United States to develop WOWTAB®-based QD products for migrane, peptic ulcers, and cold and allergy (8–10).

III. ZYDIS® TABLETS

A. Technology

R.P. Scherer Corp. has developed and commercialized various QD products based on Zydis® technology (11). The Zydis® dosage form is a freeze-dried tablet made from well-known and acceptable excipients, which does not require water to aid swallowing. When this dosage form is placed on the tongue, the tablet structure disintegrates, instantaneously releasing the drug in the mouth. The drug in Zydis® QD is physically entrapped or dissolved within the matrix of the fast-dissolving carrier material. The matrix of this fast-dissolving tablet is composed of a glassy amorphous excipient that imparts strength and resilience during handling of the tablet. Polymers such as gelatin, dextran, and alginates, and saccharides such as mannitol or sorbitol are typical examples of excipients used in Zydis® fast-dissolving tablets. The porous structure, poor crystallinity, and freeze-dried matrix are necessary attributes to achieve a fast-dissolving product (11).

The dose of water-insoluble drugs in Zydis® products is generally less than 400 mg to maintain the fast-dissolving characteristics of the product. This limitation of drug load also minimizes the taste of the drug in the mouth as the tablet dissolves in the saliva (12). To prevent sedimentation of drug and excipients during the manufacturing process, the particle size of insoluble drug and excipients should be less than 50 microns. A small particle size is also desirable for reducing the sensation of a gritty texture in the mouth and pharynx during swallowing.

The dose of water-soluble drugs is limited to about 60 mg, depending on the drug, due to incompatibility with the freezing and drying process. Some drugs may form eutectic mixtures, which might not adequately freeze or melt at the temperatures used in the freeze-drying process. The dissolved drugs may form an amorphous glassy solid on freezing that may collapse on drying due to sublimation of ice, which may lead to loss of the supporting structure of the tablet. Collapse of the tablet structure can be prevented by the addition of crystal-forming excipients. Organic solvents can be used to dissolve water-soluble drugs, which are applied on placebo Zydis® tablets. The organic solvent is then evaporated and the recrystallized drug is deposited in the pores of the Zydis® matrix (12). The chemical stability of the drug substance in the solvent system is very important and should be stable in aqueous solution or suspension for 24 hours. This time period is required for storage and filling into preformed pockets of a blister tray before the freezing stage of the manufacturing process (12).

B. Processing and Packaging

The manufacturing process for the Zydis® tablets is summarized in Figure 11.2. The drug is dispersed in the carrier matrix and dispensed by weight into preformed blister pockets. Dispensing is fully automated and typically dispensed weights are within two percent of the target weight. The dispensing system is specially designed to ensure that the homogeneity of the suspension is maintained during filling of the blister pocket. The suspension is frozen within the blister pockets by passing through a modified freeze tunnel. These frozen units are dried by subliming the ice in a freeze-drier. This process uses modified equipment and careful control of process parameters. Once dried, the blisters are inspected and sealed using an aluminum-foil paper laminate. Zydis® blister packages are loaded into magazines for bulk packaging for commercial distribution (11,13).

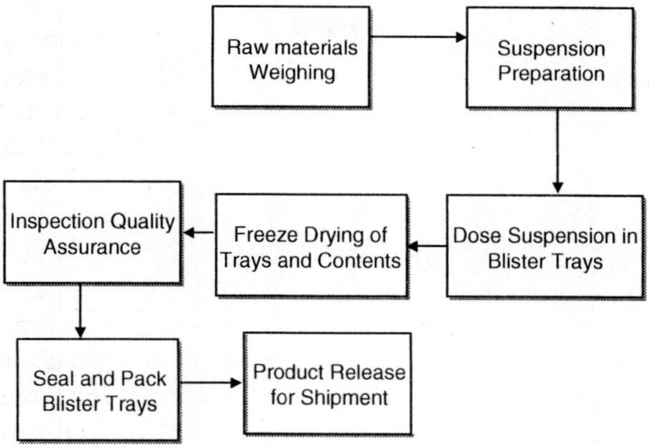

Figure 11.2 Zydis® tablet manufacturing process.

Zydis® tablets are packed in blister packs to protect them from moisture, shipping, and handling. Various materials such as polyvinylchloride (PVC), polyvinyldichloride (PVdC), or aluminum foil are used for packaging Zydis®. The influence of storage conditions on Zydis® tablets containing a poorly water-soluble drug packed in several types of blister packs is shown in Table 11.4. It is apparent that increasing the moisture protection barrier by increasing the thickness of PVdC improves the physical stability of the Zydis® units and with a foil pouch or aluminum blister, protection is maximized. Zydis® units packed in blister packs have a peelable backing foil because these units are not mechanically strong and cannot be pushed out through the foil before administration (13,14).

Table 11.4 Influence of Packaging Material on Diameter of Zydis® Tablets

	Packaging Material[b]			
Storage Conditions[a]	200 Micron PVC/40 g^{-2} PVdC	200 Micron PVC/90 g^{-2} PVdC	200 Micron PVC/40 g^{-2} PVdC in foil	Aluminum Blister Pack
Initial Diameter	15.0	14.8	15.0	16.3
1 M, 37°C, 75% R	12.2	14.0	14.3	16.3
2 M, 37°C, 75% R	12.5	12.5	14.4	16.3
3 M, 37°C, 75%R	11.1	12.2	14.3	16.4
3 M, 37°C, 70%R	14.3	14.2	14.4	—

[a] M = month; R = Relative humidity.
[b] PVC = polyvinylchloride; PVdC = polyvinyldichloride.
Source: References 11 and 13.

C. Formulations

The following examples are provided to illustrate the application of Zydis® formulations and process methods and are summarized from the patent literature (14,15).

Example 1

Hydrolyzed gelatin solution was prepared by mixing 30 g of gelatin in 1 liter of water by constant stirring and then autoclaved at 121°C, under 15 psi for 1 hour. The solution was allowed to cool to room temperature and 1 g lorazepam, colorant, and flavoring agents were mixed in the gelatin solution. This drug–gelatin solution was poured into aluminum molds containing 75 cylindrical cavities having a diameter of 0.5 cm and cooled to about –129° C in liquid nitrogen. The drug–gelatin solution was dispensed into aluminum molds from a hypodermic needle and placed into a vacuum chamber and held at 0.3 mm hg overnight. The freeze-dried Zydis® tablets each containing 0.5 mg of lorazepam were placed into an airtight container.

Example 2

The method of Example 1 was repeated substituting 2 g digoxin for 1 g lorazepam to give Zydis® tablets each containing 1 mg of digoxin.

Example 3

A formulation was prepared by placing 20 g acacia in a dry 1 liter flask and 10 ml absolute alcohol was added to wet the acacia powder. 500 ml water was added and mixed to yield a homogenous solution. 30 g sucrose, 30 g polyvinyl pyrrolidine, and 3.33 g of lorazepam were dispersed into solution with an ultrasonic mixer. Additional water was added to adjust the volume to 1 liter. Then 0.75 ml of drug–gelatin solution was dispensed from a hypodermic needle into each cylinder of a cooled aluminum molds (–129°C), placed into a vacuum chamber, and held at 0.3 mm hg overnight. The freeze-dried Zydis® tablets each containing 0.5 mg of lorazepam were taken from the mold and stored in an airtight container.

Example 4

A formulation was prepared by mixing 150 mg famotidine, 765 g mannitol, 578 g gelatin, and 6 g xantham gum in a dry mixing bowl for 5 minutes. Water was added to the mixing bowl to form a uniform paste and 14 kg of water was added to prepare a suspension. A partial vacuum of 0.8 bar was applied to the mixing bowl for 15 minutes. The mixture was heated to 40°C for 60 minutes while maintaining the partial vacuum of about 0.8 bar. The mixture was cooled to room temperature (23°C) and filtered through a 0.38 micron filter. The coloring and flavoring agents were added and the mixture was freeze-dried under a partial vacuum of about 0.8 bar for about 5 minutes. The freeze-dried Zydis® tablets each containing 10 mg of famotidine were stored in an airtight container.

D. Performance and Clinical Testing

The following in vivo and in vitro tests have been developed to study the performance of Zydis®-based products (14,16).

1. Uniformity Test

Drug content uniformity in tablets was characterized by dissolving ten tablets in water or other suitable solvent. The drug concentration in solution was measured using HPLC. The result of famotidine content uniformity is shown in Table 11.5. The result demonstrates excellent content uniformity of famotidine in Zydis® tablets (14).

2. In Vitro Disintegration Test

The disintegration of Zydis® tablets in water was studied using the following method. Five beakers filled with water were placed in a water bath at 37°C. Five tablets were secured individually in a wire clip and placed in a gauze-covered basket. The basket was then lowered at a constant rate into beakers, one basket to a beaker. Disintegration was complete when the wetted mass passed through the gauze or the gauze was visible through the remaining mass. The in vitro disintegration of typical Zydis® tablets is shown in Table 11.6. Disintegration of each Zydis® tablet occurs within five seconds in water at 37°C (14,15).

3. Bioavailability Study

The bioavailability of a water-insoluble drug such as piroxicam in Zydis® tablets is similar to a standard capsule dosage form as shown in Table 11.7 (16,17).

Table 11.5 Drug Uniformity in Zydis® Famotidine Tablets

	Content Uniformity (mg/tablet)		
Dose Strength[a] (mg)	Potency (mg)	Range (mg)	Average (mg)
Famotidine 10	10	9.7 to 10.4	10.1
Famotidine 20	20	20.0 to 21.0	20.5
Famotidine 40	40	40.7 to 41.8	40.1

Source: Reference 14.

Table 11.6 Drug Dissolution of Zydis® Famotidine Tablets

	Drug Dissolution (% Dissolved)		
Dose Strength (mg)	2 Minutes	5 Minutes	8 Minutes
Famotidine 10	79	87	101
Famotidine 20	83	95	100
Famotidine 40	83	95	99

Source: References 14 and 15.

Table 11.7 Clinical Performance of Piroxicam from Capsule and Zydis® Tablets

Time (Hours)	Piroxicam Plasma Concentration (ng/ml)	
	Piroxicam Capsule (20 mg)	Piroxicam Zydis® (20 mg)
0	0	0
2	2000	1700
25	1600	1500
50	1050	1000
100	500	500
125	240	250
175	<100	<100

Source: References 16 and 17.

Table 11.8 Clinical Performance of Selegiline Tablets and Zydis® Tablets

Time (Hours)	Selegiline Plasma Concentration (ng/ml)	
	Selegiline Tablet (10 mg)	Selegiline Zydis® (10 mg)
0	0	0
0.5	0.8	4.9
1.0	0.3	2.9
2.0	0.1	0.8
4.0	0.0	0.3
6.0	0.0	0.1
8.0	0.0	0.0
12.0	0.0	0.0

Source: References 16 and 17.

Bioavailability of a water-soluble, low-dose, and low-molecular-weight drug such as selegiline is improved with the Zydis® formulation, as shown in Table 11.8 (16,17). Lower doses of selegiline Zydis® tablets can provide blood concentration and therapeutic activities equivalent to standard oral tablets, as shown in Table 11.8 (16,17).

E. Commercial and Future Products

R.P. Scherer Corp. has developed and commercialized various QD products based on Zydis® technology. Scherer has developed the following products for the world market (except the United States): oxazepam, lorazepam, piroxicam, loperamide, famotidine, enalapril, and pheylpropanolamine/brompheniramine (16,17). In the U.S. market, the company has launched loratidine and phenylpropanolamine/ brompheniramine for the treatment of cold and allergy (18). A faster-acting thin wafer form of Viagra® using Scherer's Zydis® drug delivery system is being developed. Viagra® must be taken about an hour before sexual activity, whereas it is hoped a thin-wafer version might work within minutes so onset time will be reduced (19). The FDA approved Maxalt-MLT®, a fast-dissolving rizatriptan benzoate, for the treatment of migraine pain in June of 1998. This product is based on R.P. Scherer's Zydis® technology (20).

IV. ORASOLV® TABLETS

A. Technology

CIMA Labs has developed OraSolv® technology and commercialized various QD products based on this technology. OraSolv® technology is a combination of microparticles of active drug and effervescent disintegration excipients to achieve a fast-dissolving tablet (21). This technology provides an effective oral dosage form for systemic absorption of drugs with unpleasant flavor or taste. The microparticles are relatively fragile particles susceptible to release of the drug upon rupture of the microparticles. Because OraSolv® tablets disintegrate without chewing, the problem of microparticle rupture is substantially eliminated. The combination of rapid-releasing microparticles with the effervescent disintegration agents provides effective taste masking of the drug in OraSolv® (21,22).

Drug microparticles are prepared using a variety of coating techniques including spray coating, spray drying, spray congealing, melt dispersion, phase separation, or solvent evaporation methods. Coating materials are selected to prevent the active drug in OraSolv® from coming in contact with the taste buds in the mouth but provide for immediate release or sustained release of the drug in the stomach (22,23).

Drug microparticles are mixed with effervescent excipients composed of a dry acid and dry base to facilitate a mild effervescent reaction when the tablet contacts saliva in the mouth (22,23). This effervescent reaction accelerates tablet disintegration through the release of carbon dioxide. In addition, various flavoring, coloring, and sweetening agents are also added to the tablet. These agents are commonly used tablet excipients and are GRAS materials. As the OraSolv® tablet dissolves, it releases the drug microparticles, forming a microsuspension of the drug in saliva. This microparticulate drug suspension enters the stomach by normal swallowing (22,23).

OraSolv® tablets can be manufactured by a conventional tableting process (22,23). This involves depositing the granules into a die cavity and compression molding into the shape of the punch and the die cavity. The company has developed a process that allows high-speed packaging of soft friable tablets without breakage into specially designed protective packages (22,23).

B. Processing and Packaging

The manufacturing process for OraSolv® tablets is summarized in Figure 11.3. The drug, effervescent disintegration agents, and other excipients are weighed

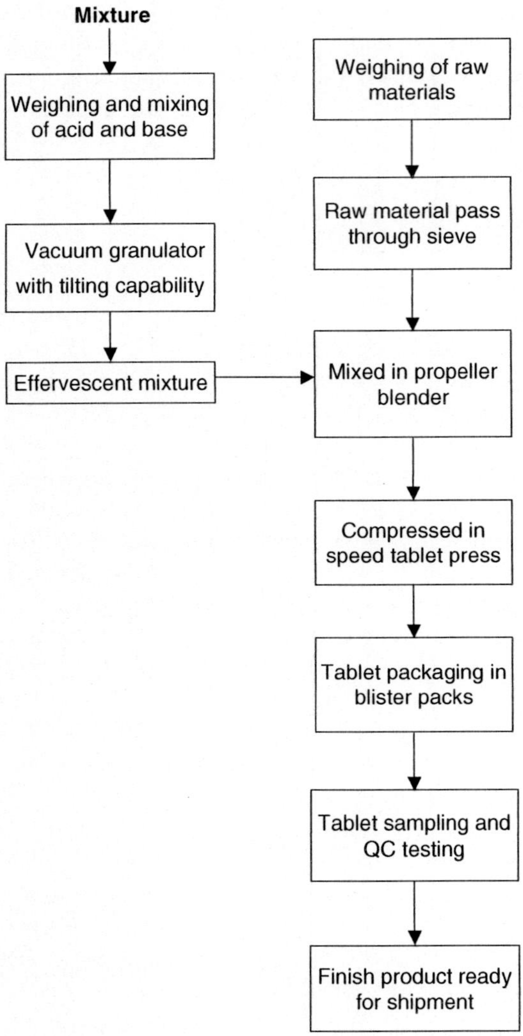

Figure 11.3 OraSolv® typical tablet manufacturing process.

individually and screened through a 10 mesh sieve. These ingredients are charged into a twin shell blender where they are mixed for 20 to 40 minutes at 24 rpm. The lubricant is then added to the blend and mixing is continued for an additional 5 to 20 minutes. The composition is then formed into tablets by compression molding using a conventional high-speed rotary tableting machine. Various studies are needed to optimize process conditions for each formulation. For example, the influence of lubricant type and blending time on disintegration of compressed calcium carbonate OraSolv® tablets is shown in Table 11.9. Based on this study,

Table 11.9 The Effect of Lubricant Blending Time on Disintegration Time of Calcium Carbonate Orasolv® Tablets

Lubricant	Lubricant Blending Time (Min)	In Vitro Disintegration Time (Min)
Talc	5	5
	10	5
	15	5
Lubritab™	5	15
	10	20
	15	25
Magnesium Stearate	5	> 60
	10	> 60
	15	> 60

Source: References 21 and 24.

Table 11.10 The Effect of Lubricant Concentration on Disintegration Time of Calcium Carbonate OraSolv® Tablets.

Lubricant	Lubricant Concentration (%)	In Vitro Disintegration Time (Minutes)
Talc	0.5	5
	1.0	5
	1.5	5
Lubritab™	0.5	5
	1.0	15
	1.5	20
Magnesium Stearate	0.5	> 60
	1.0	> 60
	1.5	> 60

Source: References 21 and 24.

the lubricant blending of 5 minutes for a 20,000 tablet batch size was selected for this process. The effect of lubricant type and concentration on disintegration time of compressed calcium carbonate OraSolv® tablets is shown in Table 11.10. Based on this study, 5 percent magnesium stearate was found to be optimal for this process (21,24).

In certain cases, the active ingredients may need taste masking by appropriate coating or microencapsulation of active drugs. The tablets produced in this process are soft, friable, and must be handled with care. A unique manufacturing process has been developed that allows high-speed packing of OraSolv® tablets without breakage into specially designed protective, child-resistant packages in a controlled, low-humidity environment. PakSolv™ is the commercial packaging system for soft and friable fast-dissolving tablets. PakSolv™ is a light and moisture-proof packaging system used for the OraSolv® products. Extensive consumer testing has been performed to confirm the integrity of packaging of OraSolv® tablets during shipping and handling.

C. Formulations

The following examples are provided to further illustrate the application of OraSolv® formulation technology and are summarized from the patent literature (21,22,25).

Example 1

A formulation of 120 g of famotidine, 1.38 kg of the effervescent mixture, 1.8 g sodium lauryl sulfate, 380 g polyethylene glycol, and 18 g magnesium stearate, along with flavor and sweetening agents, was blended and compressed into tablets. Each OraSolv® tablet weighed 1.3 g and contained 10 mg of famotidine and was packed in a sealed, water-resistant pouch. The effervescent mixture was prepared in a vacuum granulator of 50 liter capacity and capable of tilting 180°, 90° on either side of vertical, having a thermostable jacketed vessel connected to a vacuum pump and heating water source set at 80°C. 11.4 kg of granular citric acid and 9.6 kg of powdered sodium bicarbonate were added to the granulator and were mixed at 250 rpm for 2 minutes. 150 ml of water was sprayed into the vessel and the mixture was allowed to react for 5 minutes. Additional water was sprayed into the vessel and allowed to react for 5 more minutes. The mixture was dried and vacuum was applied; when the product temperature reached 80°C and moisture content was less than 0.08 percent, the product was cooled below 45°C. The effervescent granule mixture was discharged from the granulator and stored in low humidity.

Example 2

A formulation of 210 g of ranitidine, 1.5 kg of the effervescent mixture (described in Example 1), 0.19 g sodium lauryl sulfate, 44 g polyethylene glycol, and 1.8 g magnesium stearate, along with flavor and sweetener, was blended and compressed into tablets. Each OraSolv® tablet weighed 1.6 g, containing 150 mg of ranitidine and sealed in a water-resistant pouch.

Example 3

A low-sodium analgesic OraSolv® tablet was prepared by blending 163 g of 80 mesh aspirin, 163 g of 40 mesh aspirin, 110 g of chlorpheniramine maleate, 15.3 g of phenylpropanolamine hydrochloride, 30 g of beta-carotene, 2.54 kg of tableting lubricant, and 60 g of sodium carbonate (anhydrous) in a mixer. The tableting lubricant was prepared by charging 6.0 kg of magnesium carbonate into a rotoprocessor and spraying with 2.78 kg mineral oil. The rotoprocessor rotated at

130 rpm and mineral oil was sprayed for eight minutes, after which an additional 2 kg of magnesium carbonate was added to the processor and mixed at 200 rpm for five minutes. The contents of the rotoprocessor were discharged into a container and stored in a low-humidity environment. This mixture was compressed into OraSolv® tablets using a conventional high-speed rotary tableting machine. The OraSolv® tablets were effervescent in water at room temperature, as exhibited by flotation to the surface after one minute and after two minutes the tablets had completely dissolved without any precipitate.

Example 4

A formulation of 1.76 kg Vitamin E (spray dried) and 5.65 kg tableting base containing potassium bicarbonate, calcium carbonate, magnesium carbonate, and citric acid (as discussed in Example 3) was added to a 50 liter processor and mixed at 20 rpm, for 20 minutes. 60 g of aspartame and 60 g lubricant were mixed for 5 minutes at 20 rpm. The mixture was discharged into a large bag and 60 g of beta-carotene and flavor were added to the bag and thoroughly mixed. This mixture was compressed into tablets using a conventional high-speed rotary tableting machine. The tablets had good physical integrity with a very fast disintegration time.

D. Performance and Clinical Testing

The following in vivo and in vitro tests have been developed to study the performance of OraSolv® tablets (22,24).

1. Disintegration Effervescent Test

This is an in vitro test to measure the effervescent disintegration of OraSolv® tablets. The disintegration time is the time from immersion to substantially complete dispersion of the tablet as determined by visual observation. The complete disintegration of the tablet does not require dissolution of the particles. This test is conducted by immersing the tablets in water in a beaker at 37°C without agitation. The tablet remains in the bottom of beaker, foam is formed, and the tablet completely dissolves. A typical OraSolv® tablet has a disintegration time between 30 to 90 seconds at 37°C in water (22).

The OraSolv® tablet produces a distinct sensation of fizzing or bubbling in the mouth of the patient. This effervescent disintegration sensation provides a positive organoleptic sensation to the patient. This property has been studied in detail by in vitro and in vivo effervescent disintegration of tablets. A typical OraSolv® tablet should provide about 20 to 60 cm^3 of gas to achieve the desired effervescent sensation in the mouth (24).

2. Acid Neutralization Capacity (ANC) Test

ANC tests are useful because the effervescent disintegrants have an acid neutralization capacity. A standard ANC test is described in 21 CFR–331.10 part 330.

According to this test, ANC is the quantity of hydrochloric acid measured in milliequivalents that an active base is capable of neutralizing. An ANC value of less than 5 is safe and highly desirable. In general, the ANC is measured by the amount of free acid remaining after reaction with the base. A typical OraSolv® tablet contains about 1300 mg of effervescent ingredients in a 1500 mg tablet to achieve an ANC value of less than 5 (22,24).

3. Clinical Testing

Extensive clinical testing of OraSolv®-based products has been carried out, especially in children, although there are few published examples. OraSolv® multivitamin tablets were compared with conventional pediatric multivitamins in children. Children were asked to state their preference. The OraSolv® multivitamin tablets were favored by 89 percent of the children (22,25). An OraSolv® cold and allergy product was compared against various chewable and Zydis® cold and allergy tablets in about 600 children in 20 cities. The results of the study are summarized in Table 11.11 (26). The result shows that the majority of children preferred OraSolv® cold and allergy tablets.

Table 11.11 OraSolv® Cold and Allergy Product Taste Test

	Children (6 to12 Years Old) Responding (Total = 550)		
Parameter	OraSolv® Tablet (%)	Zydis® Tablet (%)	Chewable Tablet (%)
Like tablet	90	72	67
Like flavor/taste	96	72	74
Bitter	20	45	74
Would take it	79	69	59
Would buy it	66	52	26

Source: Reference 26.

E. Commercial and Future Products

Cima Labs has developed and commercialized various QD products based on OraSolv® technology. In 1997 the company started commercial manufacturing of pediatric acetaminophen as Tempra® Quicklets™ for Bristol-Myers Squibb. The product was launched in the United States and Canada in 1997. The company has also signed a global marketing license agreement with Bristol-Myers Squibb to market pediatric acetaminophen and other products in other parts of the world (23). In addition, Cima Labs has signed product development agreements with various pharmaceutical companies.

In September of 1997, the company entered into a development and license option agreement with Zeneca to develop Zomig® based on OraSolv® technology for the treatment of migraine. Recently the company has completed a Phase I clinical study with satisfactory results (27). In August of 1997, the company entered into an agreement with Schering-Plough to develop various prescription

QD products based on OraSolv® technology (28). In December of 1997, the company entered into a development and license option agreement with Novartis Consumer Health to develop various consumer QD products based on OraSolv® technology (23).

V. SHEARFORM™ TABLETS

A. Technology

Fuisz Technology Ltd. has developed and commercialized Shearform™ QD technology. Shearform™ technology comprises a mixture formed from saccharides such as sucrose and dextrose, and sugar alcohol such as sorbitol and mannitol processed into an amorphous floss which is recrystallized to a predetermined level to provide uniform flow properties and thus facilitate blending of the drug and other ingredients of the tablet. The amorphous component provides good bonding properties and contributes to the physical integrity of the matrix (29). By varying the composition and properties of the matrix, QD formulations can be obtained with different dissolution properties (30).

The Shearform™ floss is blended with coated or uncoated drug or other excipients such as effervescent agents and disintegrating agents. This blend is introduced in a dosage well and the blend is tamped (i.e., compressed below 250 psi) into the well. The tamped blend is then cured at adequate conditions of heat, moisture, and pressure (29,31,32). For example, increasing the heat under constant moisture level can cure the Shearform™ tablet. Curing transforms the amorphous composition to a crystalline matrix with sufficient strength to produce a stable Shearform™ product. Because curing results in shrinkage of the tablets, the molding and curing of Shearform™ are done in the final package for commercial manufacturing (31).

B. Processing and Packaging

The Shearform™ tablet is prepared by a flash-heat process using a cotton candy-type manufacturing process (30), which involves placing a dry powder, containing either pure drug or a blend of the drug and other pharmaceutical excipients, into a precision-engineered, fiber-spinning machine that subjects the blend to flash heat (31). The specially designed centrifuge machine and unique blend of sugars produces long, cotton candy-like fibers referred to as floss. The process requires mixing additives with the uncured Shearform™ fibers. When the Shearform™ floss is produced, it is first chopped to reduce the volume of the product to form fibers. The additives and Shearform™ fibers are compressed into tablets. Additives such as surfactants are added to enhance crystallization in the products. The product is cured in the final package well; consequently, several transfer steps can be eliminated in commercial production (29,31,32).

The manufacturing process for Shearform™ tablets is summarized in Figure 11.4. The continuous inline process feeds plastic stock material from rolls to a

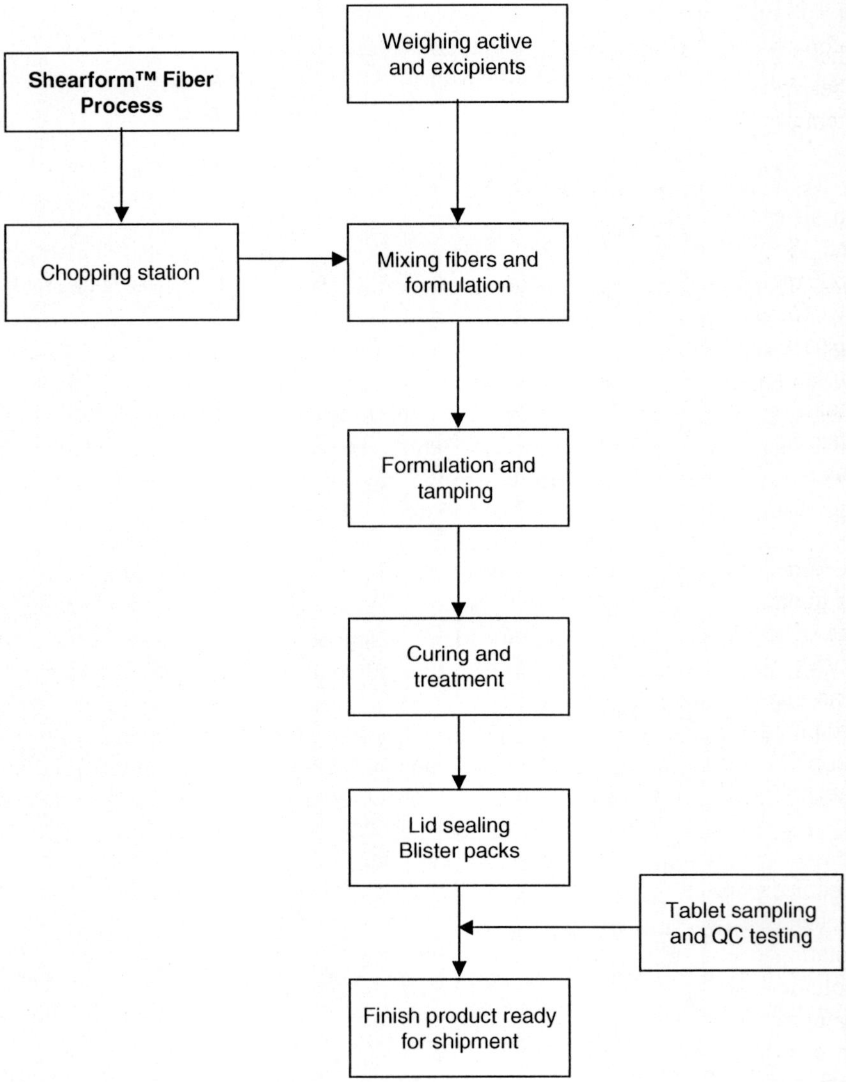

Figure 11.4 Shearform™ tablet manufacturing process.

vacuum thermoforming station where the plastic stock is formed into a tray having six wells. The web moves to a fill-and-form station where the product and additives are mixed, dispensed into each of the wells, and compressed. Proceeding along the process, the products are cured under controlled heat and moisture. The product is then covered with lid stock fed from a roll, thereby sealing the product (29,31,32). Finally, the Shearform™ products are packaged in boxes and cartons.

C. Formulations

The following examples are provided to further illustrate the application of Shearform™ technology and processing and are summarized from the patent literature (29,33).

Example 1

Shearform™ tablets were prepared by blending 4.3 kg of sucrose, 750 g of sorbitol, and 13 g of Tween 80 surfactant in a flash-flow processing apparatus rotating at 3600 rpm. The floss resulting from this process was white and was reduced in volume by chopping. 5 kg calcium carbonate, 50 g polyethylene glycol 300, and 4.7 kg of fibers were mixed with 250 g of aspartame and flavoring agent in a blender. The mixture was weighed out in 0.75 g samples and introduced into 0.75 cm diameter molds. Tablets were formed by compressing the mixture at 60 and 80 psi. The tablets were cured in an oven at 40°C and 85 percent relative humidity for about 15 minutes and resulted in fast-dissolving tablets without any chalky mouth feel.

Example 2

A formulation of 28.8 g of ibuprofen (microcap), 320 mg of lecithin, and 54.4 g of fibers (as discussed in Example 1) was mixed with 720 mg of aspartame, 230 mg of silica, and flavoring agent in a blender for about two minutes. A portion, 0.75 g, of the mixture was weighed and introduced into 0.65 cm diameter molds. The mixture was compressed at 60 and 80 psi. The formed tablets were cured for one day at room temperature and then packed in sealed pouches. The Shearform™ tablets were easy to handle and were very fast dissolving with good mouth feel.

Example 3

The floss was prepared by blending 85 g of sucrose, 12 g of sorbitol, 3 g of alpha lactose, and 0.25 g of Tween 80 surfactant 80 in a flash-flow processing apparatus rotating at 3600 rpm. The white floss resulting from this process was reduced in volume by chopping. 9 g of aspirin, 100 mg of lecithin, and 17.20 g of fibers were mixed with 250 mg of aspartame, 50 mg of flow agent, and flavoring agent in a blender for about two minutes. 0.75 g of the mixture was weighed and introduced into 0.65 cm diameter molds. The mixture was compressed at 80 psi to form tablets. These tablets were cured for one day at room temperature and then packed in sealed pouches. The resulting Shearform™ tablets were easy to handle and were very fast dissolving (i.e., less than five seconds) with good mouth feel.

Example 4

A formulation of 12.19 g of acetaminophen, 90 mg of lecithin, and 11.01 g of fibers (as described in Example 3) was mixed to coat the drug and additional Shearform™ matrix was added until a homogeneous mixture resulted. This mixture was mixed with 200 mg of aspartame and flavoring agent in a blender for about 2 minutes. The 0.90 g of the mixture was weighed and introduced into 0.65 cm diameter molds. The mixture was compressed at 60 psi and the tablets were cured at 40°C and 80 percent relative humidity for 15 minutes and then packed in sealed pouches. The Shearform™ tablets were easy to handle and were very fast dissolving with good mouth feel and taste.

D. Performance and Clinical Testing

Several in vivo and in vitro tests have been developed to study the performance of Shearform™-based tablets. Some of the published methods specific to Shearform™ technology are described below (31,32).

1. Microparticulate Size Analysis

One of the unique characteristics of this technology is the microparticulate powder mixture that can be produced with little or no dust and yet retain the excellent flow characteristic needed for tableting. This powder characteristic is demonstrated by sieve analysis. A typical sieve analysis is exemplified using two formulation compositions, shown in Table 11.12. The ingredients were prepared in accordance with the Shearform™ technology and sieve analysis was conducted to measure the particle size of the powder. The result of the sieve analysis is shown in Figure 11.5 (31). This analysis indicates that 75 to 85 percent of the microparticles are less than 250 microns and 59 to 67 percent of the microparticles are less than 150 microns. These microparticle formulations produced fast-disintegrating and excellently textured tablets.

2. Bulk Density Test

One of the characteristics of the Shearform™-based tablet is its low density. The bulk density is measured using a standard USP method (34). The typical density

Table 11.12 Shearform™ Formulations Sieve Analysis

	Shearform™ Formulations	
Ingredient	Formulation 1 (%)	Formulation 2 (%)
Shearform matrix	58.5	58.9
Calcium carbonate	36.5	36.6
Sweetener	0.1	—
Flavor	0.4	0.2
Vegetable oil	0.5	0.5
Flow agent	1.4	1.4
Starch	2.0	2.0
Color	0.1	—
Lubricant	0.5	0.5
Total	100	100

Source: Reference 31.

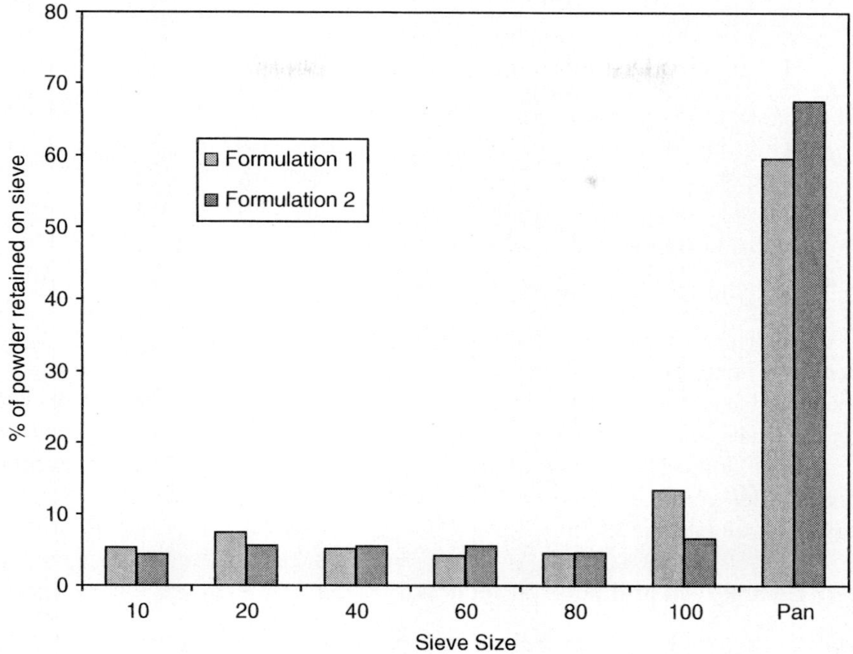

Figure 11.5 Sieve analysis of Shearform™ formulations.

of Shearform™-based tablets after low-pressure compaction is maintained below 1.2 grams per cubic centimeter. The low-density formulation results in a sufficient hardness and structural integrity to be handled manually and machine processed without degradation of the tablet.

3. Calcium Release Test

Various antacid QD products have been developed based on Shearform™ technology. These products are used as calcium supplements because of their high calcium content and release characteristics. An assay was developed to determine the calcium release based on the protocol set forth in the USP. Several tablets were powdered and 550 mg of powder was introduced into water containing 10 ml of 1N hydrochloric acid. This mixture was boiled for 30 minutes, allowed to cool, and then transferred to a 100 ml volumetric flask. This was diluted to volume with water, mixed, and filtered. This solution was further diluted with 100 ml water. Sodium hydroxide, triethanolamine, and hydroxy naphthol blue were added to the solution. The solution was then titrated with disodium ethylenediaminetetraacetate until the solution was a dark color. Each ml of disodium ethylenediamine tetraacetate is equivalent to 5.0 mg of calcium carbonate. The result of this study indicated that calcium release ranged from 87 to 96 percent depending on the formulation.

4. In Vivo Disintegration Test

The disintegration of Shearform™ tablets in healthy volunteers was conducted by placing the tablet in the buccal cavity. The time required for the complete disintegration or dissolution of the tablet without water (not holding water in the mouth) in the buccal cavity was recorded. The typical Shearform™ tablet has an average disintegration of 5 to 10 seconds depending on the individual (35).

5. Clinical Evaluation

Local delivery of the drug from Shearform™ tablets to the buccal mucosa in healthy volunteers was conducted. Antiulcer Shearform™ tablets were placed on the ulcer-bearing oral cavity tissue of each volunteer. Once the tablet was placed on the tissue, the saccharide portion of the matrix quickly dissolved and active drug in gel affixed to the oral cavity. The hydrogel with drug provided instantaneous relief from the discomfort associated with the ulcerated tissue in the oral cavity of volunteers (33,35).

E. Commercial and Future Products

Fuisz Technology Ltd. has worked with various pharmaceutical companies to develop and commercialize QD products based on the Shearform™ technology. Overall, the company has about 20 agreements with 14 multinational pharmaceutical companies for the development of new formulations of prescription and over-the-counter products (17). Selected drugs under development include fexofenadine, paracetamol, and ibuprofen.

VI. SUMMARY

Quick-dispersing oral drug delivery systems are defined as oral drug delivery systems that, when placed in the mouth, dissolve or disintegrate within a few seconds to a few minutes and do not require water to aid swallowing. The QD dosage forms are ideal for many groups of patients including geriatrics, pediatrics, and those people who have difficulty swallowing. An important benefit of QD dosage forms is the ability to provide the advantages of a liquid medication in the form of a solid preparation. This feature enables the patient to take the dose as directed at any time without water and inconvenience. There is a clear medical need and clinical benefits provided by these technologies and products.

However, QD products and technologies face various challenges, as listed in Table 11.13. These challenges are related to new technologies and products. As these technologies mature and new products are developed, some of these challenges will be addressed by various companies. Several technologies are used to achieve quick dispersion and drug delivery to the oral cavity. The four QD

Table 11.13 Challenges for Quick-Dissolving Intraoral Dosage Form Technology and Products

Challenge	Comments
Taste masking	Most drugs need taste masking.
Cost	Due to process, material, packaging, and taste masking.
Special packaging	Tablets are fragile and must be protected from water.
Novel manufacturing process	Due to new equipment, technology, and process.
Limited drug loading	Technology limitation, taste masking, and tablet size.
Clinical/medical benefits	Need more clinical trials to study medical benefits.
Older patient acceptance	Do not like taste, flavor, dissolve too fast, and cost.
Patient compliance	Frequency, may not remember, taste, take too many.

technologies reviewed in this chapter include WOWTAB®, Zydis®, OraSolv®, and Shearform™. The formulation considerations including selection of drugs, excipients, packaging, and manufacturing process have been contrasted among these different QD technologies. Each QD technology has its own advantages and disadvantages but common to all are their rapid disintegration and convenience of dosing to patients. Special in vitro and in vivo test methods to study the performance of these products are required. Although QD technology and products face many challenges as they are fairly new in the marketplace, these technologies are rapidly evolving and continue to undergo improvement which will address the future challenges and changing patient and healthcare needs. Overall, QD products have enormous commercial potential, which will be realized in the next decade as more effective QD products are being developed to address the unmet needs of the patients.

REFERENCES

1. Lindgren S, Janzon L. Dysphagia: Prevalence of swallowing complaints and clinical findings. Medical Clinics of North America, 1993; 77:3–5.
2. Andersen O, Zweidorff OK, Hjelde T, Rodland EA. Problems when swallowing tablets. Tidsskr Nor Laegeforen, 1995; 115 (8):947–949.
3. Yamanouchi Pharmaceutical Co. Press release: Open Yamanouchi Shaklee Pharma Research Center at Stanford Research Park, November 24,1997.
4. Mizumoto T, Masuda Y, Fukui M. Intrabuccally dissolving compressed molding and production process thereof. U.S. Patent 5,576,014.
5. Mizumoto T, Masuda Y, Nyshadham J, Fukui M. Intrabuccally dissolving compressed molding and production process thereof. 1998, U.S. Patent application filed.
6. JP disintegration test, The Japanese Pharmacopoeia page 32, 13th edition, Japan: The Society of Japanese Pharmacopoeia, 1996.
7. Yamanouchi Quarterly Newsletter Nucleus, Autumn 46, page 1–2, 1998.
8. Company press release: Yamanouchi Shaklee Pharma has entered into a Feasibility and Option agreement with the Glaxo Wellcome Group. Shaklee, May 1998.

9. Company press release: Yamanouchi Shaklee Pharma has entered into an agreement with Johnson & Johnson-Merck Consumer Pharmaceuticals Co. Shaklee, June 1998.
10. Company press release: Yamanouchi Shaklee Pharma and Warner-Lambert Company enter WOWTAB® commercialization phase. Shaklee, July 1998.
11. Virley P, Yarwood RJ. Zydis: A novel fast dissolving dosage form. Chemist 1990; 61 (2) 36–37.
12. Gregory GK, Peach JM, Mayne JD. Articles for carrying chemicals. U.S. Patent 4,371,516.
13. Thompson AR, Yarwood RJ, Kearney P. Method for identifying freeze-dried dosage forms. U.S. Patent 5,457,895.
14. Kearney P, Wong SK. Methods for making freeze dried drug dosage forms. U.S. Patent 5,631,023.
15. Buxton IR, Feldman. Solid shaped particles. U.S. Patent 4,754,597.
16. Seager H. Drug delivery products and Zydis® fast dissolving dosage form presented at the British Pharmaceutical Conference, Scarborough, September 1997.
17. Yarwood R. Oral fast dissolving dosage form technologies presented at IBC symposium on drug delivery systems. San Francisco, July 17–18, 1997.
18. Physician's Desk Reference (PDR), 1999, page 2825, 53d edition, Medical Economics Company, Des Moines, IA 1999.
19. Company press release: Pfizer R.P. Scherer Corp. had signed a deal to develop a faster-acting thin wafer form of Viagra® using Scherer's Zydis® drug delivery system. Pfizer, August 1998.
20. Company press release: FDA approves Maxalt™ and Maxalt-Mlt™. Merck & Co., June 1998.
21. Wehling F, Schuehle S, Madamala N. Effervescent dosage forms with microparticles. U.S. Patent 5,178,878.
22. Wehling F, Schuehle S. Base coated acid particles and effervescent formulations incorporating same. U.S. Patent 5,503,846.
23. Cima Labs Inc. Form 10-K for Cima Labs Inc. filed on March 26, 1998.
24. Wehling F, Schuehle S, Madamala N. Magnesium carbonate and oil tableting aid and flavoring additive. U.S. Patent 5,401,513.
25. Wehling F, Schuehle S, Madamala N. Pediatric effervescent dosage. U.S. Patent 5,223,264
26. Khattar JA. The role of fast-dissolving dosage forms in improving patient compliance presented at the international conference on drug delivery technologies: Capitalize on the newest scientific and commercial opportunities. Princeton, NJ, July 13–14, 1998.
27. Company press release: Cima announces completion of clinical study on Zomig™ fast dissolving tablet. Cima Labs Inc., October 1998.
28. Company press release: Cima entered into a development and license option agreement with Schering-Plough. Cima Labs Inc., August 1997.
29. Myers GL, Battist GE, Fuisz RC. Delivery of controlled release systems. U.S. Patent 5,567,439.
30. Fuisz RC. Biodegradable controlled release flash melt-spun delivery system. U.S. Patent 5,518,730.
31. Cherukuri SR, Myers GL, Battist GE, Fuisz RC. Process for forming quickly dispersing comestible unit and production therefrom. U.S. Patent 5,587,172.

32. Myers GL, Battist GE, Fuisz RC. Process and apparatus for making rapidly dissolving dosage unit and production therefrom. U.S. Patent 5,622,719.
33. Fuisz RC. Ulcer prevention method using a melt-spun hydrogel. U.S. Patent 5, 622,717.
34. USP bulk density test, The U.S. Pharmacopoeia 23, page 1830, 23rd edition, United States Pharmacopoeia, Rockville, MD, 1995.
35. Myers M. A new era in morphological drug change presented at IBC symposium on drug delivery systems. San Francisco, July 17–18, 1997.
36. Klancke J. Dissolution testing of orally disintegrating tablets, Dissolution Technology, 10(2):6–8, May 2003.
37. Parakh SR and Gothoskar AV. A review of mouth dissolving tablet technologies, Pharmaceutical Technology, 92–100, November 2003.
38. Brown D. Orally disintegrating tablets—Taste over speed, Drug Delivery Technology, 3(6):58–61, September 2000.

12

Quick-Dissolving Intraoral Tablets

S. Indiran Pather, Rajendra Khankari, and John Siebert

CONTENTS

I. INTRODUCTION

This chapter provides a comprehensive review of the intraorally dissolving tablet technology and products manufactured by Cima Labs, Inc. DuraSolv® and OraSolv® are technologies for the production of quick-dissolving tablets in which the taste of the active ingredient is masked by the use of an appropriate coating over the drug particles and by the incorporation of effective flavoring agents and an artificial sweetener.

A key attribute of the OraSolv® technology is the fact that the relatively soft tablets rapidly disintegrate, and the constituents partially dissolve in the mouth by the action of saliva. The partially dissolved excipients and powder are swallowed with the saliva. No chewing is necessary to obtain rapid breakdown of the tablet. An essential feature is the incorporation of effervescent excipients that act together to form an effervescent couple promoting rapid disintegration while also assisting with taste masking. The DuraSolv® tablets, on the other hand, dissolve rapidly without pronounced disintegration. This is due to the presence of a large proportion of fast-dissolving excipients, in fine particle form. This process is aided by the incorporation of wicking agents, which promote the rapid uptake of solvent (saliva) into the body of the tablet. Swelling disintegrants, if used at all, are kept to a minimum. The overall effect of these formulation approaches is a product that appears to melt in the mouth without a gritty feel. Small amounts of effervescence may, optionally, be incorporated to assist taste masking but the concentrations are not high enough to promote rapid disintegration.

This chapter includes, additionally, a brief discussion of the OraVescent™ technology, which shares some formulation similarities with the former technologies. The OraVescent™ technologies differ, however, from the former technologies with respect to the end use of the products. OraSolv® and DuraSolv® dosage forms are pleasant to take and offer convenience of administration and, thus, promote compliance with the dosing regimen. Enhanced drug delivery is not a feature of these technologies. The OraVescent™ SL (sublingual) and OraVescent™ BL (buccal) technologies improve the intraoral absorption of certain drugs and are, thus, fundamentally different in their scope of application. The OraVescent™ SS (site specific) technology that is designed to target drugs to certain sites within the gastrointestinal tract is beyond the scope of this chapter and is not discussed.

Other quick-dispersing technologies related to OraSolv® and DuraSolv® are covered in Chapter 11 and the reader is referred to this chapter if comparisons are desired. No direct comparisons with other quick-dissolving technologies are made in this chapter, except in the description of certain organoleptic tests in which a head-to-head comparison with a related product was one of the objectives of the test. Throughout this chapter, the terms "quick-dissolve," "fast-dissolve," "quick-melt," "fast-melt," "rapidly dispersing," and "quick-dispersing" are used interchangeably.

II. RATIONALE FOR FORMULATING QUICK-DISSOLVING TABLETS

A basic tenet of pharmaceutical formulation science is that the dosage form exists to optimize the delivery of a pharmaceutical active to its site of action in the most effective and safe manner. To this end, various sophisticated and often ingenious delivery systems have been developed including needleless injections, transdermal patches, oral transmucosal delivery devices, nasal delivery formulations, pulmonary devices, site-specific gastrointestinal delivery systems, various types of sustained-release formulations, and quick-dissolving intraoral tablets. Attainment of a suitable blood concentration/time profile is often regarded as validation of the effectiveness of the specialized dosage form. However, it should be remembered that the desirable blood-level profile is attainable with a slow intravenous injection or, in many instances, even with conventional dosage forms if they are administered sufficiently frequently. It becomes evident that the aim of all these formulation efforts, in the ultimate sense, is directed to improve patient convenience of self-medication. Patient convenience leads to improved compliance with the prescribed dosing regimen and, as a consequence, to optimal therapy.

To be effective, the drug must be delivered to the site of action. Although this statement usually implies special efforts to enhance drug permeability through a biomembrane barrier, "delivery" also encompasses the role of the patient actually taking the medication. If the patient does not take the medication, there can be no delivery.

Compliance has become a major problem, particularly for children and for senior citizens. In a study of otitis media in children, it was found that only 5 percent of the patients fully complied with the treatment regimen (1). In a study of giardiasis treatments (metronidazole, tinidazole, and furazolidone) it was found that 50 percent of the pediatric study population resisted taking the medication and spillage was common (32 percent) and, furthermore, the noncompliance was attributable to the side effects of the drug (2). A review of several studies of compliance in the elderly reveals varying rates of noncompliance, up to 60 percent, depending on the study and the definition of noncompliance (3). When intentional under-medication is considered, the rate may be as high as 90 percent. It was found that functional limitations in the elderly, such as the ability to open child-resistant containers, to distinguish colors, and to interpret labels, increased noncompliance (4). Compliance is particularly important in the case of anti-infective agents because of the need to maintain therapeutic blood concentrations (2). Antibiotics often have a bad taste that is likely to compromise compliance in children (5). The emergence of resistant bacterial strains is an ominous outcome of antibiotic overuse and noncompliance (5). Compliance is also an issue in the treatment of chronic conditions such as asthma (6) and the prevention of rejection after transplants (7).

Turning to another trend, in recent years there has been a paradigm shift in pharmaceutical marketing, whereas companies previously content to sell drugs to physicians, hospitals, and health maintenance organizations are now marketing to the patient as well. The direct-to-consumer (DTC) marketing of pharmaceuticals, including prescription drugs, has created a new "empowered" patient whose needs must be addressed in the development of pharmaceuticals. Patient-friendly dosage forms that offer ease of administration and convenience are needed to satisfy the empowered patient.

The scientific need for compliance (to ensure that the drug is delivered to the site of action at the appropriate rate) and the demands of the empowered patient are the two requirements that dictate that convenient, patient-friendly dosage forms be available. Such a dosage form, in some instances, may take the form of a transdermal patch that provides continuous medication and thus obviates the requirement for repeated doses. In other instances, particularly where constant blood levels are not essential, the need for a patient-friendly dosage form may be satisfied by a quick-dissolving intraoral tablet that has a pleasant taste and mouth feel. It can be taken at any time or place without regard to the availability of water or the need to swallow a large tablet.

III. FORMULATION AND MANUFACTURING TECHNOLOGY

In this section, a brief overview is provided on the formulation, production, and packaging of OraSolv® and DuraSolv®, the two technologies that have been patented and commercialized by Cima Labs, Inc. The OraSolv® product is a soft tablet that needs special packaging. The DuraSolv® tablet is a comparatively harder tablet that, nevertheless, is able to rapidly dissolve in the mouth. The advantage of the latter is that it can be packaged in conventional tablet containers.

A. OraSolv®

The ideal attributes of an orally disintegrating tablet formulation include: (1) disintegrates quickly in the oral cavity, (2) releases 100 percent of the active ingredient in the gastrointestinal tract, and (3) has a pleasant taste and creamy mouth feel (8). Fast disintegration is achieved by compressing water-soluble excipients using a lower range of compression forces than is normally used in conventional tableting. The time for the disintegration of OraSolv® tablets within the oral cavity varies from 6 to 40 sec, depending largely on tablet size and the compression force (within the lower range) that was used to form the tablet. The low compression force leads to high tablet porosity that, in turn, accelerates the rate of disintegration of the tablet and dissolution of the water-soluble excipients. Disintegrating agents further facilitate the process. An effervescent couple is used as a water-soluble disintegrating agent (9). Thus, the OraSolv® tablet comprises the following components:

Taste-masked active(s)
Filler
Sweetener
Disintegrating agent
Lubricant
Glidant
Flavor
Color

The active ingredients can be taste masked using a variety of techniques such as fluid bed coating, microencapsulation, or spray congealing. The type of taste-masking system used is dictated by the physicochemical properties of the active ingredient and its physical form. For example, an active ingredient with a mean particle diameter of 200 to 300 μm may lend itself to direct particle coating. The ideal orally disintegrating tablet imparts no unpleasant taste. This requirement may be met by minimizing drug dissolution of unpleasant-tasting drugs within the oral cavity. However, the product should display an acceptable bioavailability which means, ideally, that there should be 100 percent dissolution in the gastrointestinal tract. Consequently, the taste-masking process is a balancing act between these opposing requirements: insignificant dissolution in the oral cavity and maximum dissolution in the postoral cavity section of the gastrointestinal tract. When utilizing particle coating for taste masking, one may incorporate polymers that are conventionally used for sustained-release coating. The proper balance is achieved by incorporating the correct combination, and amounts, of the coating polymers.

The tablets are manufactured by a direct compression technique using conventional blending equipment and high-speed tablet presses. Because the OraSolv® tablets are produced at low compression forces, they are soft and friable. To reduce handling of the tablets, the tableting and packaging processes are integrated and a specially designed packaging system is used. This system consists of a robot that picks up and places the tablets in dome-shaped depressions in the four-layered aluminum foil blister card. A layer of top foil is heat sealed over the bottom foil to form the primary blister card package. The integrated OraSolv® manufacturing line is equipped with a printing assembly that enables each blister card to be printed individually during the manufacturing process. The automated system then cuts the foil into strips of, usually, six tablets. An automated camera visual recognition system is employed for 100 percent inspection to detect depressions that do not contain tablets and reject product with incomplete tablet count. An image of the cards is also shown on a monitor, so that the operator may also observe unfilled cards.

This specially designed package and processing system (PakSolv®) protects the OraSolv® tablets from breaking and attrition during the rigors of shipping. In particular, the dome-shaped depressions limit the vertical movement of the tablet within the package because the diameter of the lower portion of the dome is too narrow to accommodate the tablet. Thus the tablet remains in the upper part of the dome adjacent to the top foil. This is in contrast to a regular blister package in which the sides of the depression are vertical and the bottom is flat, allowing a greater range of vertical movement. The OraSolv® package also offers light, moisture, and child resistance. Moisture resistance is important when packaging an effervescent formulation or moisture-sensitive drugs. A typical OraSolv® blister package is shown in Figure 12.1.

B. DuraSolv®

DuraSolv® is Cima's second-generation fast-dissolving tablet technology. This technology provides robust yet quick-dissolving tablets. Like OraSolv®, the

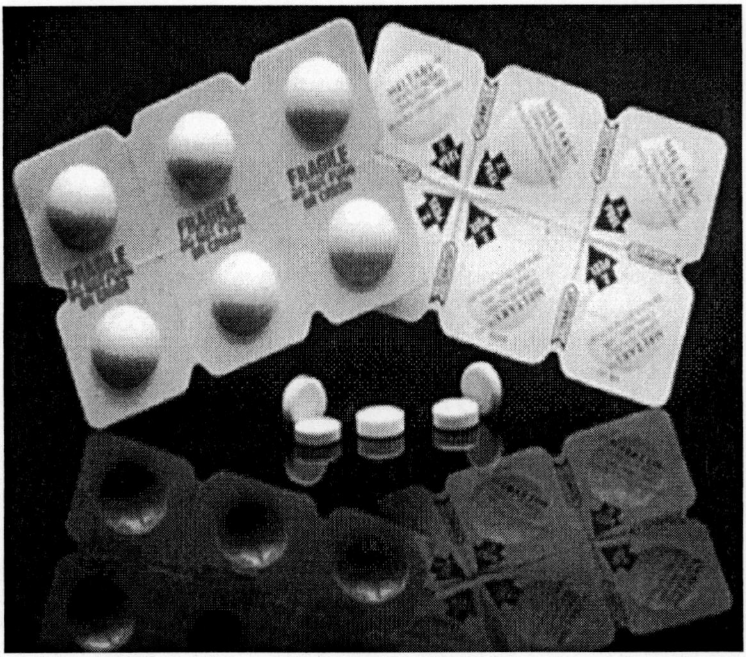

Figure 12.1 Typical OraSolv® package.

DuraSolv® tablets consist of water-soluble excipients and are manufactured using direct compression techniques. However, DuraSolv® utilizes nondirectly compressible fillers in fine particle form (10). These fillers have a high surface area, which increases their dissolution rate. The incorporation of a high proportion of such fillers causes the tablet to "melt" or dissolve rather than disintegrate. Wicking agents assist the entry of water into the body of the tablet whereas swelling disintegrants are avoided or used in small proportions. Because extensive disintegration is to be avoided, only small amounts of effervescent agents may be incorporated if they are to be included at all. The limited disintegration contributes to the nongritty mouth feel conferred to the product by the use of fine particle fillers. The DuraSolv® tablets are robust and pass the USP friability test. The reduction in tablet porosity, due to the higher compression forces, is compensated by the increased dissolution rate of the filler. DuraSolv® tablets can be packaged in bottles, blisters, or pouches. The manufacturing process utilizes conventional blenders, high-speed tablet presses, and conventional packaging equipment.

Currently available commercial products manufactured by Cima Labs, Inc. are listed in Table 12.1. Several additional products are in various stages of development or registration with the FDA.

Table 12.1 Currently Marketed Intraorally Disintegrating Tablets Manufactured by Cima Labs, Inc.

Product	Active Ingredients
Tempra® FirsTabs	Acetominophen 160 mg
TRIAMINIC® Softchews™ Cold and Allergy	Pseudoephedrine HCl 15mg, Chlorpheniramine Maleate 1 mg
TRIAMINIC® Softchews™ Cold and Cough	Pseudoephedrine HCl 15 mg, Dextromethorphan HBr Monohydrate 5mg, Chlorpheniramine Maleate 1 mg
TRIAMINIC® Softchews™ Throat Pain and Cough	Acetominophen 160 mg, Pseudoephedrine HCl 15 mg, Dextromethorphan HBr Monohydrate 5 mg
TRIAMINIC® Softchews™ Cough	Dextromethorphan HBr Monohydrate 7.5 mg
ZOMIG® Fastmelt	Zolmitriptan 2.5 mg

IV. SHIPPING TESTS

To illustrate the robustness of the OraSolv®–Paksolv® combination, a series of shipping tests have been conducted by Cima Labs. These tests fall into two broad categories: (1) laboratory tests that simulate shipping conditions or (2) tests in which the product is actually shipped to a particular destination. In tests of the first type, the packaged product is dropped from a specified distance, or vibrated at a specific frequency according to a fixed protocol. Such tests are performed on Cima's behalf by companies specializing in the performance of shipping tests on packaged goods. The latter type of test subjects the product to a real-world shipping experience. In either case, a number of cards are then opened and examined for broken tablets and for dust. A total of approximately 11 such tests have been performed by Cima. One of these tests, which involved international shipping, is described in more detail.

This test was conducted during the period December 1997 to January 1998. It involved the transcontinental shipping of 5/8 inch, OraSolv® tablets of 16 N breaking strength that were debossed on one side. The trip from Minneapolis to Prague in the Czech Republic utilized U.S. tractor-trailer transport, an ocean voyage, European tractor-trailer road transportation, and interim storage.

The tablets were packaged in cards of 6, with 4 cards per carton. The shipping cases had 24 cartons per case. The shipping cases were palletized on two standard stretch-wrapped pallets with 120 cases per pallet. The cases were packed in 6 layers with 20 cases per layer. The total number of tablets shipped thus equaled $6 \times 4 \times 24 \times 120 \times 2 = 138{,}240$. Using statistical sampling methods, 3648 tablets were removed from the pallets and observed for deterioration in their physical condition.

The tablets were rated on a scale of 0 to 3 using predetermined criteria as described in Table 12.2. The rating scale was developed prior to the test and the scores were documented by means of photographs kept on record at Cima Labs.

Table 12.2 Tablet Rating System for Shipping Tests

Evaluation	Score
No dust	0
Little dust	1
Moderate dust	2
Major dust	3
Chipped or broken tablets	3

Table 12.3 Results of International Shipping Test

Factor	Percentage Rating
Dust level 0	98.3
Dust level 1	1.7
Chipped or broken tablets	0

The results are shown in Table 12.3 and confirm that the OraSolv® tablets in their special packaging were able to withstand the rigors of shipping.

V. ORGANOLEPTIC TESTS

Two tests are described in this section: one in which an OraSolv® tablet is compared to a conventional tablet with respect to taste and subject perceptions of ease of use. Because there are several different types of quick-dissolve technologies available, in the second test an OraSolv® product is compared to a quick-dissolving freeze-dried wafer. The studies were conducted by Bases, a company specializing in the conduct of consumer preference and taste tests.

A. OraSolv® versus Standard Tablet

In this test, an OraSolv® 40 mg famotidine tablet was compared to a standard, commercially available famotidine tablet. The concept of a fast-dissolving tablet and the method of its use were explained to the subjects. Information regarding the standard tablet was also given to the subjects, using the product labeling as the source. A protomonadic design was followed in the dosing of the 206 subjects.

The subjects rated the products by means of their agreement, or disagreement, with a series of statements with reference to each product. The results are summarized in Table 12.4, in which the overall preference scores are reflected, and in Table 12.5, in which a number of subjective ratings are presented. Table 12.4 shows that three times as many subjects preferred the OraSolv® tablet compared to the standard tablet. Table 12.5 shows that a statistically significant number of subjects thought that it was convenient to use and a unique way of taking medication. After using the product, the subjects' perceptions were that it

Table 12.4 OraSolv® versus Standard Famotidine Tablet Preference Ratings

Product Preference	Respondents (%)
Preferred OraSolv® tablet	75
Preferred standard tablet	24
Liked both equally	1

Table 12.5 OraSolv® versus Standard Famotidine Tablet Attribute Ratings

Description	Swallowable Tablet (A)	OraSolv® Tablet (B)
Is safe to use	88	89
Is convenient to use	74	97[a]
Is a unique way of taking medication	54	90[a]
Dissolves quickly	52	97[a]
Does not require water to take it	27	92[a]
Total Respondents	204	202

[a] Indicates a significant difference at the 90 percent confidence level or greater.

dissolves quickly and that it did not require water to take. These perceptions were significantly different compared to those for the swallowed tablet. The subjects perceived the OraSolv® tablet to be as safe as the standard tablet (no significant difference).

B. OraSolv® versus Zydis®

In another test, OraSolv® and Zydis® formulations each containing 6.25 mg of chlorpheniramine maleate and 1 mg of brompheniramine maleate with a grape flavor were compared. The Zydis® formulation is sold under the tradename Dimetapp® Cold and Allergy tablets. The testing involved 450 interviews of children (6 to 12 years) and adults. The children were interviewed in 21 U.S. cities and the adults were interviewed in 8 U.S. cities. A monadic approach were used (i.e., each subject tested only one product). The perceptions of the subjects with respect to the length of time it took for the tablets to melt in the mouth, the taste of the products, as well as an overall preference for the products were determined. As in the previously described test, the subjects were given a series of statements for each section of the test and responded to the one with which they identified.

The results of the test measuring melting time perceptions are shown in Table 12.6. A significantly greater number of children felt that the OraSolv®

Table 12.6 OraSolv® versus Zydis®: Length of Time for Tablet to Melt

	Children (%)		Adults (%)	
	OraSolv®	Zydis®	OraSolv®	Zydis®
Too long	9	3	5	4
Just the right amount of time	71[a]	57	80	85
Melt too quickly	16	37[a]	14	10
Don't know	4	3	1	1

[a] Indicates a significant difference at the 90 percent confidence level or greater.

product melted in just the right amount of time and a significantly greater number of children felt that the Zydis® product melted too quickly. The adults did not perceive a difference. It is important to remember that it is the perceptions of the subjects that were determined in these tests. The Zydis® product is a freeze-dried wafer and, in keeping with the nature of the product, any objective testing in a laboratory will show that it dissolves more rapidly than the OraSolv® product. Yet, the adults did not perceive a difference. In a world where people are conditioned with the idea of "faster is better," it is surprising that the children felt that the Zydis® product melted too quickly. The solid form of the freeze-dried wafer breaks down almost instantaneously when placed in the mouth. Although the OraSolv® tablet disintegrates much faster than a conventional tablet, it is, nevertheless, a compressed tablet and its disintegration time is considerable longer than that of the Zydis® formulation. It is possible that the children did not like the "suddenness" of the Zydis® reaction.

A significantly greater number of adults and children liked the OraSolv® taste whereas a significantly greater number of subjects in each age group hated the Zydis® taste (Table 12.7). A significantly greater number of children were also not sure whether they liked or disliked the Zydis® taste.

The OraSolv® tablet is manufactured using drug particles that are coated with a material that prevents immediate dissolution of the drug. As a result, the drug does not dissolve appreciably in the mouth and therefore the taste is not perceived. In contrast to this, the freeze-drying process by which the Zydis®

Table 12.7 OraSolv® versus Zydis®: Intensity of Liking or Disliking the Taste

	Children (%)		Adults (%)	
	OraSolv®	Zydis®	OraSolv®	Zydis®
I liked it	96[a]	72	84[a]	70
Not sure	1	15[a]	6	4
I hated it	3	12[a]	10	26[a]

[a] Indicates a significant difference at the 90 percent confidence level or greater.

product is made (11) does not allow for a similar coating process. Both products contain a grape flavor. However, the flavor can be overwhelmed if the drugs have a stronger taste. With the OraSolv® formulation, it is possible to taste only a very small fraction of the dose (the portion that dissolves) and this taste is masked with flavor. On the other hand, the Zydis® formulation attempts to disguise the taste of a major portion of the drugs (because a large amount is dissolved and can contact the taste buds). The speed with which the wafer dissolves, however, may allow faster swallowing and removal from the taste zone. The results of the test indicate that the subjects perceived the former (OraSolv®) method to be more effective in disguising the taste of this drug combination.

With respect to their overall perceptions of the products, it is clear from the data that both products were liked by the subjects. However, a significantly greater number of adults and children liked the OraSolv® product and a significantly greater number of adults hated the Zydis® product (Table 12.8). In addition, a significantly greater number of children were unsure about their overall view of the Zydis® tablet. It is clear that the better taste masking of the OraSolv® product was preferred by the subjects and, unexpectedly, that the faster melting of the Zydis® product was not necessarily preferred by the subjects.

Table 12.8 OraSolv® versus Zydis®: Intensity of Liking Tablet Overall

	Children (%)		Adults (%)	
	OraSolv®	Zydis®	OraSolv®	Zydis®
I liked it	90[a]	72	87[a]	75
Not sure	6	19[a]	10	9
I hated it	4	9	3	16[a]

[a] Indicates a significant difference at the 90 percent confidence level or greater.

VI. PHARMACOKINETIC TESTS ON ORASOLV®

OraSolv® formulations are designed for convenience of administration, not for faster drug absorption. The coated drug particles are swallowed together with the portion of the excipients still in powder form, as well as the dissolved excipients (and possibly a minor portion of dissolved drug). Most of the coating around the drug-containing particles dissolves distal to the oral cavity, releasing the drug for absorption. Given that the drug's pharmacokinetics after OraSolv® administration are not intended to be different from those observed after administration of a conventional dosage form, the OraSolv® formulation may be used to substitute the latter. It then becomes incumbent on the marketer to show bioequivalence to a marketed product. Several studies have shown this to be the case. A brief description of a clinical study with pseudoephedrine follows.

A randomized, single-dose, three-way crossover study was performed to compare the bioavailability of pseudoephedrine hydrochloride from an OraSolv® formulation, Sudafed® tablets, and Children's Sudafed® Liquid. The dose was either one tablet or 5 ml of liquid, thus providing 30 mg of drug in each case. After an overnight fast, each of the six adult male volunteers was given one of the formulations according to a randomization schedule. After seven-day washout periods, the alternate doses were administered. Blood samples were drawn at predetermined times for 12 hours postdosing. The blood was centrifuged and the plasma separated and frozen until assayed by GC/MS. The limit of quantification for the validated assay method was 10 ng/ml. The mean plasma level versus time plots are shown in Figure 12.2 and a summary of the pharmacokinetic parameters for OraSolv® and Sudafed® tablets is provided in Table 12.9. There was no significant difference between the pharmacokinetic parameters (i.e., C_{max}, T_{max}, and AUC_0) for the two formulations ($p > 0.05$). These results indicate that the OraSolv® formulation of pseudoephedrine is bioequivalent to Sudafed® tablets.

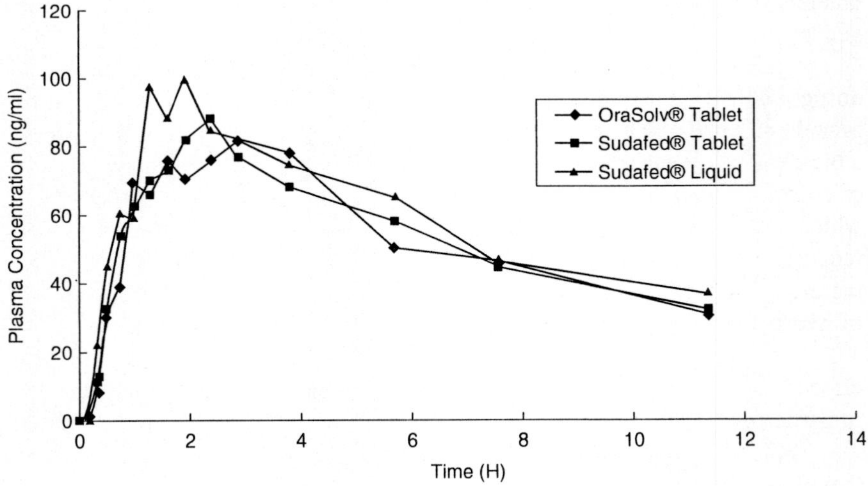

Figure 12.2 Comparative pseudoephedrine bioavailability.

Table 12.9 Pseudoephedrine Pharmacokinetic Parameters for OraSolv® and Sudafed® Tablets

PK Parameter[a]	OraSolv®	Sudafed®
C_{max} (ng/ml)	94.7 (17.1)	94.5 (26.2)
T_{max} (hr)	2.07 (1.2)	2.03 (0.6)
AUC_0 (ng. hr/ml)	948.3 (328.0)	989.2 (307.7)

[a] Each value is the mean ± S.D.

VII. THE ORAVESCENT™ ORAL CAVITY DRUG DELIVERY SYSTEMS

In contrast to the OraSolv® formulations, the OraVescent™ oral cavity drug delivery systems have been designed to promote drug absorption through the oral mucosa. This may enable more rapid absorption of drugs that have a long T_{max}. In other instances, this route of administration may be desirable in order to avoid the first-pass effect and so improve the bioavailability of the drug. Some drugs are not absorbed to a significant extent when administered orally and delivery through the oral cavity mucosa may represent a convenient method of administration for such drugs. Formulation of drugs of this type into intraorally dissolving tablets for transmucosal absorption may permit the substitution of a transmucosal delivery system for an injection. In some instances, the disease condition of the patient or side effect of the medication may compromise efficient absorption of a drug that would otherwise be well absorbed. An obvious example is the case where the patient is vomiting frequently. In addition, gastrointestinal transit may be so severely compromised in migraine as to render efficient drug absorption unlikely. Under these circumstances, a transmucosal system may be advantageous.

The effervescence reaction has been known and utilized in pharmaceutical dosage forms for a long time, a patent having been issued in 1872 (12). Carbon dioxide is released as a result of the interaction of an acid with a carbonate, or a bicarbonate, salt. More recently, the potential for CO_2 to promote the transport of a drug across a biological membrane was explored (13). The mechanisms by which CO_2 acts as an absorption promoter can be summarized as follows: (1) reduces the thickness of the mucous layer, (2) opens tight junctions, and (3) increases the hydrophobicity of the cell membrane, thus promoting the absorption of hydrophobic drugs.

This absorption enhancement may be improved by several additional effects, such as the use of mucoadhesion or additional penetration enhancers and, in particular, by pH effects. The latter are examined in more detail.

From a consideration of the Henderson–Hasselbach equation, it is known that pH values lower than the pK_a of a weak base promote its ionization in aqueous solution. On the other hand, when the pH of the solution is above the pK_a, the ionization of the weak base is repressed and the nonionized form predominates. The ionized form of the drug is much more water soluble than the nonionized form, whereas the latter is usually much more permeable to biological tissues. Therefore, conventional wisdom directs that a pH lower than the pK_a of the drug be used when the solubility of the drug is limited. Either the solution to be administered has this pH or the local environment of a solid or semisolid dosage form develops this pH as a result of the dissolution of formulation adjuvants incorporated into the dosage form. An analogous situation is encountered in the packaging of weakly basic alkaloids used as eye drops. Because of poor aqueous solubility, it is a challenge to keep the drug in solution and to prevent precipitation during storage. Hence, the solution is buffered to a pH below the pK_a of the drug to allow predominance of the more soluble ionized form of the drug.

Where absorption of the weakly basic drug into a biological system is slow or incomplete, a pH that is somewhat higher than the pK_a of the drug may be chosen for a system to be administered. This ensures that a significant proportion of the dissolved drug is in the nonionized form and hence is more readily absorbed. Standard texts dealing with physical pharmacy usually portray these as opposite requirements: a low pH for dissolution and a higher pH for absorption. Often, the pH that is chosen is a compromise between that which is desirable for optimizing the solubility of the drug and that which promotes absorption (14). The compromise implies a decrease in one potential in order to facilitate the other.

A variable pH would give the best of both worlds. If we could have a system that develops a low pH initially, the solubility of the weak base will be good. The drug from a solid dosage form, such as a sublingual tablet, would dissolve adequately in the limited volume of aqueous medium surrounding the dosage form. If the pH of this system could then be caused to slowly rise, the ionized form of the drug in solution would slowly change to the nonionized form and, hence, absorption would be promoted. It is important that the pH change slowly, so that the concentration of the nonionized form does not exceed its limited water solubility. Absorption of the drug into the biological tissues effectively reduces its concentration in solution and further helps to prevent precipitation from solution.

When the effervescence reaction occurs, CO_2 is liberated and dissolves in the aqueous medium. If a buccal or sublingual tablet contains the effervescent couple, the CO_2 that is released will dissolve in the saliva. The saliva becomes more acidic due to the formation of carbonic acid (Equation (12.1)). Due to the later loss of CO_2 from solution, the pH of the solution gradually rises (Equation (12.2)).

$$CO_2 + H_2O \longrightarrow H_2CO_3 \qquad (12.1)$$

$$H_2CO_3 \longrightarrow CO_2 + H_2O \qquad (12.2)$$

The liberated CO_2 is either absorbed by the mucosal tissues where it might promote drug absorption, or it is lost to the airspace in the oral cavity. The overall pH variation with approximately equimolar amounts of citric acid and sodium bicarbonate is from about 6 to about 8.4 (i.e., a variation of about one pH unit in either direction from normal salivary pH). This reaction, therefore, conveniently produces a changing pH profile that can be exploited for drug delivery. The upper and lower limits of the pH values that are covered by this dynamic system can be modified by the use of additional pH-adjusting substances, while maintaining a pH range of more than two units. For example, the pH range can be changed from ~5.5 to ~7.5 for use with a drug that has a pK_a of 6.5. A limitation in the pH range that can be achieved, for practical purposes, is the amount of pH-adjusting substance that must be used and its taste and, hence, the patient acceptability of the final product.

By substitution of the appropriate pH and pK_a values into the Henderson–Hasselbach equation, it can be shown that pH changes greater than one pH unit from the pK_a produce only small additional changes in the proportion of the desired species. Hence, it usually suffices to produce pH variations of one unit from the pK_a. In fact it may be impractical, in some instances, to formulate for pH variations of one unit in either direction and the formulator would have to settle for a smaller enhancement of dissolution and absorption.

The formulation of these products differs from that of the OraSolv® or DuraSolv® products first, in the presence of the pH-adjusting substances as additional components and, second, in the fact that the product does not contain any mechanism that will inhibit drug absorption in the oral cavity. In the OraSolv® and DuraSolv® products, the drug particles are usually coated with a material to retard dissolution in the oral cavity, so that the taste of the drug will not be perceived. The inhibition of dissolution effectively prevents consequential drug absorption in the oral cavity.

The hypothesis that effervescence and the resulting pH transition can be utilized to enhance the delivery of poorly soluble weak bases was tested in human subjects with fentanyl as the model drug. Fentanyl is an anesthetic agent that may be used for induction of anesthesia and, in combination with other anesthetic agents, in the postinduction phase (15). It is also used for the treatment of moderate to severe chronic pain, such as that experienced by cancer patients. For this purpose, it is often administered as a transdermal drug delivery system (16). This system provides continuous delivery of the medication for 72 hours and its efficacy and safety have been established (17). To provide the sustained delivery effect, the drug is absorbed slowly from the system. The rate of fentanyl absorption in the initial hours may be too slow to provide pain relief and additional short-acting opioids, such as morphine, may be needed during this period (18). Recently a fentanyl oral transmucosal system has become available (19) and may be useful to treat the severe breakthrough pain sometimes experienced by cancer patients on a transdermal fentanyl regimen (20). The breakthrough pain is of relatively short duration and this product (Actiq, marketed by Anesta Corporation in the United States) is intended to create a spike in the fentanyl blood concentration to provide transient relief.

Fentanyl was selected for the present study because its pK_a is 7.3 and its nonionized form is highly lipophilic and poorly water soluble. Absorption of fentanyl from the gastrointestinal tract is slow and the drug also undergoes gut wall, and extensive hepatic, metabolism. Only about one third of a swallowed dose is absorbed. Fentanyl is thus an ideal drug to test the hypothesis outlined above because its physicochemical characteristics fit the model well and there is a reduced tendency for gastrointestinal absorption (from swallowed drug) to confuse the assessment of oral transmucosal absorption.

The study was conducted in Belfast, Northern Ireland and the protocol was approved by the Ethics Committee of Queen's University of Belfast. Twelve normal, healthy male volunteers aged 18 to 55 years participated in the study.

Each subject's body weight was not more than 15 percent above or 15 percent below the ideal weight for his height and estimated frame, adapted from the 1983 Metropolitan Life Table. After an overnight fast, the subjects were given one of the following three dosing regimens according to a randomization schedule:

A. An OraVescent™ fentanyl bucccal tablet (200 μg)
B. A tablet that was similar to A in size, shape, and drug content but which contained lactose in place of the absorption-enhancing components
C. An Actiq® oral transmucosal delivery system, which is marketed by Anesta Corporation in the United States

Water was allowed ad lib except for the period from dosing to four hours postdose. Subjects followed a menu and meal schedule as determined by the clinic with the first meal approximately five hours after dosing.

Subjects taking treatments A and B were asked to place the tablet between the upper gum and inside of the cheek, above a premolar tooth. The tablets were left in place for 10 minutes. If, at this time, a subject felt that a portion of the tablet remained undissolved, he was requested to gently massage the area of the outer cheek, corresponding to the tablet's placement, for a maximum of 5 minutes. A member of the clinic staff checked the subjects' mouths at 15 minutes to see if any portion of the tablet remained. The residue, if any, was allowed to dissolve on its own without further manipulation.

Administration of treatment C was according to the directions on the package insert of the product. After removal from the wrapper, the candy-based delivery system was placed between the lower gum and cheek with the handle protruding from the subject's mouth. From time to time, the unit was moved to the other side of his mouth. Subjects were instructed to suck and not to chew the unit. To assist consumption of the unit at the correct rate (completion in 15 minutes), a diagram was developed of a two-thirds dissolved unit. At 10 minutes, the unit was removed from the subject's mouth and a clinic staff member compared the profile of the residual candy to the diagram. If it was observed that the subject was consuming the unit too slowly, he was asked to increase his rate of consumption (and vice versa). Using this technique it was possible to get the subjects to complete the consumption of the unit in the 15 minutes recommended in the package insert. The diagram was developed as follows. An Actiq unit was fully immersed in a beaker of water and stirred at a fixed rate using a mechanical stirrer. By trial and error, the stirring rate that gave complete dissolution in 15 minutes was found. A fresh Actiq unit was then immersed in a beaker of water and stirred constantly at this rate for 10 minutes. The unit was removed from the water at this point and its outline drawn.

After dosing, blood samples were withdrawn at predetermined time points for 36 hours. Sampling was every 2 minutes up to 6 minutes and thereafter every 3 minutes up to 15 minutes in order to capture the anticipated rapid rise in fentanyl blood levels in the early postdosing period. Sampling was increasingly less frequent thereafter. The blood samples were centrifuged and the separated serum was frozen until assayed by an LC/MS/MS method. The assay method was sufficiently sensitive to allow a limit of quantitation of 0.05 ng/mL.

The serum level versus time plots are shown in Figure 12.3 and the pertinent pharmacokinetic parameters are summarized in Table 12.10. The OraVescent™ enhanced delivery system displayed superior pharmacokinetics when compared

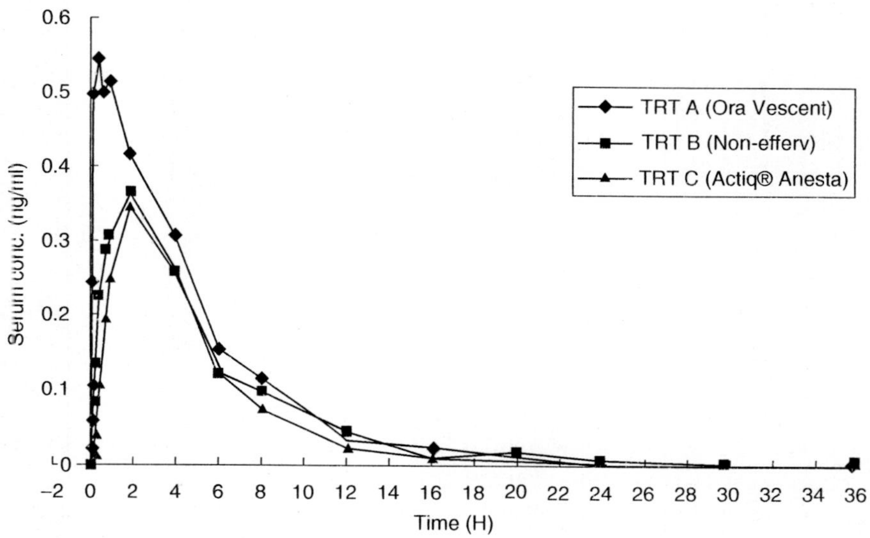

Figure 12.3 Fentanyl bioavailability after buccal administration.

Table 12.10 Fentanyl Human Pharmacokinetics

	OraVescent™ (A)			Nonproprietary Buccal Tablet[a] (B)			Actiq® (C)
PK Parameter	Result	Ratio A:C	Significance	Result	Ratio B:C	Significance	Result
T_{max} (hr)	0.696	0.362	S[b]	1.627	0.846	NS[c]	1.923
C_{max} (ng/mL)	0.641	1.575	S[d]	0.399	0.980	NS[e]	0.407
AUC(0–t) (ng. hr/mL)	2.656	1.468	S[d]	2.041	1.128	NS[e]	1.809

[a] Similar to OraVescent™ in tablet size, shape, and formulation, except that absorption-enhancing components were replaced with lactose.

[b] Difference statistically significant using nonparametric analysis ($p \leq 0.005$).

[c] Difference not statistically significant using nonparametric analysis ($p > 0.005$).

[d] Difference statistically significant using ANOVA ($p \leq 0.05$).

[e] Difference not statistically significant using ANOVA ($p > 0.05$).

to either the unenhanced tablet (B) or the commercial dosage form (C). A comparison of the enhanced and unenhanced formulations (A versus B) indicates that the enhanced delivery system functioned as intended; that is, effervescence promoted absorption. The following comparisons can be made of fentanyl pharmacokinetics after OraVescent™ versus Actiq® administration.

1. Fentanyl peak serum levels are higher (0.6 ng/mL compared to 0.4 ng/mL).
2. Systemic fentanyl bioavailability is more complete (the AUC is approximately 1.5 times as high).
3. Fentanyl is more rapidly absorbed (T_{max} is 0.7 H compared to 1.9 H).

In Figure 12.4, the serum levels obtained during the first 30 minutes are plotted on an expanded scale. These graphs clearly reveal the faster absorption of fentanyl from the OraVescent™ formulation during the initial stages. The rapid initial rise in fentanyl serum levels indicates that the delivery system has the potential to provide quicker onset of pain relief. This may be useful in the following situations.

1. Treatment of breakthrough cancer pain.
2. Adjunct therapy during the initial stages of fentanyl transdermal administration when the full effect of the transdermal system is not yet evident.
3. During dose titration of patients on fentanyl transdermal therapy.
4. Low dose fentanyl administration for the conscious sedation of patients to facilitate endoscope insertion and other procedures in which the cooperation of the patient is needed. At present, fentanyl injection in combination with other drugs, such as midazolam, may be used for this indication (21).

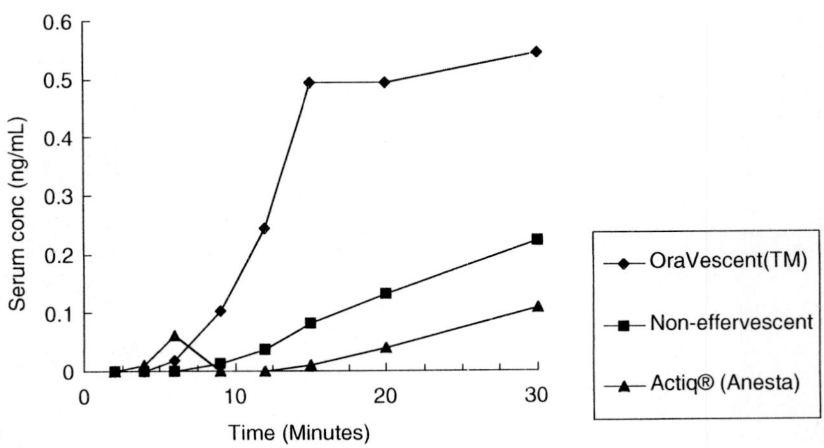

Figure 12.4 Early phase fentanyl serum levels.

A family of patent applications has been filed to cover this technology. Notices of Allowance have been received in respect to two of these applications, one pertaining to intraoral drug delivery and another to site-specific delivery in the gastrointestinal tract.

REFERENCES

1. Mattar ME, Markello J, Yaffe SJ. Pharmaceutic factors affecting pediatric compliance. Pediatrics 1975; 55:101–108.
2. Boreham PFL, Benrimoj S, Ong M, Craig M, Shepherd RW, Hill D, Singleton B. Aust J Hosp Pharm 1986; 16:138–142.
3. Fincham JE. Patient compliance in the ambulatory elderly: A review of the literature. J Geriatric Drug Therapy 1988; 2 (4):31–52.
4. Hurd PD, Butkovich SL. Compliance problems and the older patient: Assessing functional limitations. Drug Intell Clin Pharm 1986; 20:228–231.
5. Goldfarb J, Patel J, Wormser G. Antimicrobial flavors: A consideration in drug compliance. Drug Therapy 1986; 2:59–71.
6. Weinstein AG. Clinical management strategies to maintain drug compliance in asthmatic children. Ann Allergy Asthma Immunol 1995; 74:304–310.
7. Fennell RS, Foulkes L-M, Boggs SR. Family-based program to promote medication compliance in renal transplant children. Transplantation Proceedings 1994; 26:102–103.
8. Habib W, Khankari R, Hontz J. Fast-dissolve drug delivery systems. Crit Rev Thera Drug Carrier Syst 2000; 17:61–72.
9. Wehling F, Schuele S, Madamala N. Effervescent dosage form with microparticles. US Patent 5178878, 1993.
10. Khankari RK, Hontz J, Chastain SJ, Katzner, L. Rapidly dissolving robust dosage form. US Patent 6024981, 2000.
11. Seager H. Drug-delivery products and the Zydis® fast-dissolving dosage form. J Pharm Pharmacol 1998; 50:375–382.
12. Cooper WT. Improvements in preparing or making up medicated and other effervescing mixtures. British patent 3160.
13. Eichman JD, Robinson JR. Mechanistic studies on effervescent-induced permeability enhancement. Pharm Res 1998; 15:925–930.
14. Martin A. Physical Pharmacy. 4th ed. Philadelphia: Lea and Febiger, 1993.
15. Kietzmann D, Hahne D, Crozier TA, Weyland W, Groger P, Sonntag H. Comparison of sufentanil-propofol-anesthesia with fentanyl-propofol in major abdominal surgery. Anaesthesist 1996; 45:1151–1157.
16. Yee LY, Lopez JR. Transdermal fentanyl. Ann Pharmacother 1992; 26:1393–1399.
17. Southam MA. Transdermal fentanyl therapy: System design, pharmacokinetics and efficacy. Anticancer Drugs 1995; 6 (Suppl 3):29–34.
18. Jeal W, Benfield P. Transdermal fentanyl. A review of its pharmacological properties and therapeutic efficacy in pain control. Drugs 1997; 53:109–138.
19. Streisand JB, Busch MA, Egan TD, Smith BG, Gay M, Pace NL. Dose proportionality and pharmacokinetics of oral transmucosal fentanyl citrate. Anesthesiology 1998; 88:305–309.

20. Payne R. Factors influencing quality of life in cancer patients: The role of transdermal fentanyl in the management of pain. Semin Oncol 1998; 25 (No 3, Suppl 7): 47–53.
21. Claussen DS. Fentanyl citrate (fentanyl sublimaze). Gastroenterol Nurs 1993; 16:85–87.

13

Process Development and Scale-Up of Oral Fast-Dissolving Tablets

Srikonda V. Sastry and Janakiram Nyshadham

CONTENTS

I. INTRODUCTION TO ODTS

Oral administration is by far the most popular route of drug administration due to ease of ingestion, no or minimal pain in swallowing, versatility to accommodate drug molecules with varied physicochemical characteristics, and most important, patient compliance (1). Also solid oral dosage forms do not need to be sterilized, therefore, they are less expensive to manufacture compared to parenteral dosage forms.

Several novel oral drug delivery systems are currently available and many others are in various stages of development useful in:

- Maximizing drug bioavailability
- Increasing patent life
- Conveniently delivering the drug
- Enhancing patient compliance (2)

As a result, several drug delivery companies have products on the market with a projected value of several billion dollars annually, reaching a revenue of \$38 billion in 2002 (3). Orally disintegrating tablets (ODT) are one such drug delivery technology that has become a popular dosage form, capturing a market value of \$1.1 billion worldwide (4). This chapter focuses on ODTs with specific emphasis on reviewing the technological and innovative concepts, manufacturing process, and scale-up considerations of selected commercialized ODT technologies with a brief overview of the regulatory considerations for such dosage forms.

II. TECHNOLOGY OVERVIEW

Oral fast disintegrating tablet dosage forms are also known as fast-dissolve, rapid-dissolve, rapid-melt, and quick-disintegrating tablets. The common definition of these dosage forms is: "a solid dosage form that dissolves or disintegrates quickly in the oral cavity in the presence of saliva resulting in a solution or suspension without the need of administration of water" (2).

The center for Drug Evaluation and Research states an ODT to be "a solid dosage form containing medicinal substances, which disintegrates rapidly, usually within a matter of seconds, when placed upon the tongue" (5).

The ODTs possess unique characteristics addressing several patient compliance issues and in particular offer improved alternatives to administration of conventional oral dosage forms (i.e., tablets and capsules) in the following conditions.

- Dysphagia, difficulty in swallowing, which is common among all age groups and in swallowing conventional tablets and capsules (6). As much as 35 percent of the general population, and an additional 30 to 40 percent of elderly institutionalized patients, and 18 to 22 percent of patients in long-term care facilities suffer from dysphagia (7).

- Medical conditions such as stroke, Parkinson's disease, autoimmune deficiency disease (AIDS), gastrointestinal reflux disease (GERD), surgical conditions such as thyroidectomy, head and neck injury, and some neurological disorders such as cerebral palsy are associated with dysphagia (8,9).
- In a study designed to understand problems associated with dosage form administration, 26 percent out of 1576 patients experienced difficulty in swallowing tablets and a common complaint was tablet size, surface area, form, and taste (9).
- Problem of swallowing is more evident in geriatric and pediatric patients and patients who are traveling and may not have access to water (9).

ODTs have also been shown to provide the following advantages.

- Suitable for administration of drug to those patients who cannot swallow (e.g., elderly, stroke victims, healthcare facility and bedridden patients), patients who should not swallow (e.g., renal failure situations), and patients who will not swallow (e.g., pediatric, geriatric, and psychiatric patients) (10,11).
- Useful in situations where rapid drug therapy intervention is required (10).
- Offer convenience and patient compliance (e.g., disabled, bedridden patients) and for active and busy people who do not have access to water. They also can be swallowed without the need for water intake (11).
- No capsule or tablet to swallow because ODTs form a solution/suspension within several seconds when placed on the tongue and accurate dosing as compared to liquid dosage forms (10).
- Improved and rapid drug absorption upon administration. This is evident in at least one bioequivalency study (e.g., selegiline) through pregastric absorption from the mouth, pharynx, and esophagus (11,12).

ODT as a novel drug delivery technology also offers new business opportunities such as product differentiation, line extension with added product value, exclusivity of product promotion, and patent life extension of the pharmaceutical active ingredient (11,12).

A. Commercial ODT Technologies

Although many ODT technologies are currently available, only a few have reached the commercial market. The available ODT technologies are summarized in Table 13.1, and can be classified according to differences in their manufacturing process. The excipients of choice for the preparation of several types of ODTs share some commonalities in characteristics such as water solubility, sweet taste, low viscosity even at high concentrations, and compressibility. However, each technology requires distinct manufacturing methods to achieve fast disintegration of the dosage form in water (10–13).

B. ODT Concepts, Formulation, and Mechanisms

The following sections provide a brief discussion of selected ODT technologies and the underlying approach of manufacturing. The principal ODT technologies are classified based on their manufacturing process and include conventional tablets, freeze-dried tablets, and floss technology.

1. Conventional Tablet Process

ODT technologies based on conventional tablet processes include WOWTAB®, Orasolv®, Efvdas®, Flashtab®, and Advantab® and are summarized in Table 13.1.

Table 13.1 Orally Disintegrating Tablet Technologies

Technology	Company
I. Conventional Tablet Process	
WOWTAB®	Yamanouchi Pharma Technologies, USA
OraSolv®	Cima Labs, USA
Efvdas®	Elan corporation, USA
Flashtab®	Prographarm, France
Advantab®	Eurand America, USA
II. Freeze-Drying Process	
Zydis®	Cardinal Health, USA
Lyoc®	Farmalyoc, France
Quicksolv®	Janssen Pharmaceutica, USA
III. Floss Formation Process	
Flashdose®	Fuiz Technologies, USA

The WOWTAB® technology from Yamanouchi Pharma features conventional tablet characteristics for handling and packaging, and the fast-disintegrating characteristics upon contact with water are maintained (14). The technology is based on a new physically modified combination of polysaccharides possessing water dissolution characteristics aiding fast disintegration and high compressibility characteristics resulting in fast-disintegrating tablets having adequate hardness for handling purposes. The manufacturing process involves granulation of polysaccharides such as mannitol, lactose, glucose, sucrose, and erythritol that show quick-dissolution characteristics (also known as low-moldable sugars) with polysaccharides such as maltose, sorbitol, trehalose, and maltitol (also known as high-moldable sugars) resulting in a mixture of excipients possessing fast-dissolving and high-moldable characteristics The drug can be added during the granulation or subsequently during the blending process with other standard tableting excipients. The tablets are manufactured at a low compression force followed by humidity treatment to increase tablet hardness (14). The manufacturing process is shown schematically in Figure 13.1.

OraSolv® is yet another ODT technology manufactured using a conventional tableting process. The technology involves direct compression of effervescence excipient, active, and taste-masking agents (15). Quick disintegration of the tablet occurs due to the production of effervescent carbon dioxide upon contact with moisture. The effervescence excipient known as effervescence couple in this technology is prepared by coating the organic acid crystals using a stoichiometrically less amount of base material. The particle size of the organic acid crystals is carefully chosen to be larger than the base excipient for uniform coating of the base excipient onto the acid crystals. The coating process is initiated by the addition of a reaction initiator, in this case purified water. The reaction is allowed to proceed only to an extent to completion of base coating on organic acid crystals. The required endpoint for the reaction termination is determined by measuring CO_2 evolution. This excipient is then mixed with active ingredient or active microparticles along with other standard tableting excipients and compressed into tablets (16–18). The schematic representation of the manufacturing process is shown in Figure 13.2 and a typical process is shown in Figure 13.3.

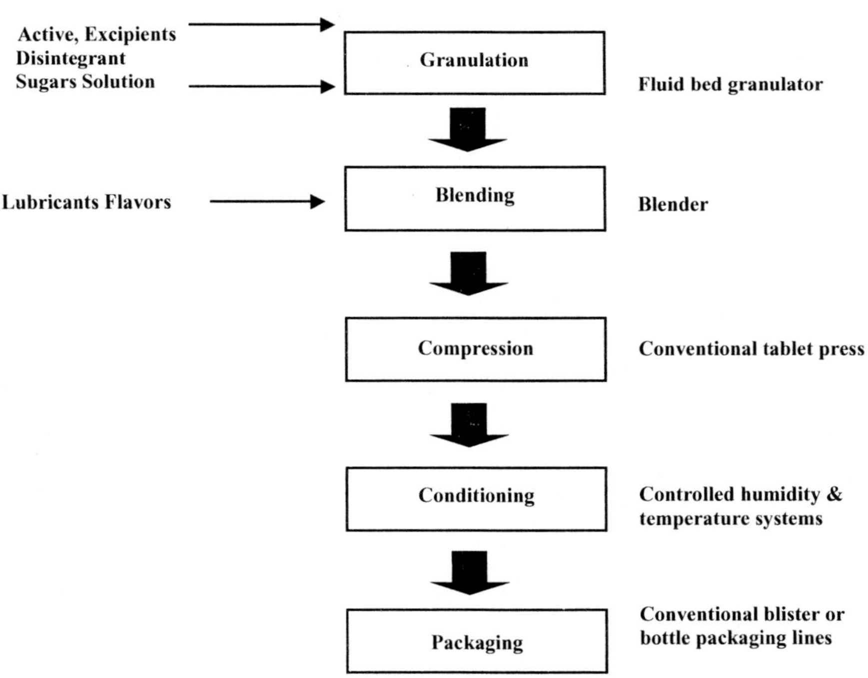

Figure 13.1 WOWTAB® schematic manufacturing process.

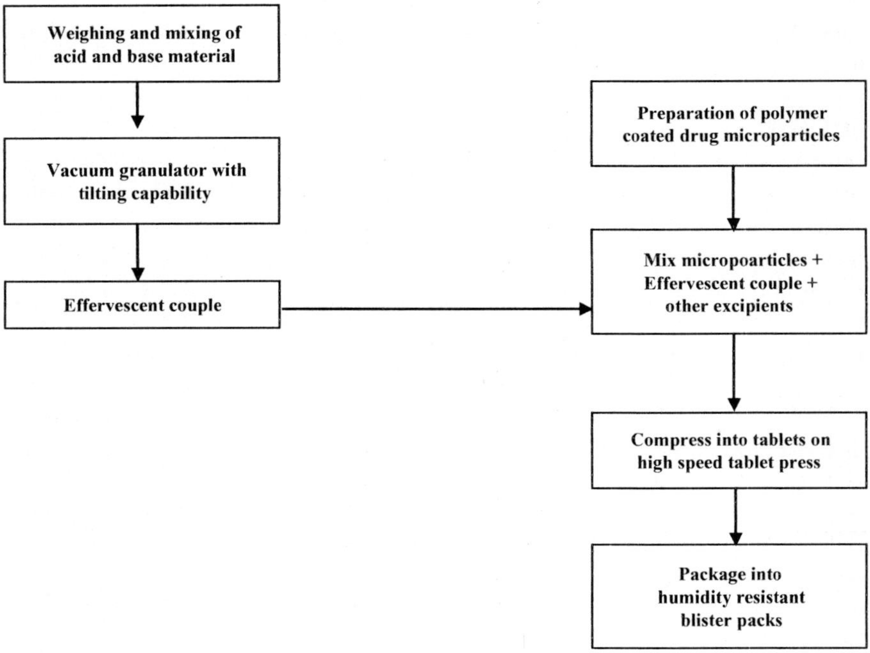

Figure 13.2 OraSolv® schematic manufacturing process.

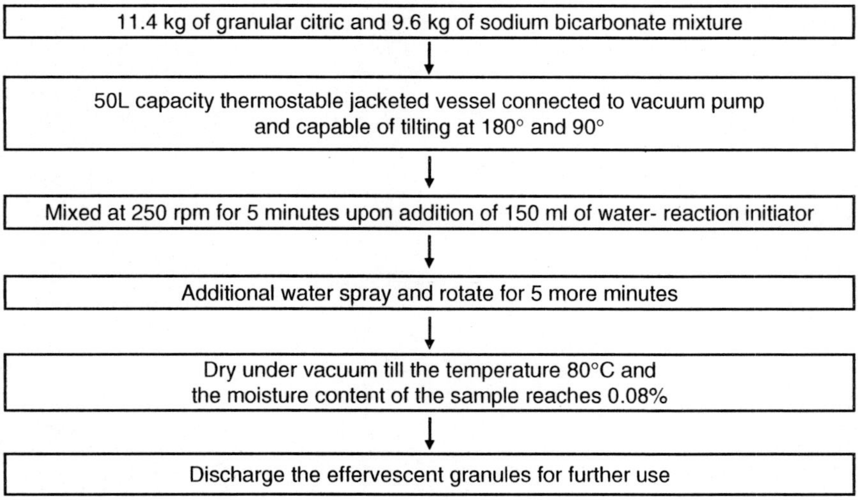

Figure 13.3 OraSolv® schematic manufacturing process for effervescent couple.

2. Freeze-Drying Process

Zydis® from Cardinal Health is one of the first generation of ODTs that uses a freeze-drying manufacturing process. The Zydis® technology is the most successful among all ODT technologies having numerous commercially marketed products including lorazepam, piroxicam, loperamide, loratidine, enalpril, and selegiline (11,13). The freeze-drying process involves removal of water by sublimation upon freeze-drying from the liquid mixture of drug, matrix former, and other excipients filled into preformed blister pockets. The formed matrix structure is very porous in nature and rapidly dissolves or disintegrates upon contact with saliva (11,19). The Zydis® manufacturing process is presented in Figure 13.4. This basic process has undergone several modifications to accommodate drugs with different physicochemical characteristics, drug loading, and particle size, and matrix modifications to result in an acceptable dosage form (20–26).

The Zydis® technology requires specific characteristics for drug candidates and matrix-forming material. Drug loading for water-insoluble drugs approaches 400 mg and the upper limit for water-soluble drugs is approximately 60 mg. The ideal drug characteristics for insoluble drugs are relative water insolubility with a fine particle size and aqueous stability in the suspension. The water-soluble drugs pose different formulation challenges as they form eutectic mixtures, resulting in freezing-point depression and the formation of a glassy solid that might collapse on drying because of loss of the supporting structure during the sublimation process (11,13,23). Matrix-forming excipients such as mannitol can induce crystallinity and hence impart rigidity into the amorphous composite. This characteristic may help prevent the collapse of the matrix structure during the

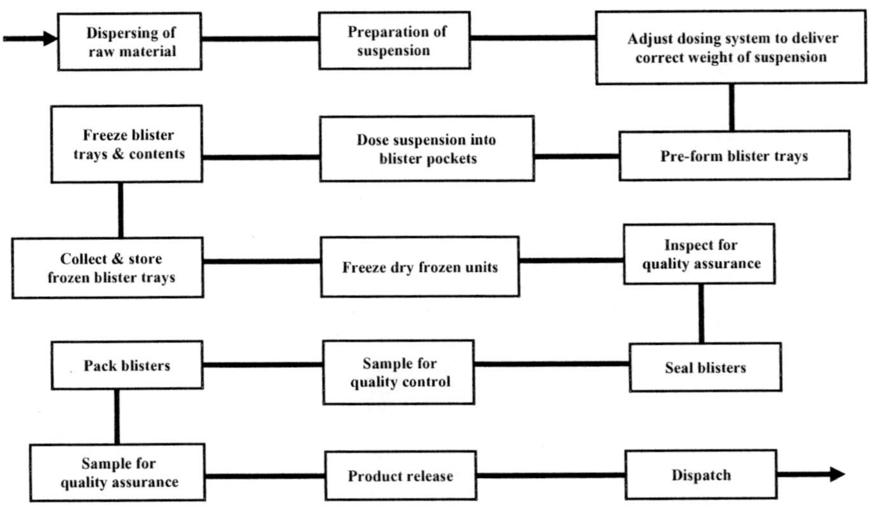

Figure 13.4 Zydis® schematic manufacturing process.

freeze-drying process. The formulation of highly water-soluble drugs into freeze-dried wafers was achieved by complexing them onto ion exchange resins. This technique also helps achieve taste masking of medicaments in this technology (13,27). The appropriate particle size for insoluble drugs is about 50 microns, however, drugs with larger particle size can also be formulated using suitable suspending agents such as gelatin, and flocculating agents such as xanthan gum (13,26). In yet another modification, the solutions of soluble drug can be sprayed onto a preformed matrix, with subsequent evaporation of solvent (13,21).

Matrix formation and its characteristics are equally important for freeze-drying technology. The common matrix-forming agents are gelatin, dexran, or alginates as glassy amorphous mixtures providing structural strength; saccharides such as mannitol or sorbitol to provide crystallinity, hardness, and elegance; and water as a manufacturing process medium to induce the porous structure upon sublimation during the freeze-drying step. Besides, the formulation can contain taste-masking agents such as sweeteners, flavorants, pH-adjusting agents such as citric acid, and preservatives to ensure the aqueous stability of the suspended drug in media prior to sublimation. Finally, the freeze-dried formulations are manufactured and packaged in PVC or PVDC plastic packs, or may be packed into Aclar laminates or aluminum foil preparations to protect the product from external moisture (11,13). A typical processing example using Zydis® technology is presented in Figure 13.5.

3. Floss Technology

The floss technology, also known as cotton candy process, forms the basis of the Flashdose® technology (Biovail, Mississauga, ON, Canada). A fast-dissolving

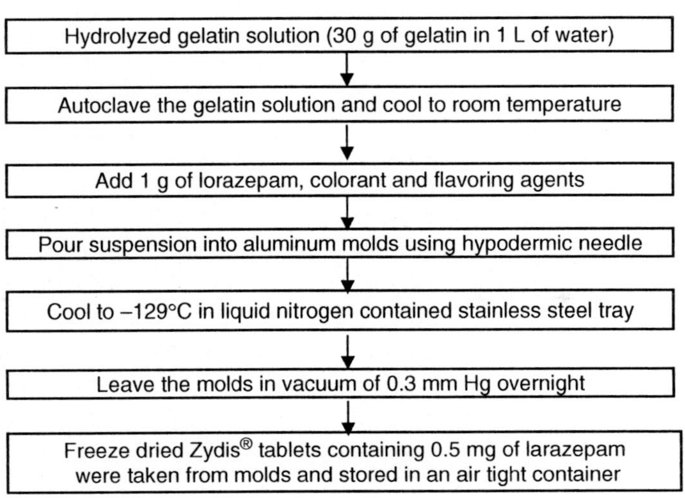

Figure 13.5 Zydis® typical manufacturing process example.

tablet is formed using a cotton candy known as floss or shearform matrix (3,28). The floss is commonly made of saccharides such as sucrose, dextrose, lactose, and fructose. The saccharide is converted into floss by a simultaneous action of flash-melting and centrifugal force in a heat-processing machine similar to a cotton candy machine (29–35). The produced fibers are usually amorphous in nature. These fibers are then partially recrystallized which finally results in a free-flowing floss (3,28). The floss is then mixed with an active ingredient and other excipients followed by compression into a tablet possessing fast-dissolving characteristics (29–35).

III. PRODUCT QUALITY AND DESIGN APPROACHES IN DEVELOPING ODT

A. PQD Overview

1. Philosophy behind PQD

Quality is often defined as "the totality of features and characteristics of product or service that bear on its ability to satisfy stated or implied needs." In other words, the definition of quality is "fitness for use" or how well the user's (i.e., customer's) needs are met. Two aspects of product quality that are critical to the understanding of important product and process performance factors are as follows.

- Quaiity of Design (QOD): QOD pertains to product functionality, the characteristics of the product, and its potential for achieving highest quality of conformance. This aspect of quality addresses product suitability and product consistency that the user demands or expects.
- Quality of Conformance (QOC): QOC pertains to the uniformity of product characteristics with respect to user demands.

Table 13.2 explains the functional areas and application of QOD and QOC for achieving, maintaining, and improving product quality.

2. Benefits of PQD

Typically both QOD and QOC tools are used in design and conformance, respectively. QOD tools can be creatively used during product design and development to build quality and quantify the quality and product robustness. QOC tools can be applied to recommend manufacturing limits and in-process controls for future manufacturing of Phase 3 batches, primary stability batches, or process validation batches using historical process development data. Additionally, QOC can be used to assess and quantify the ability of the manufactured product to meet preestablished and preapproved controls and product specifications. QOD and QOC can be combined to develop and implement a systematic approach for engineering and quantifying quality into product. This overall PQD approach in product development programs would enable a cross-functional, broad-based, and multidisciplinary team to:

- Achieve better product attributes
- Reduce development times
- Establish product robustness
- Enable high quality prediction
- Quantify quality
- Maximize benefit-to-cost ratio for experimental effort

Table 13.2 Functional and Application Areas of QOD and QOC

Quality of Design (QOD)		Quality of Conformance (QOC)	
Functional Area	Examples	Functional Area	Examples
Basic research	Identify critical factors	Process control	Identifying and controlling control knobs, applying statistical process control (SPC)
Product development	Optimizing product design	Sampling and testing	Design of sampling protocols, OC curves, AQLS
Process design	Optimizing process design, determining control adjustments	Measurement methods	Optimizing instrument and measurement procedures
Product/process improvements	Trouble-shooting experiments, process characterization studies	Product release	SPC and process capability

B. Quality of Design

1. Input–Output Relationships

The science- and technology-driven knowledge base and understanding of input–output relationships in product performance to meet the intended design are integral parts of QOD. The schematic representation shown in Figure 13.6 depicts a systems approach to identifying the effect of variables on output product response.

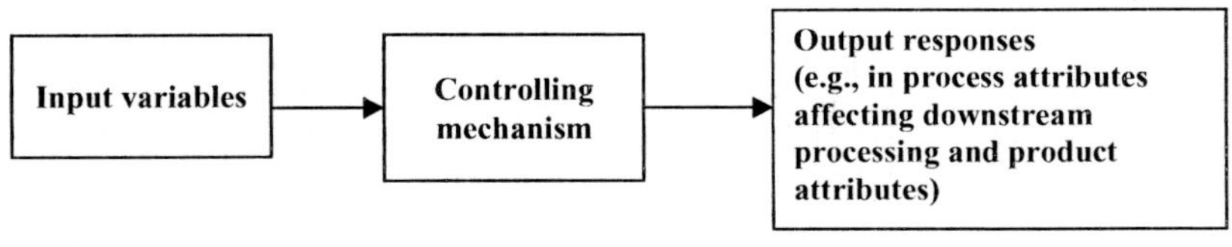

Figure 13.6 Input–output relationship: systems approach.

This systems approach of learning input–output relationships, and thereby engineering quality to characterize and maximize the design features for the product to meet the intended needs, can be reduced to practice in critical process stages of solid oral products such as granulation, drying, milling, blending, compression, film coating, and so on. Specifically, Figure 13.7 illustrates the application in film coating. It is evident that to achieve appropriate coating attributes such as coating elegance, surface roughness, and the like, input variables such as thermodynamics of film-coating conditions are required to be controlled to result in appropriate mechanical characteristics of the film-forming material. Understanding the relationship between input and output variables is critical to process development and product optimization.

2. Design of Experiments

The design of experiments (DOE) process provides quality and productivity thereby achieving better product characteristics with decreased development time and sensitivity to manufacturing conditions. The input variables, also known as independent variables (*X*s), can be altered and controlled to achieve a desired change of an output variable, also known as dependent variables (*Y*s), to form the basis for the DOE. A deliberate process of experimentation follows where data is collected in order to identify the relationship between dependent and independent variables and to develop mathematical models to predict/estimate system behavior. The polynomial equations generated during this process also help confirm the model validity.

The design usually consists of diagnosis of the experimental environment, which includes identification of controllable variables, size of the variable effects to be detected, properties of variables, the variable settings, and the number of experimental runs. Following the design and execution of experimentation the

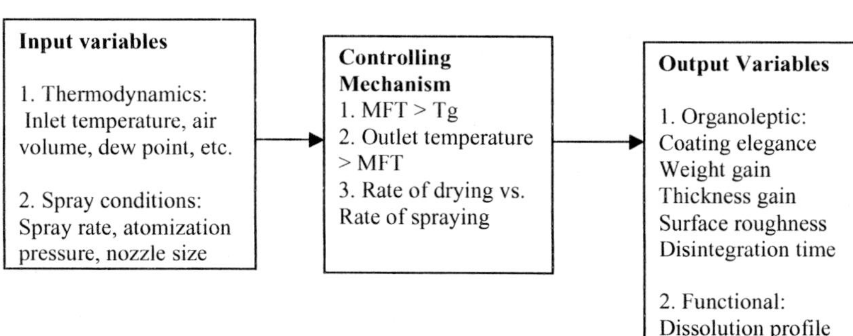

Figure 13.7 Input–output relationship: pharmaceutical coating.

data collection and data analysis are done to generate the models. Finally, the analyzed data is then optimized using appropriate optimization methods. The experimental designs can be first-order designs such as simplex and Plackett–Burman screening designs, and second-order designs such as factorial and Box–Behken designs. Optimization can be done either by simple inspection, Lagrangian method, or by optimizing (i.e., maximizing or minimizing) the polynomial equations. In the design of WOWTAB® ODT the DOE is employed to achieve product quality attributes.

C. Quality of Conformance

Although several statistical tools are available to characterize, understand, and implement QOC as part of PQD approaches, this particular section addresses process capability due to limited availability of references for process capability in the pharmaceutical literature. Specifically, philosophy, definitions, terminology, computational approach, assumptions, and applications of process capability are covered.

1. Overview of Statistical Process Capability Analysis

Statistical process control (SPC) has been utilized in a variety of industries for many years for establishing that a manufacturing process is in a state of control (36). Pharmaceutical manufacturing environments have utilized control charts to ensure that product process parameters and outputs are uniform and in control (37). However, the use of statistical process capability analysis in quantifying manufacturing process performance and in demonstrating capability of the process to meet manufacturing limits and in-process controls for critical attributes have been fairly limited in the pharmaceutical industry.

Statistical process capability measures the inherent reproducibility of the product produced by the manufacturing process (36). It is a measure of a process being adequate and capable of meeting product specifications. The quality of a manufacturing process can be understood by quantifying the process performance and variation. The process mean (μ) and standard deviation (σ), as estimated by x (sample mean) and s (sample standard deviation), are not unitless and sometimes are not convenient summary statistics (38).

Capability indices can be used to relate the process parameters μ and σ to specification requirements. In addition, process capability indices are unitless and provide a basis for quantifying the performance of a process (38). Although a standard definition for process capability analysis has not emerged, the primary objective is to determine how well the output from a process meets specification limits. Capability indices can be used to effectively summarize process information in a succinct manner. The indices C_p, C_{pu}, C_{pl}, K, and $C_{pk,}$ which are defined below, form a group of complementary measures that comprise a convenient unitless system and collectively determine whether the process has sufficiently low variability to meet product/process specifications (38).

2. Definitions, Terminologies, and Computational Approaches

Process Range (6σ): The range of the process is computed by 6σ where σ is standard deviation. For $\pm 3\sigma$ limits, based on a normal distribution, it can be estimated that 99.73 percent of the data will fall within the limits (38).

Specification Range (d): The specification range (d) is the difference between the upper specification limit (USL) and the lower specification limit (LSL) (38).

Potential Capability (C_p): The potential capability (C_p) is the simplest and most straightforward indicator of process capability. It is defined as the ratio of the specification range to the process range. Computationally,

$$C_p = (\text{USL} - \text{LSL})/6^* \sigma.$$

This ratio expresses the proportion of the range of the normal curve that falls within the specification limits. A C_p of 1.0 indicates that a process is judged to be capable. A capable process with an underlying stable normal distribution will in theory result in 0.27 percent of product beyond the specification limits (38). Due to sampling variation and machine testing limitations, C_p = 1.0 is generally not used as a minimally acceptable value. A minimum value of C_p = 1.33 resulting in a very low rejection rate of 0.007 percent is generally used for determining the capability of the process (38,39).

Upper and Lower Capability Index (C_{pl} and C_{pu}): A major shortcoming of the C_p index is that it may yield erroneous information if the process is not on target, that is, if it is not centered. C_{pl} and C_{pu} are the standardized values of normal distribution for the deviation of the observed mean from the LSL and USL (38,40). Computationally,

$$C_{pl} = (\text{Mean} - \text{LSL})/3\sigma \quad \text{and} \quad C_{pu} = (\text{USL} - \text{Mean})/3\sigma.$$

Noncentering Correction (K): The noncentering correction factor (K) expresses the noncentering of the process. It is a reflection of the deviation of the process mean from the midpoint of specification limits (38,40). Computationally,

$$K = \text{Abs (Target-Mean)}/1/2 \text{ (USL–LSL)}.$$

Performance Index (C_{pk}): The performance index (C_{pk}) is also referred to as demonstrated excellence. The C_p index involves only the process spread as related to the specification limits; the location of process mean is not considered. The C_{pk} index is related to the C_p index, but utilizes the process mean and can be considered a measure of process performance (38). It can be computed in two ways with the same results as follows:

$$C_{pk} = \text{Min } (C_{pu}, C_{pl}) \quad \text{or} \quad C_{pk} = (1\text{-K}) * C_p.$$

If the process is perfectly centered, then $K = 0$ and $C_{pk} = C_p$. However, as the process drifts from the target specification, K increases and C_{pk} becomes smaller than C_p (40–42).

3. Interpretation of C_{pk}

As shown in Table 13.3 (43), at a C_{pk} value of 1.33 only 30,000 units/billion units (0.003 percent) would fall outside the specified acceptable limits.

C_{pk} for product attributes such as assay, content uniformity, weight, hardness, and so on should be targeted at a minimum value of 1.33. To consistently achieve a C_{pk} of 1.33 during routine production, C_{pk} values of greater than 1.33 should be obtained in process validation studies.

Table 13.3 Interpreting C_{pk}

	Units Outside of Specifications	
C_{pk}	Billion	Percentage
0.5	70,000,000	7
1.0	1,300,000	0.13
1.33	30,000	0.003
1.67	1,000	0.0001
2.0	1	0.0000001

4. Assumptions for Process Capability

As in any methodology, process capability also has some associated assumptions. These assumptions can be summarized as follows.

- The manufacturing process is in a state of statistical process control.
- The data is normally distributed.
- The data is collected from independent random samples.
- The data is truly representative of the manufacturing process.

5. Applications of Process Capability in Pharmaceutical Development

A few areas in the process development of pharmaceutical products, where the process capability tool can be applied to characterize and quantify quality, can be summarized as follows.

- Generate rationale for and establish in-process control limits based in historical data.
- For those attributes, such as content infirmity, where the limits are governed by pharmacopies, one can develop design criteria during the development phase. For example, for stage 1 content uniformity per USP, tablets should conform to 85 to 115 percent limits in order to achieve $C_{pk} > 1.33$.

- Standard deviation (SD) < (115–85)/(1.33*6) or less than 3.76.
- Because a formulation is expected to provide 100 percent of label claim a priori, a formulation scientist or process engineer should aim to get an RSD that is far less than 3.76 percent during development trials.
- Process validation: Because process capability is highly statistical in nature, a high degree of assurance would be achieved in ensuring the product attribute conforms to its preapproved limit. This "high degree of assurance" is a key requirement of process validation mandated by the FDA and necessitated by the development objectives of the development team.
- Assess process performance, predictability, and process improvements.

D. Case Studies for PQD: Developing WOWTAB® ODT

Use of QOD and QOC as part of overall PQD features in the development, optimization, and scale-up for WOWTAB® ODT formulation is dealt with in detail in this section. The overall product development involves use of QOD and QOC features in the development and product with optimal and consistent quality.

1. General Formulation Considerations in WOWTAB® ODT

According to WOWTAB® technology, saccharides either possessed fast dissolution characteristics or good hardness upon compaction (44,45). Table 13.4 shows the compression and hardness characteristics of commonly available saccharides.

Therefore, by granulating low-moldable sugar with high-moldable sugar a new excipient possessing both fast dissolution and high-moldable characteristics is produced. When the material is compressed at low hardness followed by moisture conditioning the result is a tablet with increased hardness but possesses fast oral disintegrating characteristics.

Table 13.4 Compression[a] and Hardness Characteristics of Saccharides

Ingredients	Hardness (kp)	In Vivo Disintegration Time (sec)
Mannitol	0.0	<10
Lactose	0.0	<10
Glucose	0.2	<10
Sucrose	0.5	<10
Erythritol	0.0	<10
Xylitol	0.0	<10
Maltose	6.8	>30
Sorbitol	2.2	>30
Trehalose	3.4	>30
Maltitol	2.5	>30

[a] Compression force: 10 kg/cm^2.

A typical formulation composition of a placebo tablet and the processing stages at different scales are are shown in Table 13.5 and Figure 13.8.

In a typical manufacturing process of WOWTAB®, high-moldable sugar such as maltose dissolved in water at 0.5 to 1 percent w/w concentration is sprayed onto a low-moldable sugar such as mannitol in a fluid bed granulator. The granulation is dried to a percent loss on drying (LOD) value of 1 to 3 percent. The dried granulation is then blended with drug, colorants, flavorants, and other standard tableting excipients. The drug can also be incorporated into the formulation during the granulation stage. The resulting final blend is then compressed into soft tablets followed by moisture conditioning resulting in tablets with increased hardness.

2. Case Study for Applying QOD via DOE

The overall objective of the study was to optimize the moisture-conditioning process in the product and further to scale up to commercial scale. Studies performed to develop, optimize, and scale up other unit operations such as

Table 13.5 Placebo WOWTAB® ODT Composition and Tablet Characteristics

Ingredients	Quantity/Tablet (mg)
Mannitol	141.8
Maltose	7.5
Magnesium stearate	0.75
Total weight	150.0

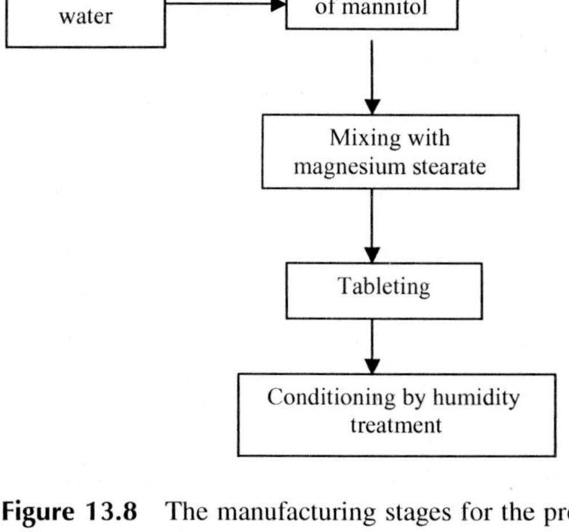

Figure 13.8 The manufacturing stages for the process.

granulation, drying, blending, and compression are not covered in this chapter as these process steps are typical of any solid oral dosage forms.

3. DOE Application at Pilot Scale

To achieve moisture-treatment optimization, in order to maximize the output response of tablet hardness, a fractional factorial (two-level) experimental design was employed. The design included optimization of seven process variables including three variables at humidification stage ($X_1 - X_3$) and four variables at the drying stage ($X_4 - X_7$), which are summarized in Table 13.6. The dependent variables were final hardness (Y_1) and final percent moisture (Y_2).

Table 13.6 Independent and Dependent Variables Used in Fraction Factorial Design

Independent Variables
Humidification State
X_1 = Humidity (%)
X_2 = Airflow (CFM)
X_3 = Time (min)
Drying Stage
X_4 = Temperature (°C)
X_5 = Humidity (%)
X_6 = Airflow (CFM)
X_7 = Time (min)
Dependent Variables
Y_1 = Final Hardness (Kp)

A 16-run, two-level fractional factorial design for the above variables was employed to obtain the statistical polynomial equation shown below.

Polynomial Equation for Final Hardness (Y_1):

$$Y_1 = 4.5 + 0.53X_1 - 0.10X_2 - 0.07X_3 - 0.49X_4 + 0.03X_5 - 0.19X_6 - 0.09X_7$$

$$R^2 = 0.98$$

The polynomial equation can be used to calculate the predicted response values. The model was then optimized to identify the optimum experimental domain that would maximize the hardness. The optimal conditions were then employed in experiments to confirm and validate the statistical model. Table 13.7 shows the results from a triplicate confirmation study and the corresponding predicted and actual values for tablet hardness.

The final hardness for the model confirmation studies was 5.3 kp to 5.6 kp and found to be within the predicted value of 5.54 kp, therefore establishing the validity of the statistical model employed (44).

Table 13.7 Optimization Confirmation

	Final Hardness (kp) (Y_1)		
Trial	Predicted	Observed	Residual
Trial 1	5.54	5.60	−0.06
Trial 2	5.54	5.39	0.15
Trial 3	5.54	5.30	0.24

IV. COMMERCIAL SCALE-UP

Subsequently, the above process was scaled up from 15 kg to 120 kg scale, a commercial scale for this product. Scale-up of several critical process parameters such as spray rates, granulation conditions, blending times, and compression parameters are not discussed in this chapter as they are typical of any solid oral dosage form.

For the moisture-treatment process, the process equipment consists of a chamber that has a bottom plate with perforations. This is similar to the bowl of the fluid bed granulators that are equipped with bottom plates or meshes. The tablets to be subjected to moisture treatment are placed on top of the bottom plate inside the chamber of the treatment equipment. The schematic of bottom plate diagrams is presented in Figure 13.9 along with parameters used in the calculation of air volume and process times.

During scale-up, only air volumes and process times were needed to be calculated to reflect larger batch sizes and bigger process equipment at commercial scales. Other input variables for the moisture treatment steps, such as inlet temperature, inlet humidity in each of the two phases—humidification and drying—were identical to the optimized values identified via DOE at pilot scale.

Parameters such as cross-sectional surface area of the openings and size and shape of the opening at each of the two scales—pilot and commercial—for the calculation of air volumes and process times were constant.

The process-engineering basis for calculating air volumes and process times is described as follows.

$$\text{Air Volume}_c = \text{cfm}_p \times (D_c/2)2/(D_p/2)2$$

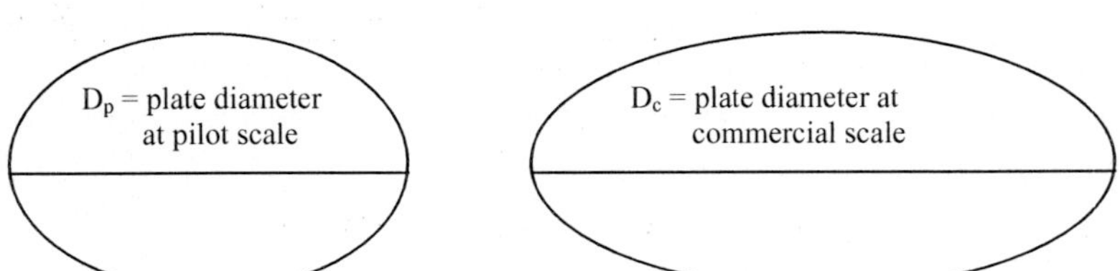

Figure 13.9 Schematic diagram for bottom plate.

$$\text{Process Time}_c = (\text{cfm} \times \text{time/BS, in kg})_p \times (\text{BS in kg/cfm})_c$$

Where cfm_p = air volume in cubic feet
D_p = bottom plate diameter at pilot scale
D_c = bottom plate diameter at commercial scale
BS = batch size
p and c refer to pilot and commercial scales.

Commercial process scale-up trials were undertaken for the moisture treatment step using the calculated values for air volumes and process times from these process-engineering equations. Tablets were obtained at random from several locations of the treatment chamber and tested for hardness. The obtained final hardness values (sample size of $n = 150$/batch) from the commercial scale-up batches are as follows.

Batch 1: 6.1 ± 0.22 kp;
Batch 2: 5.9 ± 0.23 kp;
Batch 3: 6.0 ± 0.13 kp.

These hardness values were congruent to the maximized predicted hardness values from pilot scale (Table13.7) thus validating the process-engineering basis for scaling up the moisture treatment process.

Hence the DOE approach, in an attempt to engineer quality:

- Provided the basis for developing high-quality predictive power though generation of polynomial equations. These equations enabled understanding of input–output relationships.
- Identified the experimental domain that led to maximum tablet hardness at pilot scale.
- Incorporated the optimal experimental domain when combined with process engineering that led to successful scale-up efforts.
- Assisted in scaling up of optimal process parameters from pilot scale to commercial scale.

A. Case Study for QOC via Process Capability

The data for the process capability evaluation was obtained at batch size of 15 kg (pilot scale) and 120 kg (commercial scale). Pilot scale data came from 3 batches with 5 samples per batch and 6 tablets per sample (overall $n = 90$). Commercial scale data came from 4 batches, 10 samples, and 3 tablets per sample (overall $n = 120$). Process capability was applied for content uniformity stage 1 USP limits (i.e., 85 to 115 percent) for both scales. Figures 13.10 through 13.12 depict the successful process scale-up showing the process capability parameters at required levels as explained in previous sections (44).

Generally the C_{pk} value for assay, content uniformity, weights, and other attributes should be targeted at 1.33. However, to achieve a C_{pk} value of 1.33

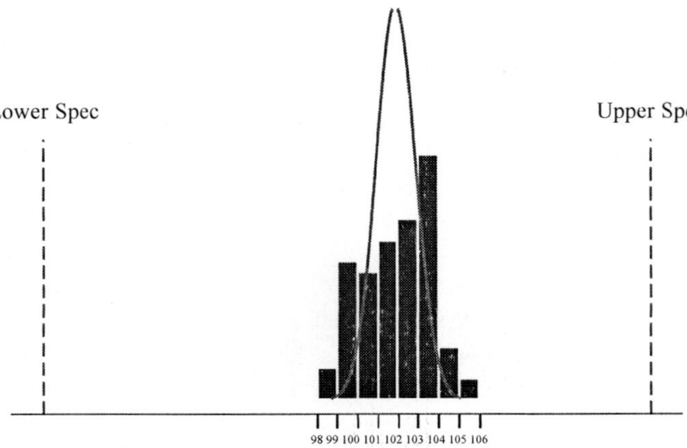

Cp	5.11	Targ	100.000	Mean	102.393	%>USL Exp	0.00	PPM>USL Exp	0
CPU	4.30	USL	115.000	Mean+3s	105.327	Obs	0.00	Obs	0
CPL	5.93	LSL	85.000	Mean-3s	99.459	%<LSL Exp	0.00	PPM<LSL Exp	0
Cpk	4.30	k	0.160	s	0.978	Obs	0.00	Obs	0
Cpm	1.68	n	90.000						

Figure 13.10 Pilot scale: content uniformity for compression run.

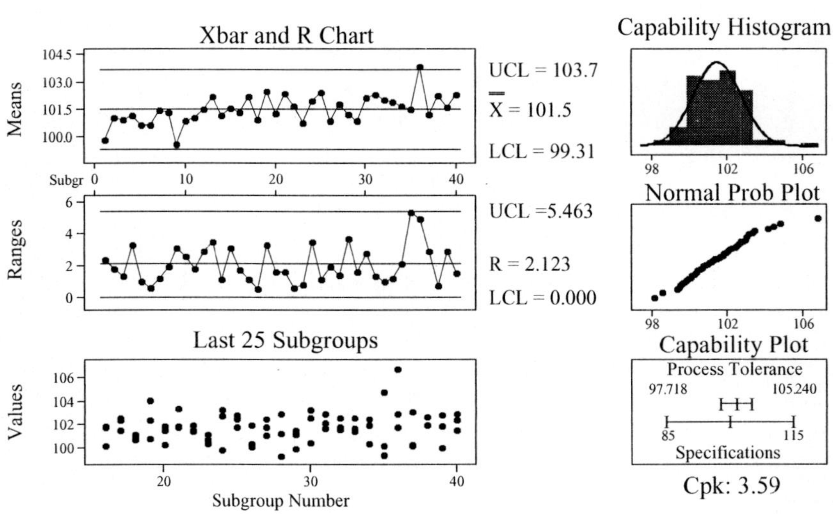

Figure 13.11 Commercial scale: content uniformity for compression run.

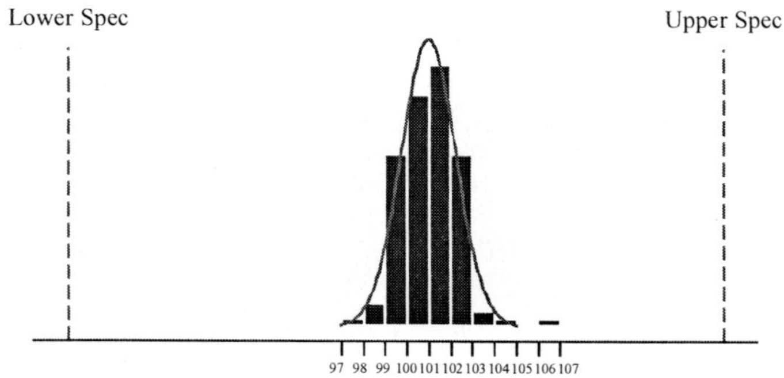

Cp	3.99	Targ	100.000	Mean	101.479	%>USL Exp	0.00	PPM>USL Exp	0
CPU	3.59	USL	115.000	Mean+3s	105.240	Obs	0.00	Obs	0
CPL	4.38	LSL	85.000	Mean-3s	97.718	%<LSL Exp	0.00	PPM<LSL Exp	0
Cpk	3.59	k	0.099	s	1.254	Obs	0.00	Obs	0
Cpm	2.55	n	120.000						

Figure 13.12 Commercial scale: content uniformity for compression run.

consistently during routine production, $C_{pk} > 1.33$ should be obtained during process development. In Figures 13.10 to 13.12 it is shown that the process is capable of meeting the Stage 1 content uniformity requirements (85 to 115 percent) at both pilot and commercial scales. In Figure 13.10 it is shown that the C_{pk} value for the content uniformity at pilot scale was 4.30 and greater than the required value of 1.33. Similarly, a greater C_{pk} value of 3.59 (Figure 13.12) was achieved during commercial scale-up operation. The capability histogram and the normal probability plots show that the content uniformity values are normally distributed.

B. Stability Testing of Scale-Up Product

The scaled-up product, from the previous case study, was also tested for stability in bottles and blisters. The product was analyzed for drug content along with critical product attributes including disintegration time and hardness. For the commercial scale product Table 13.8 shows the stability data.

As shown in Table 13.8, the critical product attributes such as drug assay, in vitro disintegration times, and final hardness did not show changes at ambient and accelerated stability conditions over a time period of six months. The stability analysis establishes that the quality of product design conformed to the expected product quality throughout product development including pilot scale and commercial scale development stages. This establishes the validity of the QOC and QOD concepts (11,45).

Table 13.8 Commercial Scale Drug Product Attributes in Stability Studies

Configuration	Month 0	Month 2	Month 4	Month 6
Drug Assay (% Label Claim) at Ambient Conditions (25°C/60%RH)				
Bottle	100.0	100.1	100.5	100.0
Blister	100.0	101.2	101.1	100.4
Drug Assay (% Label Claim) at Accelerated Conditions (40°C/75%RH)				
Bottle	100.0	100.1	100.9	100.1
Blister	100.0	100.7	101.8	100.7
Tablet Disintegration (sec) at Ambient Conditions (25°C/60%RH)				
Bottle	33	28	27	27
Blister	33	28	29	26
Tablet Disintegration (sec) at Accelerated Conditions (40°C/75%RH)				
Bottle	33	27	29	27
Blister	33	30	31	27
Tablet Hardness (kp) at Ambient Conditions (25°C/60%RH)				
Bottle	5.5	5.4	5.1	5.3
Blister	5.5	5.3	5.0	5.6
Tablet Hardness (kp) at Accelerated Conditions (40°C/75%RH)				
Bottle	5.5	5.3	4.9	6.4
Blister	5.5	5.0	5.0	6.5

V. REGULATORY CONSIDERATIONS AND GUIDANCE IN ODT PRODUCT DEVELOPMENT

In vivo studies are performed on ODTs to establish their behavior in the GIT, their pharmacokinetic specifications, therapeutic efficacy, and patient acceptability (46). In general, ODTs showed similar or improved pharmacokinetic behavior in comparison to conventional tablet dosage forms (47). Chapter 14 provides a detailed discussion on the clinical and biopharmaceutical aspects of ODT dosage forms.

For quick-dissolving ODTs a disintegration test may be used in lieu of a dissolution test if it is shown to be a discriminating method (46). If it is decided that a disintegration test alone is adequate from a biopharmaceutics point of view, then the issues of disintegration media, temperature, and pH have to be carefully evaluated. If it is decided that a disintegration test alone is inadequate from a biopharmaceutics point of view, then a dissolution test is also needed with careful consideration of dissolution method selection. Data generation and evaluation for both in vitro and in vivo disintegration times may be useful as an objective measure to classify the dosage form as an ODT. One such method may be use of a modified dissolution apparatus in evaluation of ODT dosage forms (48). In another case, a texture analyzer apparatus may be used to estimate tablet disintegration to compare against in vivo disintegration times (49).

It was also observed that use of a typical dissolution method conducted according to monograph conditions resulted in very fast dissolution of ODTs, hence slower paddle speeds are recommended. For the ODTs having a weight in excess of one gram and slow dissolution characteristics, a higher paddle speed may be necessary to keep the particles from agglomerating. It is also often encountered that taste-masked drug particles used in manufacture of ODTs to improve patient compliance greatly influenced dissolution method development. The incorporation of taste-masked active into ODTs mandates a suitable dissolution method development for the evaluation of ODT dosage forms. In this situation dissolution testing of ODTs becomes very critical for both product development and quality control. USP *Q*-type specifications may not be applicable as the *Q*-type criterion is not part of the finished dosage unit but is a homogeneous sample of the dispersed medium (50).

The following regulatory guidance can be used either for NCEs or approved drugs to support ODT product marketing applications. For an NCE, if an ODT is the first dosage form developed, all the general clinical pharmacology and biopharmaceutics (CPB) requirements to support the development and approval of the NCE should be followed as covered under 21 CFR 320 in a 505(b)(1) application. In the case of an ODT as a new dosage form for an already approved conventional oral product, the primary CPB as well as overall regulatory requirement will be to evaluate comparative BA or BE between the ODT and the conventional oral tablet under 505(b)(1), 505(b)(2), or ANDA applications. However, currently, there is no guidance specifically aimed at ODTs. Nonetheless, regulatory guidance for food effect BA studies, BA–BE studies for orally administered drug products, and others should be consulted (47).

VI. CONCLUSIONS

Clearly ODTs are one of the more popular orally administered dosage forms. The main advantages of the ODT lie in improved dosing, convenience, and product and line extensions. Even though numerous technologies have been developed to achieve oral fast disintegration when placed on the tongue or in the mouth, the product development efforts still fall short in answering critical issues such as biopharmaceutical optimization and development of QC tools. Despite the several ODT products and technologies such as Zydis®, WOWTAB®, and Orasolv®, the science and engineering of commercial scale-up for ODT technologies are not generally available in the published literature. The large commercial competitive nature of the pharmaceutical drug delivery business explains the proprietary nature of these technologies. With continued innovations in pharmaceutical technology such as process analytical technology (PAT) one can expect improvements in process control and new methods for optimization and measurement of product quality to support the development of ODTs in the future.

QOD and QOC use statistical methods in product design and improvements, manufacturing process, and life-cycle testing. This process compares materials and components and determines system and component tolerance during the product design phase leading to lower development costs and reduced product development time. In product manufacturing, this process systematically improves overall productivity by reducing variability, increasing the production quality, and reducing production costs. Finally, the QOD and QOC provide reliable data on the product quality leading to new and improved designs during life-cycle testing of the product. However, the statistical process capability must be designed in, not just inspected in. Also, the practitioners of these methods usually estimate capability without verifying the process stability. Therefore, applied correctly, these measures enable effective communication of process potential and performance information during product design and continuing on to manufacturing.

ACKNOWLEDGMENTS

We would like to thank the following people for their help in bringing out this work. Dr. William Pfister of Nexmed Inc. and co-editor of this book for his valuable suggestions and input; Dr. Ian Muir and Kevin Ostrander of Cardinal Health; and Brian McClellan and James Chu of Yamanouchi Pharma Technologies for providing the input in the Zydis® and WOWTAB® technologies area.

REFERENCES

1. Annual report on drug delivery—Controlling their destiny (1996) Med. Ad. News. 15, 1–32.
2. Chang, R.K. et al. (2000) Fast dissolving tablets. Pharm. Tech., June, 52–58.
3. Drug Delivery 2003: New technologies in the global market. Med. Ad. News. 27–30.
4. Brown, D. (2003) Orally disintegrating tablets—Taste over speed. Drug Deliv. Technol. Vol. 3, No. 6, 58–61.
5. Ostrander, K. (2003) Oral fast dispersing systems: Recent advances—Advances in Zydis® fast dispersing technology, October, AAPS, National Meeting.
6. Fasano, A. (1998) Novel approaches in oral delivery of macromolecules. J. Pharm. Sci. 1351–1356.
7. Avery, S.W. and Dellarosa, D.M. (1994) Approaches to treating dysphagia in patients with brain injury. Am. J. Occup. Ther. 48, 235–239.
8. Gisel, E.G. (1994) Oral motor skills following sensorimotor intervention in the moderately eating impaired child with cerebral palsy. Dysphagia 9, 180–192.
9. Anderson, O. et al. (1995) Problems when swallowing tablets. Tidsskr Nor Laegeforen. 115, 947–949.
10. Wilson, C.G. et al. (1987) The behavior of a fast dissolving dosage form (Expidet) followed by γ-scintigraphy. Int. J. Pharm. 40, 119–123.
11. Fix, J.A. (1998) Advances in quick-dissolving tablets technology employing WOWTAB. In IIR Conference on Drug Delivery Systems, October, Washington, DC.

12. Virely. P. and Yarwood, R. (1990) Zydis—A novel, fast dissolving dosage form. Manuf. Chem. February, 36–37.
13. Seager, H. (1998) Drug delivery products and the Zydis® fast dissolving dosage form. J. Pharm. Pharmacol. 50, 375–382.
14. Takao. M. et al. (1996) Intrabuccally dissolving compressed molding and production process thereof. US Patent 5,576,014.
15. Hontz. J. (1999) Development of a rapidly dissolving famotidine tablet. In 41st Annual International Industrial Pharmaceutical Research Conference, June, Madison, WI.
16. Wehling, F. et al. (1993) Pediatric effervescent dosage form. US Patent 5,223,264.
17. Wehling, F. et al. (1993) Effervescent dosage form with microparticles. US Patent 5,178,878.
18. Wehling, F. and Schuehle, S. (1996) Base coated acid particles and effervescent formulation incorporating same. US Patent 5,503,846.
19. Gregory, G.K.E. et al. (1983) Articles for carrying chemicals. US Patent 4,371,516.
20. Gregory, G.K.E. and Ho, D.S.S. (1981) Pharmaceutical dosage form packages. US Patent 4,305,502.
21. Gregory, G.K.E. (1988) Solid shaped articles. US Patent 4,758,598.
22. Buxton, I.R. and Feldman, H. (1988) Solid shaped articles. US Patent 4,754,597.
23. Iles, M.C. et al. (1993) Freeze-dried dosage forms and methods for preparing the same. US Patent 5,188,825.
24. Courteille, F. and Vanhoeve, M. (1993) Porous pharmaceutical form and its preparation. US Patent 5,206,025.
25. Cynthia, B.M. and Shah, M.N. (1993) Aqueous pharmaceutical suspension for pharmaceutical actives. US Patent 5,272,137.
26. Kearney, P. and Wong, S.K. (1997) Method for making freeze dried drug dosage form. US Patent 5,631,023.
27. Deasy, P.B. (1984) Release of drug from microcapsules and microparticles. In Microencapsulation and Related Drug Processes (Deasy, P.B., ed.), pp. 289–319, New York: Marcel Dekker.
28. Parakh, S.R. and Gothoskar, A.V. (2003) Mouth dissolving tablet technologies. Pharm. Technol. November, 92–100.
29. Cherukuri, S.R. et al. (1996) Process for forming quickly dispersing comestible unit and product therefrom. US Patent 5,587,172.
30. Fuisz. R. (1997) Ulcer prevention method using a melt-spun hydrogel. US Patent 5,622,717.
31. Cherukruri, S.R. and Fuisz, R. (1995) Process and apparatus for making tablets and tablets made therefrom. European Patent 0677,147 A2.
32. Myers, G.L. et al. (1996) Delivery of controlled-release systems. US Patent 5,567,439.
33. Myers, G.L. et al. (1999) Apparatus for making rapidly dissolving dosage units. US Patent 5,871,781.
34. Cherukuri, S.R. and Fuisz, R. (1997) Process and apparatus for making tablets and tablets made therefrom. US Patent 5,654,003.
35. Myers, G.L. et al. (1997) Ulcer prevention method using a melt-spun hydrogel. US Patent 5,622,719.
36. Juran, J.M. (1974) Quality Control Hand Book, 3rd edition. New York: McGraw-Hill.

37. Berry, I.R and Nash, R. (1993) Pharmaceutical Process Validation, 2nd edition, New York: Marcel Dekker Series, Volume 57, xiii–xxxix.
38. Kane, V.E. (1986) Process capability indices. J. Qual. Technol., 18, pp 41–52.
39. King, P. (2001) Process validation—Establishing the minimum process capability for drug product manufacturing—I & II, Pharmaceutical Engineering, 11/12, 1999 and 3/4, 2001.
40. Gunter, H. (1989) Manual for Software STATISTICAB. "The Use and Abuse of C_{pk}," part 2 and 3. Quality Progress, March & May 1989.
41. Rodriguez, R.N. (1992) Recent developments in process capability analysis, J. Qual. Technol., 24, pp 176–187.
42. Kotz, S. (2002) Process capability indices—A review, 1992–2000, J. Qual. Technol., 34, pp 2–19.
43. Torbeck, L. (1998) Pharm Technol., June, 66–76.
44. Nyshadham, R. (2001) Process development and scale-up of oral fast disintegrating tablet (FDT) technology—WOWTAB®. In Advanced Practices in Pharmaceutical Tablet and Capsule Technology Conference, May, Las Vegas.
45. Chu, J. (2003) Product development, scale-up and technology aspects of WOWTAB® technology, Oct, AAPS, National Meeting.
46. Dobetti, L. (2001) Fast melt tablets: Developments and technologies, Pharm. Technol: Drug Delivery, 44–50.
47. Ghosh, T. (2003) Quick dissolving oral dosage forms: Clinical pharmacology, biopharmaceutics and regulatory considerations, October, AAPS, National Meeting.
48. Bi, Y.X. et al. (1996) Preparation and evaluation of a compressed tablet rapidly disintegrating in the oral cavity, Chem. Pharm. Bull. 44 (11), 2121–2127.
49. Dor, J.M., Fix, J.A., and Johnson, M.I. (1999) In vitro method to measure the disintegration time of a fast-disintegration tablet, Proc. Intl Symp. Control. Rel. Bioact. Mater. 26, 939–940.
50. Klanke, J. (2003) Dissolution testing of orally disintegrating tablets, Diss. Technol. Vol. 10(2), May, 6–8.

14

Quick Dissolving Oral Dosage Forms: Scientific and Regulatory Considerations from a Clinical Pharmacology and Biopharmaceutics Perspective*

Tapash K. Ghosh, Dhruba J. Chatterjee, William R. Pfister, Venkateswar R. Jarugula, Emmanuel O. Fadiran, John P. Hunt, Lawrence J. Lesko, Vijaya K. Tammara, and Donald B. Hare

CONTENTS

*Disclaimer: No official support or endorsement of this chapter by the Food and Drug Administration is intended or should be inferred.

I. INTRODUCTION

Over the last three decades, intraoral dosage forms (IODFs) have been evolving as an acceptable, and in some cases as the preferred, alternative to conventional tablets (CTs) and capsules. Quick-dissolving intraoral dosage forms (QODs) are a type of IODFs that have gained much attention recently due to improved patient compliance and ease of administration compared to conventional oral dosage forms (e.g., tablets and capsules). QODs include orally disintegrating tablets (ODTs), the only dosage form of this nature recognized by the FDA listed in *Approved Drug Products with Therapeutic Equivalence Evaluations* (i.e., the *Orange Book*)[1] [e.g., Claritin® RediTabs® 24 Hour Non-Drowsy Orally Disintegrating Tablets (loratadine orally disintegrating tablets)].[2] The European Pharmacopoeia, however, defines a similar term, "orodisperse," as a tablet that can be placed in the mouth where it disperses rapidly before swallowing.[3] The simplest definition of an ODT is a unit dose that disintegrates in the oral cavity. The Center for Drug Evaluation and Research (CDER) defines an ODT to be "a solid dosage form containing medicinal substances, which disintegrates rapidly, usually within a matter of seconds, when placed upon the tongue."[4] As the majority of the QODs on the market or in different stages of development are in the category of ODTs, the discussion in this chapter primarily focuses on this particular type of dosage form.

Conventional oral dosage forms are administered to the mouth and immediately swallowed intact with water to release the drug for absorption in the stomach and/or intestine (i.e., gastric and postgastric absorption). Potential problems associated with bioavailability (BA) of drug molecules (e.g., stability issues associated within the gastrointestinal tract (GIT) as a chemically hostile environment, gastrointestinal wall, hepatic and gastrointestinal flora metabolism, or variable absorption) can be overcome with a formulation intended for pregastric delivery. To this end, targeted drug delivery to the oral cavity may offer several advantages over the GI and other alternative routes of administration in terms of ease of dosing, enhanced permeability, avoidance of first-pass clearance, better patient acceptability, product life extension, patentability, and increased systemic absorption with certain drugs.[5-10] However, unless the dosage form is a targeted delivery system such as a buccal tablet or buccal patch, only a minor fraction of the dose is absorbed through the buccal mucosa limiting the main advantage of orally disintegrating dosage forms to convenience of administration and better patient acceptance.

It has long been evident to drug delivery scientists that achieving therapeutically effective plasma levels with intraoral drug delivery systems can be a major challenge. However, nitroglycerin was the first drug shown to be successfully delivered and absorbed from the oral cavity a number of decades ago. Since that time, many other drug delivery dosage forms administered to the oral mucosa have been commercialized. The development of oral mucosal delivery systems can be distinguished into two periods of innovation. The first period focused on the development of sublingual tablets and solutions. More recently, delivery systems have been refined for oral mucosal delivery by using new technologies and delivery systems such as sprays, mucoadhesive patches, or quick-dissolving solid matrices using advanced manufacturing processes (i.e., lyophilized wafers, solvent cast films, etc.). The second period of drug delivery innovation was spawned over the last decade by the renewed interest in using the highly permeable oral mucosa for targeted delivery of drugs.

Within the United States, the research and development of new unconventional IODFs and their marketing are regulated by legislation or law enacted by the U.S. Congress that is promulgated and enforced by the U.S. Food and Drug Administration (FDA). To assist in the management of the new drug approval process, FDA proposes rules or regulations via the Federal Register,[11] which upon finalization are compiled in the U.S. Code of Federal Regulations (CFR) which are also available on the Internet.[12] The relevant CFR sections that address different aspects of clinical pharmacology and biopharmaceutics (CPB) informational needs for supporting a new product's approval are Part 201 of Title 21 (21 CFR 201)[13] and Part 320 of Title 21 (21 CFR 320)[14] entitled *Labeling* and *Bioavailability and Bioequivalence Requirements*, respectively.

The FDA also publishes guidances that are not legally binding but are intended to provide sponsors with informal guidance and the FDA's current thinking on how regulatory requirements can be satisfied. Currently there are numerous FDA guidances that have been finalized or are in draft that are accessible to the pharmaceutical industry at the FDA's Web site.[15] Official and draft guidances related to this chapter may be found under the general headings of *Clinical Pharmacology, Biopharmaceutics, and Chemistry.* However, it is noted that currently there is no specific FDA guidance for the types of dosage forms discussed in this chapter. In addition to FDA guidances, there are International Conference on Harmonization (ICH) guidelines prepared jointly by regulatory authorities and pharmaceutical representatives from the European Union, Japan, and the United States.[16] The ICH guidance documents also provide insight as to informational needs (i.e., including the areas of clinical pharmacology and biopharmaceutics) for the drug development process. The ICH guidelines and FDA regulations and guidances complement each other.

The overall approval cycle for any product encompasses multiple disciplines. The intent of this chapter is to review general and specific CPB considerations when developing a new QOD. However, relevant issues pertaining to any IODF in general are also discussed. As some CPB issues overlap with chemistry, manufacturing, and control (CMC) issues mainly in the area of product specifications and in vitro disintegration/dissolution, these considerations are also discussed in this chapter.

II. QUICK-DISSOLVING ORAL DOSAGE FORMS

It is believed that routes of administration that avoid first-pass metabolism may lead to better clinical safety and efficacy predominantly due to lower doses leading to production of a lower amount of potentially toxic metabolites. It has been demonstrated that administration of morphine by routes that avoided first-pass metabolism resulted in lower metabolite production.[17] An increase in the area under the plasma drug concentration time curve (AUC) and C_{max} for an ODT may be explained in part by avoidance of first-pass metabolism and significant buccal absorption as it has been reported that a lower dose of Zelpar® Zydis® (1.25 mg Zydis® selegiline) is equivalent to a 10 mg selegeline CT.[18] Moreover, because of the lower dose, Zelpar® Zydis® produces less selegiline metabolites (N-desmethylselegiline, methamphetamine, and amphetamine), especially amphetamine metabolites (approximately thirteenfold less) which are linked to side effects of insomnia and cardiac arrythmias.

Difference in metabolism via the buccal route compared to metabolism in the GIT has also been demonstrated for verapamil. Observed higher C_{max} and AUC for verapamil and lower C_{max} and AUC for norverapamil (major metabolite) from a buccal formulation with half a dose of verapamil than that in the CT shows that the buccal drug product might be more desirable than the conventional one in a clinical setting.[19] However, not all compounds can be absorbed through the oral mucosa because pregastric absorption depends on the appropriate physicochemical characteristics of the drug. For example, it has been found that sublingual administration of salbutamol tablets has no clinical benefit over conventional oral delivery and buccal absorption of salbutamol was found to be negligible.[20]

The ideal characteristics of a drug for pregastric absorption include: less than 20 mg dose; small to moderate MW; good solubility in water/saliva; partially nonionized at the pH of oral cavity (approximate pH 6.8); ability to diffuse and partition into the epithelium of the upper GI tract (log $P > 1$, or preferably >2); and ability to diffuse and partition into the blood.[18,21]

Table 14.1 provides a description of the various IODFs compiled from the approved labeling of a variety of products based on different drug delivery technologies and the *CDER Data Standards Manual*.[4] Some of these dosage forms are synonymous. The majority of the products mentioned in Table 14.1 are designed to rapidly disintegrate or dissolve in the oral cavity, and other dosage forms are slow- or controlled-release in nature (i.e., sublingual tablets, gums, and patches). Rapidly disintegrating dosage forms release the drug substance for immediate absorption through the buccal mucosa (although insignificant in most cases) followed by subsequent absorption from different segments of the gastrointestinal tract.

However, there is at least one product (Prevacid® SoluTab®) that disintegrates rapidly in the oral cavity and releases coated microgranules from which the drug substance is not absorbed at all from the buccal mucosa, and upon swallowing releases the drug substance in the GIT in a delayed fashion. However,

Table 14.1 Description of Different Intraoral Dosage Forms

Dosage Form	Description of Dosage Form
Freeze-dried wafer	A quick-dissolving thin matrix containing a medicinal agent that needs no water to aid in swallowing. This is a unique freeze-dried dosage form that when placed in the mouth disintegrates instantaneously, releasing the drug which dissolves or disperses in the saliva. The saliva is then swallowed and the drug is absorbed across the gastrointestinal (GI) tract.
Orally disintegrating tablet (ODT)	A solid dosage form containing medicinal substances that disintegrates rapidly, usually within a matter of minutes to seconds, when placed upon the tongue, releasing the drug which dissolves or disperses in the saliva. The saliva is then swallowed and the drug is absorbed across the GI tract.
Quick-dissolving tablet Fast-dissolving tablet Rapid-dissolving tablet Mouth-dissolving tablet Fast-melting tablet Quick-melting tablet	An oral tablet that requires no intake of liquid. The dosage form dissolves in less than a few minutes when placed in the mouth (i.e., faster than a lozenge). The active ingredients are absorbed into the body through the mucous membranes of the mouth into the bloodstream, bypassing the digestive system where they can be inactivated by the stomach acids and enzymes.
Orodispersing tablet	A tablet that can be placed in the mouth where it disperses rapidly before swallowing.
Rapid disintegrating tablet Rapid dispersing tablet	This dosage form is placed in the patient's mouth and the saliva rapidly dissolves the tablet, releasing the active ingredient (either as coated granules or as solubilized drug), which is swallowed in a liquid form.
Buccal gum Chewing gum	A nondissolving polymer matrix modified release dosage form containing the drug and other excipients that must be chewed but not swallowed to promote release of the drug from the dosage form in the oral cavity. The gum is removed from the mouth and disposed of following use.
Buccal patch Gingival patch Odontal patch	A nondissolving thin matrix modified-release dosage form composed of one or more polymer films or layers containing the drug and/or other excipients. The patch may contain a mucoadhesive polymer layer which bonds to the oral mucosa, gingiva, or teeth for controlled release of the drug into the oral mucosa (unidirectional release), oral cavity (unidirectional release), or both (bidirectional release). The patch is removed from the mouth and disposed of after a specified time.

Table 14.1 Description of Different Intraoral Dosage Forms (continued)

Dosage Form	Description of Dosage Form
Buccal tablet	A modified-release dosage form that dissolves in the mouth, releasing the drug which is intended to be absorbed through the mucosal lining of the mouth without the aid of water, and should not be swallowed whole. The tablet slowly dissolves once placed in the upper or lower cheek pouch (buccal pouch) between the gum and the side of the cheek. It is recommended not to eat, drink, chew, or smoke while the tablet is dissolving. Patients are also instructed to rinse the mouth and brush the teeth after dissolution to remove the taste, if necessary.
Sublingual tablet	A fast-dissolving, usually small flat tablet intended to be inserted beneath the tongue, where the tablet slowly dissolves and active ingredient is absorbed directly through the oral mucosa. The tablet must not be chewed.
Spray	A unit actuation pump or aerosol spray in a gas or solvent carrier vehicle for rapid drug absorption by the buccal mucosa.
Quick-dissolving film Quick-dissolving wafer	A fast-dissolving polymer film embedded with drug that melts and dissolves in the saliva of the oral cavity quickly and completely, releasing the drug for absorption through the oral mucosa. A fraction of the drug will be swallowed with the saliva and absorbed along the length of the GI tract.

both of these types of dosage forms are considered as immediate-release (IR) dosage forms from the standpoint of releasing the drug substance or coated microgranules instantaneously in the oral cavity. There are dosage forms mentioned in Table 14.1 (such as buccal tablets, patches, and gums) that are designed to release the drug in a slow and controlled manner when first placed in the oral cavity, and therefore they are categorized as modified-release (MR) dosage forms. However, MR dosage forms are not within the scope of discussion of this chapter.

A list of selected ODT products marketed in the United States along with the associated technologies upon which they are based are summarized in Table 14.2. Most of these ODTs are over-the-counter (OTC) line extensions of CTs developed for life-cycle management with improved patient convenience, which can be taken with or without water.[22] A human pharmacokinetic (PK) study comparing apomorphine Zydis® versus subcutaneous injection showed that the Zydis® formulation provides comparable plasma levels of apomorphine and therefore may even be used as an alternate route of administration for apomorphine.[18] Market research studies have shown that a significant number of consumers prefer

Table 14.2 Summary of Selected ODT Products

Technologies (Innovator)	Type of Dosage Form	Marketer	Drug	Indication	Products on the U.S. Market
DuraSolv®	Compressed tablet	Wyeth	Loratadine	Allergy	Alavert™
OraSolv®		Schwarz Pharma	Hyoscyamine Sulfate	Irritable bowel	NuLev®
		Organon	Mirtazapine	Depression	Remeron® SolTabs™
(CIMA)		Bristol-Myers Squibb	Acetaminophen	Analgesic	Tempra® Quicklets/ Tempra® FirsTabs
		Novartis	Pseudoephedrine HCl (other actives)	Pediatric cough & cold	Triaminic® Softchews® (several formulations)
		AstraZeneca	Zolmitriptan	Migraine	Zomig Rapidmelt™
Flashtab® (Ethypharm)	Compressed tablet	Bristol-Myers Squibb	Acetaminophen Caffeine	Headache	Excedrin® QuickTabs™
WOWTAB® (Yamanouchi)	Compression molded tablet	Pfizer	Diphenhydramine Citrate Pseudoephedrine HCl	Allergy & sinus	Benadryl® Fastmelt®
Zydis® (Cardinal Health)	Freeze-dried wafers	Schering	Loratadine	Allergy	Claritin® Reditabs®
			Rizatriptan Benzoate	Migraine	Maxalt-MLT®
		GlaxoSmithKline	Ondansetron	Nausea & vomiting	Zofran ODT®
		Eli Lilly	Olanzapine	Schizophrenia	Zyprexa® Zydis®
		Roche	Clonazepam	Seizures	Klonopin® Wafers
Other	Tablet	TAP Pharm. Inc.	Lansoprazole	Duodenal ulcers	Prevacid® SoluTab®

ODTs to the CT or liquid counterpart products.[23] This may, in part, be due to the pleasant taste and variations of flavors available with the ODTs.

III. ADVANTAGES AND LIMITATIONS OF ORALLY DISINTEGRATING TABLETS

The administration of ODTs may not inherently result in a faster therapeutic onset, but can circumvent problems such as difficulty in swallowing traditional solid oral dosage forms such as tablets and capsules, particularly by pediatric and geriatric patients. The driving force in developing an ODT or other QODs may include one or more of the following advantages.

Patient compliance: To allow patients to easily swallow the dosage form anytime, anywhere, for systemic absorption via rapid dissolution or disintegration in the oral cavity

Patient convenience: To enhance convenience to the patients in carrying and administering these dosage forms especially when traveling by designing them to be taken without water

Rapid absorption and onset of action: To produce rapid absorption and faster onset of therapeutic efficacy mainly from the rapid disintegrating/dissolving dosage forms, presumably due to rapid disintegration, dissolution, and absorption (i.e., antianginal therapy)

Avoidance of first-pass effect: To improve bioavailability owing to partial avoidance of first-pass metabolism resulting from at least partial absorption through the buccal mucosa

Elimination of water/improved stability: To provide an alternative to liquid dosage forms (i.e., syrups, solutions, dispersions) and thereby improve physical/chemical stability of the drug

Product life-cycle management: To extend product life cycle by product differentiation by providing alternative formulations of the conventional oral products

Although one or more of the advantages mentioned above may be desired in the product labeling of a particular ODT, the planned clinical studies should support each claim. In contrast to the advantages, many ODTs have limitations in terms of the amount of drug that can be incorporated in each unit dose. For lyophilized dosage forms, the drug dose must generally be less than 400 mg for insoluble drugs and less than 60 mg for soluble drugs. Also, due to the nature of ODTs, special packaging is needed for products that are fragile, which may add to the cost.[24]

IV. CLINICAL PHARMACOLOGY CONSIDERATIONS

Dosage forms intended to dissolve in the oral cavity may also be designed to favorably alter delivery of a drug. Avoidance of first-pass metabolism may be demonstrated by information comparing a difference in the drug and metabolite profiles when given as ODTs versus a conventional dosage form. Standard single and multiple dose pharmacokinetic studies may help to characterize absorption of active and metabolites as well as the bioavailability. Studies to evaluate taste, tolerability to the oral cavity, and toxicity (i.e., mucosal irritation, stinging, etc.) should be carried out. Depending on the proposed indication, clinical safety and efficacy data may also be needed to support the approval of a QOD.

The published information on clinical pharmacology requirements for approval of ODTs is limited at this time for products that are either a pharmaceutical alternative (PA) or pharmaceutical equivalent (PE) to a CT. Based on available information, the following CP information appears relevant depending upon the label claims made by the sponsor in the application for approval of an ODT, and should be considered for characterization of the product.

A. In Vivo Assessment

If the safety and efficacy of a drug in a CT has already been established and the product is already approved for a specific indication, and the purpose of the ODT is to make another dosage form with equal dosage strength to improve patient compliance only, then the primary CP requirement will be to establish bioequivalence (BE) between the ODT and the CT. The design of a BE study should consider sample size in view of variability in absorption of the ODT through the oral mucosa due to the inter- and intraindividual differences in salivary output and variable physiologic conditions of the oral cavity. However, considering the nature of the dosage form, time to reach maximum plasma levels (T_{max}) may be an important PK parameter. On the other hand, if the ODT is being developed with a different dosage strength for similar exposure, bioavailability, exposure-response, and/or clinical studies should be conducted. Also, if the ODT is targeted for a new molecular entity (NME), all of the CP studies for developing a new drug will be required. The following examples are provided to address the general BE issues. [25-28]

1. ODT is generally bioequivalent with the CD as exemplified by the fact that 4 and 8 mg doses of either Zofran® (ondansetron) Oral Solution or Zofran® ODT Orally Disintegrating Tablets are bioequivalent to corresponding doses of Zofran® Tablets.[25] Therefore, in this case, no clinical safety or efficacy study was conducted on this new ODT.
2. Contrary to the usual expectation that drugs from an ODT will be absorbed faster than a CT with a shorter T_{max}, it has been found that the AUC and maximum plasma concentration (C_{max}) of rizatriptan were similar following administration of Maxalt® Tablets and Maxalt-MLT® Orally Disintegrating Tablets, but the rate of absorption was somewhat slower with Maxalt-MLT® (T_{max} ranged from 1.6 to 2.5 hours) as compared to Maxalt® Tablets (T_{max} ranged between 1 and 1.5 hours).[26] Another example is zolmitriptan in which the AUC and C_{max} of zolmitriptan were similar following administration of Zomig® Tablets and

Zomig-ZMT® Orally Disintegrating Tablets, but the T_{max} was longer with Zomig-ZMT® (median T_{max} of 3 hours for the ODT compared with 1.5 hours for the Zomig® Tablets).[27] In both of these examples, the apparent rate of absorption was found to be slower with a longer T_{max} with the ODT as compared to the conventional tablet.

3. AUC, C_{max}, and T_{max} may differ between a CT and an ODT dosage form. If these parameters are different for an ODT compared to the CT, the ODT dose may need to be adjusted accordingly. For example, Zelapar™ (selegiline HCl) ODT (not approved in the United States) dissolves instantly in the mouth and provides therapeutic levels of selegiline in Parkinson patients with one eighth of the usual daily dose.[28]

In light of the preceding examples and general understanding of these dosage forms, the following section describes general CP considerations for development of an ODT.

1. Bioequivalence

Generally the PK profiles between CT and ODT with the same dosage strength are similar and may demonstrate pharmaceutical equivalents (PE). However, if there is any change in PK parameters such as AUC, C_{max}, trough plasma concentration (C_{trough}), and T_{max} with the ODT compared to CT, the sponsor may need to demonstrate or justify that there are no safety- and efficacy-related issues associated with these factors for the proposed indication(s). A hypothetical comparison of PK profiles between a CT and an ODT is shown in Figure 14.1.

2. Water Effect

As for patient convenience, ODTs can be administered without water; the effect of water intake on bioavailability should be studied. In light of the directions given to patients for most of the approved ODTs, which do not require the patients to take the dosage form with water, the following study design may be considered to evaluate BE as well as the effect of water: three-way design where ODT is administered with and without water and the reference listed drug (RLD) is administered with water under fasting conditions. The hypothesis of the above study design is that the ODT administered with and without water will be bioequivalent to each other and to the RLD administered with water. A typical profile of a BA study comparing a CT and ODT when taken with or without water is shown in Figure 14.2. Generally the profiles are similar. However, if the results show any difference, the clinical relevance of the differences will need to be evaluated.

3. Food Effect

Food generally has a similar effect on drug absorption following administration of an ODT or a CT. However, the effect of food intake may need to be addressed

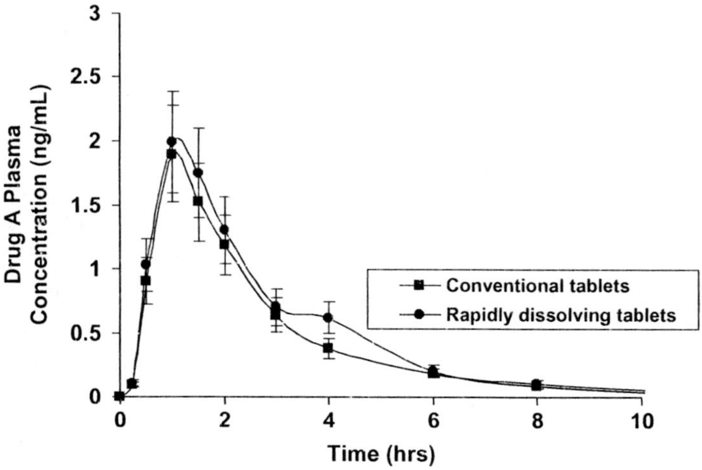

Figure 14.1 Comparative human plasma pharmacokinetic profiles of drug A following administration of either conventional or orally disintegrating tablets.

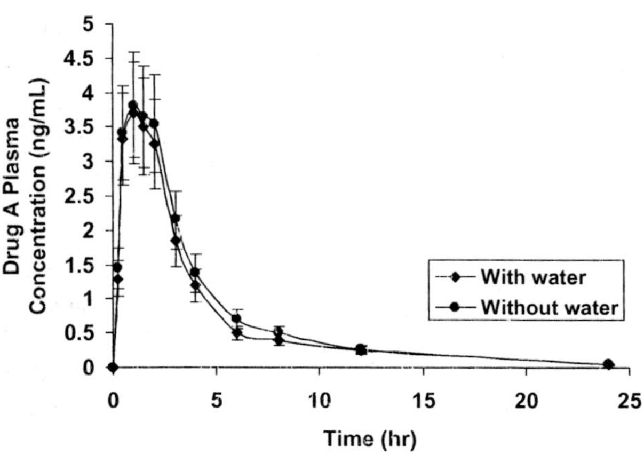

Figure 14.2 Comparative human plasma pharmacokinetic profiles of drug A following administration of orally disintegrating-tablet with or without water.

particularly when the drug in an ODT disintegrates in the mouth and the resulting solution/suspension is swallowed rendering a significant fraction of the drug available for absorption in the GIT. This is particularly important for drugs having a documented significant effect of food on absorption, efficacy, or safety. A typical profile showing the effect of food on retarding absorption and altering the bioavailability of a drug in an ODT is shown in Figure 14.3. The following scenarios may be worthwhile to consider in this context.

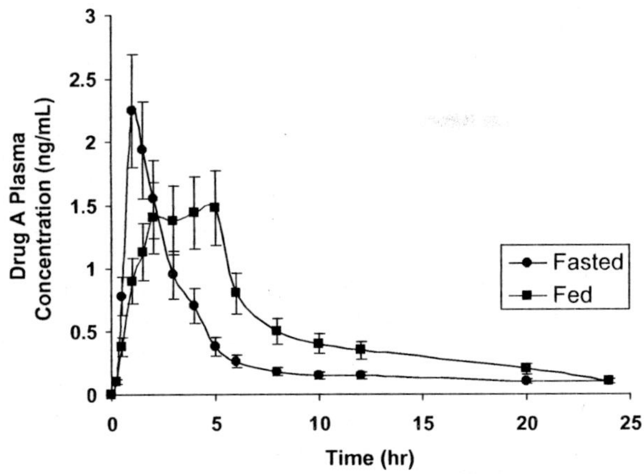

Figure 14.3 Comparative human plasma pharmacokinetic profiles of drug A following administration of an ODT with or without food.

- If food has not affected the BA of the active drug in the CT, similar lack of food effect may be expected with an ODT. Therefore, the sponsor may choose not to conduct any food effect study with the ODT as long as absorption and metabolism remain unchanged from the CT.
- If food does affect BA of the active drug in the CT, then the sponsor should conduct a food effect study on the ODT as per FDA guidance on food effect studies.[29] Given the nature and label instructions of the dosage form, ODTs should be administered without water in the food effect study.

In addition, the following CP information may need to be addressed in the development of an ODT on a case-by-case basis. However, not all are needed or applicable to every ODT.

- Contribution of oral (i.e., local site within the oral cavity, e.g., buccal, sublingual, etc.) absorption to drug bioavailability.
- Taste characteristics of the dosage form (compliance/adherence).
- Interactions with drugs that influence production of saliva (e.g., anticholinergics).
- Effect of chewing or swallowing an ODT on the PK of the drug.
- Local (buccal) mucosa irritation.
- Effect of the volume and ingestion of saliva on drug bioavailability. The effects of saliva (volume, pH, etc.) on drug BA from an ODT are not well understood.

V. BIOPHARMACEUTICS CONSIDERATIONS

To be formulated as a lyophilized ODT, the drug molecules should have certain characteristics that include but are not limited to:[24]

- Relative water insolubility with fine particle size and good aqueous stability.
- Drug loading for water insoluble drugs less than 400 mg.
- Upper limit for water soluble drug loading of about 60 mg.
- Appropriate particle size is ~50 microns as sedimentation is possible with large particles.

The major challenge with ODTs lies in establishing product specifications in terms of in vitro assessments. The considerations in establishing in vitro specifications for ODTs are discussed below.

A. In Vitro Assessment

Most sponsors have used in vitro dissolution time as the release specification for ODTs similar to CTs. Given the nature and mode of action of ODTs and in line with the ICH Q6A guidance document, there is an ongoing initiative to address whether a dissolution test can be replaced by a disintegration test for these dosage forms.[30] According to this guidance, if the disintegration test is shown to be an equal or better discriminating test compared to the dissolution test, then the disintegration test may be considered in lieu of the dissolution test. However, it remains to be determined whether a disintegration test by itself is adequate from a biopharmaceutics perspective. To further address this issue, the following points should be considered.

- Disintegration time does not necessarily correlate with dissolution time, even for soluble drugs. Factors attributed to excipients, lubricants, agglomeration, particle size, and dosage form (coated or delayed release drug products) can delay drug dissolution and may have little apparent effects on disintegration.
- Short of in vivo studies, the characterization of a product's dissolution profile is the best quality control test that may correlate with the drug's in vivo bioavailability.
- A disintegration test cannot detect chemical degradation of a drug, whereas a dissolution test can.

If it is decided that a disintegration test alone is adequate from a biopharmaceutics point of view, then the following issues and questions remain to be addressed.

- What medium (e.g., water vs. simulated saliva) should be used for a disintegration test?
- At what temperature should the test be conducted?

- What volume of medium should be used?
- What type of apparatus and mechanical condition should be used for that purpose?
- What should be an appropriate disintegration time specification?

However, if it is decided that a disintegration test alone is not adequate from a biopharmaceutics point of view, and a dissolution test is also needed, then the issues similar to the above would need to be addressed for selecting a dissolution method.

Another factor that may or may not have any clinical relevance but may be a valuable tool in early development is the consideration of in vivo (on the tongue) disintegration time. However, the utility of this information is debatable.[31]

B. In Vitro Dissolution Methodology, Requirements, and Challenges

Dissolution testing is a very important tool in drug development and quality control. Initially developed for immediate-release solid oral dosage forms, the application of dissolution testing in recent years has widened to a variety of modified-release and "novel" or "special" dosage forms such as suspensions, ODTs, chewable tablets, chewing gums, transdermal patches, semisolid topical preparations, suppositories, implants, injectable microparticulate formulations, and liposomes. The ultimate goal of these in vitro tests for ODTs is analogous to that for conventional solid oral dosage forms, that is, to use the test for the biopharmaceutical characterization of the drug product, and as a tool to assure batch-to-batch consistency in product quality within a defined set of specifications.

ODTs create an in situ suspension by disintegrating typically within one minute or less in the mouth. Taste masking (i.e., drug coating) is very often an essential feature of ODTs, and thus can also be the rate-determining mechanism for dissolution/release. In vitro dissolution testing of ODT should follow the principles of conventional solid oral dosage forms (i.e., tablets, capsules, or suspensions). A suitable method using an appropriate apparatus needs to be developed before the method is finalized, validated, and accepted for developing release specifications. Due to the different characteristics of ODTs and their sites and modes of application, it is essential that apparatus selection, composition of the dissolution medium, agitation (i.e., flow rate or stirring speed), and temperature be given appropriate consideration during method development and validation. In some cases, the method used in the early phases of product/formulation development could be different from the final test procedure utilized for control of product quality.

A single-point specification is considered appropriate for ODTs with fast dissolution properties. If taste masking is a key aspect of the dosage form by use of polymer coating, a multipoint profile in a neutral pH medium with early points of analysis (e.g., ≤5 min) may be recommended. It is to be noted that this early time point in the profile is intended to address the taste-masking properties of the formulation and may not have any relevance to the product's biopharmaceutical properties. Such a dissolution criterion (typical example: ≤10 percent dissolved in 5 min) would largely depend on the taste intensity of the drug and may enable the in vitro evaluation of the taste-masking properties while avoiding organoleptic measurements.

If a separate quality control assessment is deemed necessary from the biopharmaceutics point of view, that test should attempt to mimic in vivo conditions. The purpose of this assessment is to predict the performance of the ODTs in vivo and to establish an in vitro/in vivo correlation, if possible. This test may or may not have utility in predicting batch-to-batch product uniformity. Due to instantaneous disintegration/dissolution of most of the ODTs, the sponsor may need to consider which is the more appropriate test to use (i.e., a dissolution test, a disintegration test, or both). Whichever test is used, it should take into account salivary volume, physiologic pH, and temperature of the oral cavity. Therefore, instead of traditional compendial apparatuses, this test may be performed in temperature-controlled test tubes with or without mechanical stirring to determine whether the dosage units dissolve/disintegrate in accordance to patient expectations (as in labeling). If product performance can be characterized by disintegration and dissolution, measures to capture those parameters should be addressed by appropriately designing sampling times. This will help to accurately characterize a disintegration time and/or dissolution profile for the product, and to set appropriate biopharmaceutical specifications. For intraoral products that are intended for rapid drug release, dissolution specifications (typically not less than 80 percent drug dissolved) are usually set within 15 minutes.

Compendial methods (CM) should be used as a first approach in drug development. Alternative methods should be considered only when it has been shown that compendial methods do not provide meaningful data for a new dosage form. In instances where a compendial (e.g., USP, Ph. Eur., and Ph. Jap.) method is employed for in vitro drug release testing, the experimental test conditions, qualifications, and validation steps should conform to those discussed in the FIP (International Pharmaceutical Federation) and FDA guidelines on dissolution testing.[32] A disintegration test may be used in lieu of a dissolution test for an ODT if it is shown to be an optimally discriminating method.[30] Procedures similar to dissolution should also be followed for disintegration method development. Although there is a method for determination of disintegration time in the United States Pharmacopoeia (USP) for conventional hard tablets (Method <701> for troches or lozenges),[33] it may not be appropriate for use with ODTs, and modified methods may be required.[32] Use of validated dissolution methods may then be used to demonstrate equivalence between ODTs as an alternative to conducting bioequivalence studies. Validated in vitro dissolution tests may in some instances be used as an alternative to conducting in vivo BA and BE studies for IR solid oral dosage forms based on the Biopharmaceutics Classification System (BCS), and are outlined in a Guidance Document.[34]

VI. GENERAL REGULATORY ISSUES AND CONSIDERATIONS

The developmental steps for an ODT or a QOD in general should be similar to any other new dosage form provided the active ingredient in the ODT is already approved in a CD (e.g., tablets or capsules) for a specific indication. In the absence of any separate therapeutic claim, a comparative BA or BE study between the ODT and the reference-listed CD may be required. Information on the design of a BA or BE study can be found in 21 CFR 320 and in the FDA guidance entitled *Guidance for Industry: Bioavailability and Bioequivalence Studies for Orally Administered Drug Products—General Considerations*.[35] Bioequivalence by definition is established if the maximum plasma concentration and area under the plasma concentration time curve (systemic exposure) are comparable between the two dosage forms as per the criteria set forth in the general BA/BE guidance (90 percent confidence interval of geometric mean ratio of AUC and C_{max} of test/reference should fall within 80 to 125 percent). Sometimes ODT may result in a different exposure than CD. In this case, it would be important to know what the exposure–response relationships are in order to understand the meaning of differences in exposure. However, if a different therapeutic claim is made for an ODT over the reference-listed conventional product, the new formulation needs to be supported by clinical safety and efficacy information. The exposure–response relationship should be considered in finding the optimal dose.

In addition to a comparative BA or BE study, local and systemic tolerability and irritation/toxicity in the oral cavity with the use of an ODT may need to be adequately evaluated in preclinical and clinical studies.

VII. SUMMARY

The pediatric, geriatric, and psychiatric populations are the primary targets for ODTs. The future potential for ODTs is promising because of the availability of technologies and strong patient demand. Several ODT products have been commercialized, and the market size for ODTs will surely continue to expand. By paying close attention to the advances in technologies, pharmaceutical professionals should take advantage of this new dosage form.[36]

To develop an ODT, the following clinical pharmacology and biopharmaceutics areas may need to be addressed: site of absorption (buccal, sublingual, etc.), extent of absorption through oral cavity, involvement of metabolizing enzyme systems (including transporters), net effect of all these on BA, effects of intrinsic and extrinsic factors on drug PK and PK/PD, physicochemical characteristics of the drug, formulation effect, and BCS classification. Taken collectively, the clinical development of an ODT should include a complete characterization of the drug in terms of absorption characteristics, metabolism pathways, and formulation effects. Furthermore, consideration should be given to explore parameters that can be used as predictors of in vivo performance of the ODT and also parameters that can be used to determine/ensure product quality. Categorizations of ODTs and QODs, in general, into a new dosage form along with guidance documents may assist sponsors in designing the appropriate preclinical, clinical, and biopharmaceutical studies to support marketing applications. Many of these issues were recently discussed in a symposium on this topic.[18]

An important issue that remains to be addressed is harmonization of definitions of ODT. It remains to be decided whether it is necessary to perform in vivo disintegration tests in addition to in vitro dissolution or disintegration tests. As the disintegration time of an ODT is generally less than 60 seconds, it is expected that the in vivo disintegration time will not correlate with the in vivo drug plasma concentration profiles. However, both in vitro as well as in vivo disintegration times may be useful as objective measures to determine whether a dosage form can be classified as an ODT, and this area needs to be explored in the future. Given the rapid dissolution characteristics of the ODTs and the recent developments of the BCS for drug molecules, the BCS may be useful in defining regulatory requirements for developing an ODT or any QOD in the future.

ACKNOWLEDGMENTS

The comments provided by the following individuals are gratefully acknowledged: John A. Lazor, Arzu Selen, Paul Hepp, Ramana Uppoor, and Edward D. Bashaw of the Office of Clinical Pharmacology and Biopharmaceutics, CDER, FDA.

REFERENCES

1. Electronic Orange Book, 2003, www.fda.gov/cder/ob/default.htm.
2. Claritin Reditabs 24 hour Non-drowsy Orally Disintegrating Tablets (Schering-Plough) Brand of Loratadine (PPI). PDR.Net, Thomson Healthcare, Montvale, NJ, 2003, http://www.pdr.net/HomePage_template.jsp.
3. European Directorate for the Quality of Medicines, Pharmeuropa 10 (4):547, 1988, http://www. pheur.org.
4. CDER Data Standards Manual, 2003, www.fda.gov/cder/dsm/DRG/drg00201.htm.
5. M.E. de Vries et al., Developments in buccal drug delivery, Crit. Rev. Ther. Drug Carr. Sys., 8:271, 1991.
6. M.J. Rathbone and J. Hadgraft, Absorption of drugs from the human oral cavity, Int. J. Pharm., 74:9, 1991.
7. C.A. Squier, The permeability of oral mucosa, Crit. Rev. Oral Biol. Med. 2:13, 1991.
8. H.P. Merkle and G.J.M. Wolnay, Muco-adhesive patches for buccal peptide administration, Minutes, European Symposium, Buccal and nasal administration as an alternative to parenteral administration (D. Duchne, ed.), Editions de Santé, Paris, p. 110, 1992.
9. D. Harris and J.R. Robinson, Drug delivery via the mucous membranes of the oral cavity, J. Pharm. Sci., 81:1, 1992.
10. M.J. Rathbone et al., The oral cavity as a site for systemic drug delivery, Adv. Drug Del. Rev., 13:1, 1994.

11. Federal Register, Supt. of Documents, U.S. Government Printing Office, Washington, DC, 2002, http://www.access.gpo.gov/su_docs/index.htm.
12. Code of Federal Regulations, Title 21, Supt. of Documents, U.S. Government Printing Office, Washington, DC, 2002, http://www.access.gpo.gov/su_docs/index.htm.
13. Code of Federal Regulations, U.S. Government Printing Office, Washington, DC, http://www.gpoaccess.gov/cfr/index.html, Title, 21, Volume 4, Revised April 1, 2003, Chapter I—Food and Drug Administration, Department of Health and Human Services, Part 201—Labeling.
14. Code of Federal Regulations, U.S. Government Printing Office, Washington, DC, http://www.gpoaccess.gov/cfr/index.html, Title, 21, Volume 5, Revised April 1, 2003, Chapter I—Food and Drug Administration, Department of Health and Human Services, Part 320—Bioavailability and Bioequivalence Requirements.
15. U.S. Food and Drug Administration, Center for Drug Evaluation and Research (CDER), Guidance Documents, http://www.fda.gov/cder/guidance.
16. International Conference on Harmonization of Technical Requirements for Registration of Pharmaceuticals for Human Use (ICH), http://www.ich.org.
17. C.C. Faura et al., Systemic review of factors affecting the ratios of morphine and its major metabolites, Pain, 74:43, 1998.
18. S.V. Sastry, K. Ostrander, J.S. Chu, V. Agarwal, and T.K. Ghosh, Annual AAPS Symposium: Oral Fast Dispersing Delivery Systems—Recent Advances, Salt Lake City, Utah, October 29, 2003.
19. W. Sawicki and S. Janicki, Pharmacokinetics of verapamil and its metabolites norverapamil from a buccal drug formulation, Int. J. Pharm., 238:181, 2002.
20. B.J. Lipworth et al., Pharmacokinetics, efficacy and adverse effects of sublingual salbutamol in patients with asthma, Eur. J. Clin. Pharmacol., 37 (6):567, 1989.
21. D. Harris and J.R. Robinson, Drug delivery via the mucous membranes of the oral cavity, J. Pharm. Sci., 81(1):1, 1992.
22. D. Brown, Orally disintegrating tablets—Taste over speed, Drug Delivery Technology, 3(6):58,
23. K. Cremer, Orally disintegrating dosage forms. Industry Summary Report. Pharma Concepts GMbH and Co., 2001: www.pharma-concepts.com.
24. W. Habib et al., Fast-dissolve drug delivery systems, Crit. Rev. Ther. Drug Carrier Sys., 17(1):61, 2000.
25. Zofran® ODT Patient Package Insert (PPI), PDR.Net, Thomson Healthcare, Montvale, NJ, 2003: http://www.pdr.net/HomePage_template.jsp.
26. Maxalt-MLT® Patient Package Insert (PPI), PDR.Net, Thomson Healthcare, Montvale, NJ, 2000: http://www.pdr.net/HomePage_template.jsp.
27. Zomig-ZMT® Patient Package Insert (PPI), PDR.Net, Thomson Healthcare, Montvale, NJ, 2002: http://www.pdr.net/HomePage_template.jsp.
28. Amarin press release, Amarin Corporation Announces Acceptance for Filing of New Drug Application for Zelapar by U.S. FDA, May 23, 2002.
29. Guidance for Industry, Food-Effect Bioavailability and Fed Bioequivalence Studies, Food and Drug Administration, Center for Drug Evaluation and Research (CDER), January, 2003, http://www.fda.gov/cder/guidance/5194fnl.htm.
30. International Conference on Harmonization; Guidance on Q6A Specifications: Test Procedures and Acceptance Criteria for New Drug Substances and New Drug Products: Chemical Substances, Federal Register, 65(251):83041–83063, December 29, 2000, http://www.fda.gov/OHRMS/DOCKETS/98fr/122900d.pdf.

31. P. Chue et al., Dissolution profile, tolerability, and acceptability of the orally disintegrating olanzapine patients with schizophrenia, CJP, 47:771, 2002.
32. M. Siewert et al., FIP/AAPS Guidelines for dissolution/in vitro release testing of novel/special dosage forms, Dissolution Technologie, Feb 6, 2003.
33. USP-NF Online, USP26-NF21S2, United States Pharmacopeial Convention, Inc., http://www.uspnf.com/uspnf/usp26-nf21s2/default.htm..
34. Guidance for Industry, Waiver of in vivo Bioavailability and Bioequivalence Studies for Immediate-Release Solid Oral Dosage Forms Based on a Biopharmaceutics Classification System, Food and Drug Administration, Center for Drug Evaluation and Research (CDER), August 2000, http://www.fda.gov/cder/guidance/3618fnl.pdf.
35. Guidance for Industry, Bioavailability and Bioequivalence Studies for Orally Administered Drug Products—General Considerations, Food and Drug Administration, Center for Drug Evaluation and Research (CDER), March 2003, http://www.fda.gov/cder/guidance/5356fnl.htm.
36. R.-K. Chang et al., Fast-dissolving tablets, Pharm. Tech., June: 52, 2000.

Appendix 1

Quick Reference Guide to Worldwide Companies Developing Intraoral Drug Delivery Technologies and Products

William R. Pfister

This appendix provides an easy-to-use reference guide with the addresses of companies developing intraoral drug delivery technologies and commercial products worldwide. Each company is profiled with its trademarks, description of technology, and selected brand names of commercial products for site-specific delivery to the oral cavity. The listed brand names are a representative sampling of drug products, not a comprehensive listing. Internet sites are provided for additional information on the technology or products of selected companies. Drug delivery technology companies are indicated as innovators, if known, and cross-referenced with companies that are marketing the brand name products. Source data was obtained from: (1) the Physicians' Desk Reference, PDR 54 Edition, 2000; www.pdr.net; (2) Physicians' Desk Reference for Nonprescription Drugs and Dietary Supplements, PDR 20 Edition, 1999; (3) Nonprescription Products: Formulations & Features 98-99, Ed. L. Knodel, American Pharmaceutical Association, Washington, DC (1998); and (4) the Internet. Every effort was made to ensure that the details are correct; however, if information changes, please either fax changes to 609.208.1868, or e-mail pfister@mindspring.com so we can update this guide for the next edition. Dialing prefixes are indicated for calls originating within the United States (i.e., calling within the U.S. prefix = 1; calling outside the U.S. prefix = 011).

3M Drug Delivery Systems
275-3E-10 3M Center
PO Box 33275
St. Paul, MN 55133-3275 USA
Tel: +1 800.643.8086
Fax: +1 651.633.2072
Internet: www.3m.com/DDS
Trademarks: Cydot
Technology: Transmucosal drug delivery system (patch)
Brand Names:

3M Health Care, Ltd.
3M House
Morley Street
Loughborough, Leicestershire
LE11 1EP UK
Tel: +49 2861 95 4796
Fax: +49 2861 95 4770
Internet: www.3m.com/DDS
Trademarks: Cydot
Technology: Transmucosal drug delivery system (patch)
Brand Names:

Abbott Laboratories, Inc.
Pharmaceutical Products Division
North Chicago, IL 60064 USA
Tel: +1 800.255.5162
Internet: www.abbott.com
Trademarks: Actiq; Cylert; Ery-Ped; Uprima
Technology: Oral transmucosal delivery (innovator: Anesta Corp.); chewable tablets; chewable wafers; sublingual apomorphine (innovator: Pentech Pharmaceuticals)
Brand Names: Actiq (oral transmucosal fentanyl citrate); Cylert (pemoline) Chewable Tablet; Ery-Ped Chewable (erythromycin ethylsuccinate, USP) Wafer

Access Pharmaceuticals, Inc.
2600 Stemmons Freeway
Suite 176
Dallas, TX 75207-2107 USA
Tel: +1 214.905.5100
Fax: +1 214.905.5101
Internet: www.accesspharma.com

Trademarks: OraDisc
Technology: Oral-mucosal polymer disc delivery system
Brand Names:

ADD Advanced Drug Delivery
Technologies AG
Kriegackerstrasse 30 Muttenz
CH-4132
Switzerland
Tel: +41 61 647 47 04
Fax: +41 61 647 47 03
Internet: www.addtechnologies.com
Trademarks:
Technology: Effervescent and fast dispersible dosage forms
Brand Names:

Alkermes, Inc.
64 Sidney Street
Cambridge, MA 02139 USA
Tel: +1 617.494.0171
Fax: +1 617.494.9263
Internet: www.alkermes.com
Trademarks:
Technology: Dose sipping technology
Brand Names:

Alza Corporation
950 Page Mill Road
Palo Alto, CA 94303 USA
Tel: +1 650.494.5000
Fax: +1 650.494.5151
Internet: www.alza.com
Trademarks: Actisite
Technology: Periodontal fiber for controlled drug delivery
Brand Names: Actisite (tetracycline hydrochloride) Periodontal Fiber

Anesta Corporation
4745 Wiley Post Way
Salt Lake City, UT 84116 USA
Tel: +1 801.595.1405
Fax: +1 801.595.1406
Internet: www.anesta.com
Trademarks: Fentanyl Oralet; OT-system; Actiq

Technology: Oral transmucosal (OT) drug delivery systems; "lollipop" technology
Brand Names: Actiq (oral transmucosal fentanyl citrate)

AstraZeneca
S-151 85
Sodertalje, Sweden
Tel: +011 46 8 55326000
Fax: +011 46 8 553290001
Internet: www.astra.com
Trademarks: PerioChip; Zomig Rapidmelt
Technology: Flexible film for periodontal drug delivery
Brand Names: PerioChip (Chlorhexidine gluconate) Biodegradable Polymer; Zomig Rapidmelt (zolmitriptan) Fast Dissolving Tablets

Atrix Laboratories, Inc.
2579 Midpoint Drive
Fort Collins, CO 80525 USA
Tel: +1 970.482.5868
Fax: +1 970.482.9735
Internet: www.atrixlabs.com
Trademarks: Atridox; Atrigel; Bema; Heska Perioceutic
Technology: Local gel for periodontal disease treatment; bioerodable mucoadhesive patch
Brand Names: Atridox (doxycycline); Heska Perioceuti Gel

Bayer Corporation
Pharmaceutical Division
400 Morgan Lane
West Haven, CT 06516 USA
Tel: +1 800.288.8371
Internet: www.bayer.com
Trademarks: Mycelex; Alka-Mints; Bayer
Technology: Slow-dissolving oral lozenge (Troche); chewable tablets
Brand Names: Mycelex (clotrimazole) Troche; Alka-Mints (calcium carbonate 850 mg) Chewable Antacid and Calcium Supplement; Bayer Children's Chewable (aspirin 81 mg) Orange and Cherry Flavored Tablets

Bayer Corporation
Consumer Care Division
36 Columbia Road
P.O. Box 1910
Morristown, NJ 07962-1910
Tel: +1 800.331.4536

Internet: www.bayer.com
Trademarks: Flintstones, Bugs Bunny
Technology: Chewable multivitamin tablets
Brand Names: Flintstones Original Children's Chewable Multivitamin Supplement; Bugs Bunny Plus Iron Chewable Children's Multivitamin Plus Iron Supplement; Flintstones Plus Iron Chewable Children's Multivitamin Plus Iron Supplement; Flintstones Complete Children's Chewable Multivitamin/Multimineral Supplement; Bugs Bunny Complete Children's Chewable Multivitamin/Multimineral Supplement (sugar free); Flintstones Plus Calcium Children's Chewable Multivitamin Plus Calcium Supplement; Flintstones Plus Extra C Children's Chewable Multivitamin Supplement; Bugs Bunny With Extra C Children's Chewable Multivitamin Supplement

Bentley Pharmaceuticals, Inc.
65 Lafayette Road
North Hampton, NH 03862 USA
Tel: +1 603.964.8006
Fax: +1 603.964.6889
Internet: www.bentleypharm.com
Trademarks:
Technology: Permeation Enhancer Technology (CPE-215)
Brand Names:

Besins-Iscovesco Laboratories
5 rue du Bourg
L'Abbe 75003 Paris
France
Tel: +011 33 01 42775825
Fax: +011 33 01 42771462
Internet: www2.biam2.org/www/Lab74098.html
Trademark: Lenitral
Technology: Sublingual spray
Brand Names: Lenitral Spray (5% nitroglycerin) Sublingual Solution

BF Goodrich Chemical
9911 Brecksville Road
Brecksville, OH 44141 USA
Tel: +1 216.447.5000
Fax: +1 216.447.5770
Internet: www.bfgoodrich.com
Trademarks: Carbopol; Noveon
Technology: Mucoadhesive polymer excipients
Brand Names:

Bioglan AB
P.O. Box 50310
Borrgaten 31
Malmo
S – 20213
Sweden
Tel: +011 46 40 2875 80
Fax: +011 46 40 2919 55
Internet: www.bioglan.com
Trademarks: Crystalip
Technology: Slow-release buccal tablets; sublingual aerosol; liquid crystal dermal/mucosal delivery system
Brand Names:

Bioglan Pharma, Ltd.
One The Cam Centre
Wilbury Way
Hitchin
SG4 0TW
UK
Tel: +011 44 1462 633 286
Fax: +011 44 1462 451 400
Internet: www.bioglan.com
Trademarks:
Technology: Slow-release buccal tablets
Brand Names:

Bioject, Inc.
7620 SW Bridgeport Road
Portland, OR 97224 USA
Tel: +1 800.683.7221
Fax: +1 503.624.9002
Internet: www.bioject.com
Trademarks: Bioject; Vitaject
Technology: Needle-free jet injection technology for dental application
Brand Names: Vitajet

Block Drug Company, Inc.
257 Cornelison Avenue
Jersey City, NJ 07302 USA
Tel: +1 201.434.3000
Fax: +1 201.432.6183
Internet: www.blockdrug.com
Trademarks: Aphthasol; Atridox

Technology: Oral-mucosal paste (innovator: Access Pharmaceuticals); bioresorbable periodontal gel (innovator: Atrix Laboratories)
Brand Names: Aphthasol (5% amlexanox) Oral Paste; Atridox (doxycycline hyclate, 10%) in the Atrigel Delivery System

Bristol-Myers Squibb Company
345 Park Avenue
New York, NY 10154 USA
Tel: +1 800.468.7746
Internet: www.bms.com
Trademarks: Tempra; Videx
Technology: Quick-dissolving chewable tablets (innovator: Cima Labs); chewable tablets
Brand Names: Tempra Quicklets (80 mg acetaminophen) Quick-Dissolving Chewable Tablets; Videx (didanosine) Chewable/Dispersible Buffered Tablets

Camurus AB
Ideon Science Park
Solvegatan 41
S-223 70 Lund
Sweden
Tel: +011 46 46 286 5730
Fax: +011 46 46 286 5739
Internet: www.camurus.se
Trademarks: Elyzol
Technology: Lipid liquid crystal drug delivery technology; dental gel
Brand Names: Elyzol (metronidazole) Dental Gel

CigArrest
6361 Yarrow Drive
Ste. B
Carlsbad, CA 92009
Tel: +1 760.438.1935
Fax: +1 760.438.3212
Internet: www.cigarrest
Trademarks: Cigarrest
Technology: Homeopathic no smoking gum
Brand Names: Cigarrest No Smoking Gum Tablets and Lozenges (lobelia inflata 6X)

Cima Labs, Inc.
10000 Valley View Road
Eden Prairie, MN 55344 USA

Tel: +1 612.947.8700
Fax: +1 612.947.8700
Internet: www.cimalabs.com
Trademarks: OraSolv; OraSolv SR/CR; SuraSolv; OraVescent SL; OraVescent BL; OraVescent SS; DuraSolv
Technology: Fast-dissolving sublingual and buccal tablets; transmucosal systems
Brand Names: Tempra FirsTabs Acetaminophen 160 mg; Quicklets; Triaminic Softchews Throat Pain & Cold (acetaminophen 160 mg, pseudoephedrine HCl 15 mg, dextromethorphan HBr monohydrate 5 mg); Triaminic Softchews Cold & Cough (pseudoephedrine HCl 15 mg, dextromethorphan HBr monohydrate 5 mg, chlorpheniramine maleate 15 mg) Triaminic Softchews Cold and Allergy (pseudoephedrine HCl 15 mg, chlorpheniramine maleate 1 mg); Triaminic Softchews Cough (dextromethorphan HBr monohydrate 7.5 mg); Zomig Fastmelt (zolmitriptan 2.5 mg)

Colgate Oral Pharmaceuticals, Inc.
Colgate-Palmolive Company
One Colgate Way
Canton, MA 02021 USA
Tel: +1 800.226.5428
Internet: www.colgate.com
Trademarks: Luride; Prevident 5000 Plus
Technology: Oral rinse; oral cream; oral solutions
Brand Names: Luride Drops (sodium fluoride) Oral Solution; Luride Lozi-Tabs (sodium fluoride) Lozenge; Periogard (chlorhexidine gluconate, 0.12%) Oral Rinse; Prevident 5000 Plus (sodium fluoride, 1.1%) Dental Cream

Cygnus, Inc.
400 Penobscot Drive
Redwood City, CA 94063 USA
Tel: +1 650.599.3539
Fax: +1 650.599.2519
Internet: www.cygn.com
Trademarks:
Technology: Oral mucosal delivery technology
Brand Names:

Del Pharmaceuticals, Inc.
A Subsidiary of Del Laboratories, Inc.
178 EAB Plaza
Uniondale, NY 11556
Tel: +1 516.844.2020
Fax: +1 516.293.1515

Internet: www.dellabs.com
Trademarks: Orajel, Mouth-Aid
Technology: Cold sore and canker sore treatment, oral local anesthetics
Brand Names: Baby Orajel Teething Pain Medicine (benzocaine 7.5%); Orajel Maximum Strength Toothache Medicine (benzocaine 20%); Orajel Covermed Fever Blister/Cold Sore Treatment Cream (dyclonine HCl 1% and allantoin 0.5%); Orajel Mouth-Aid Cold/Canker Sore Medicine (benzocaine 20%, benzalkonium chloride 0.02%, zinc chloride 0.1%)

Delsys Pharmaceutical Corporation
5 Vaughn Drive
Suite 305
Princeton, NJ 08540 USA
Tel: +1 609.720.0033
Fax: +1 609.520.6692
Internet: www.delsyspharma.com
Trademarks: Acccudep™
Technology: Electrostatic deposition process for tablet manufacturing
Brand Names:

Elan Pharmaceutical Research Corp.
1300 Bould Drive
Gainesville, GA 30504 USA
Tel: +1 770.534.8239
Fax: +1 770.534.8247
Internet: www.elancorp.com
Trademarks: Zelepar
Technology: Zydis fast dissolving formulations (innovator: RP Scherer)
Brand Names: Zelepar (selegiline) Fast-Dissolving Tablets

Elan Corporation, plc
Lincoln House
Lincoln Place
Dublin 2 Ireland
Tel: +011 353 1 7094000
Fax: +011 353 1 6624949
Internet: www.elancorp.com
Trademarks: Zelepar
Technology: Zydis fast dissolving formulations (innovator: RP Scherer)
Brand Names: Zelepar (selegiline) Fast-Dissolving Tablets

Eli Lilly and Company
Lilly Corporate Center

Indianapolis, IN 46285
Tel: +1 800.545.5979
Internet: www.lilly.com
Trademarks: Zyprexa Zydis
Technology: Zydis fast dissolving formulation (innovator: RP Scherer)
Brand Names: Zyprexa Zydis (olanzapine) Orally Disintegrating Tablets

Entec Drug Delivery Technologies
Entec Corporation Pharmaceutical Division
1735 Market Street
Philadelphia, PA 19103 USA
Tel: +1 609.514.4824
Fax: +1 609.514.4824
Internet: www.entec-fmc.com
Trademarks: Ensolv, Encirc, Envel, EnTec
Technology: Taste masking chewable tablets, dissolution of hard-to-dissolve drugs, Soft chew technology
Brand Names:

Ethex Corp.
10888 Metro Court
St. Louis, MO 63043 USA
Tel: +1 314.567.3307
Fax: +1 314.567.0701
Internet: www.ethex.com
(see KV Pharmaceutical Co.)
Trademarks: Site Release; OraSert; OraSite; OraQuick; Trans-EP; MicroMask
Technology: Quick-dissolving technology; bioadhesive technologies

Eurand International SpA
Via Martin Luther King, 13
20060 Pessano con Bornago
Milan, Italy
Tel: +011 39 02 954 281
Fax: +011 39 02 957 45018
Internet: www.eurand.com
Trademarks: Opus; Ziplets; Microcaps
Technology: Buccal tablets for trans-oral-mucosal absorption; quick-dissolving tablets; chewable tablets
Brand Names:

Eurand International SpA
845 Center Drive

Vandalia, OH 45377 USA
Tel: +1 937.898.9669
Fax: +1 937.898.9529
Internet: www.eurand.com
Trademarks: Opus; Ziplets; Microcaps
Technology: Buccal tablets for trans-oral-mucosal absorption; quick-dissolving tablets; chewable tablets
Brand Names:

Ethypharm SA
194 Bureaux de la Colline
Saint-Cloud
92213 France
Tel: +011 33 1 41 12 1720
Fax: +011 33 1 41 12 1730
Internet: www.ethypharm.com
Trademarks: Flashtab
Technology: Fast dispersible tablet of soft pellets; fast dispersible tablets
Brand Names:

Faulding
1538 Main North Road
Salisbury South 5106
Australia
Tel: +011 61 88209 2406
Fax: +011 61 88281 6998
Internet: www.faulding.com.au
Trademarks:
Technology: Mucoadhesive patch technology
Brand Names:

Fertin A/S
Division of Dandy A/S
Industrivej 8
DK-7120 Vejle
Tel: +011 45 72 15 13 00
Fax: +011 45 72 15 13 01
Internet: www.fertin.com
Trademarks: Fluorette
Technology: Medicated chewing gum
Brand Names: Fluorette® (0.25 mg fluoride, peppermint) Chewing Gum; Endkay (Vitamin C) Gum, developed and manufactured for Stafford Miller; Mentadent® ActiGum, developed and produced for Unilever; Fludent (0.25 mg fluor) Chewing Gum, developed and produced for Dumex-Alpharma;

Aquafresh™ Protect Triple Action (xylitol) Chewing Gum, developed and produced for SmithKline Beecham; Nicotinell®(2 mg kaugumni mint) Chewing Gum, developed and produced for Novartis Consumer Health

Flamel Technologies
Parc Club du Moulin a Vent
33 Avenue du Dr Georges Levy
Venissieux Cedex
69200 France
Tel: +011 33 4 72 78 34 34
Fax: +011 33 4 72 78 34 35
Internet: www.flamel.com
Trademarks: Medusa
Technology: Mucosal nanoparticle protein delivery
Brand Names:

Flemington Pharmaceutical Corp.
43 Emery Avenue
Flemington, NJ 08822 USA
Tel: +1 908.782.3431
Fax: +1 908.782.2445
Internet: www.Flemington-Pharma.com
Trademarks: I^2R
Technology: Immediate-immediate release lingual sprays; gelatin-bite capsules
Brand Names:

FMC Corporation
Avenue Louise 480-B9
Brussels
1050 Belgium
Tel: +011 32 2 645 9211
Fax: +011 32 2 640 0564
Internet: www.fmc.com
Trademarks: Entec; Envel
Technology: Soft chewable gel delivery system; chewable taste masking technology
Brand Names:

FMC Corporation
Pharmaceutical Division
1735 Market Street
Philadelphia, PA 19103 USA
Tel: +1 215.299.6614

Fax: +1 215.299.6821
Internet: www.entec-fmc.com
Trademarks: Entec; Envel
Technology: Soft chewable gel delivery system; chewable taste masking technology
Brand Names:

Forest Laboratories
909 Third Avenue
New York, NY 10022 USA
Tel: +1 212.421.7850
Fax: +1 212.750.9152
Internet: www.forestlabs.com
Trademarks: Susadrin
Technology: Bioadhesive tablets
Brand Names: Susadrin (nitroglycerin) Bioadhesive Tablets

Fuisz Technologies, Ltd.
14555 Avion at Lakeside
Chantilly, VA 20151 USA
Tel: +1 703.995.2384
Fax: +1 703.803.6460
Internet: www.fuisz.com
Trademarks: Ceform; FlashDose; Soft Chew; EZ Chew
Technology: Chewable tablets; quick-dissolving tablets
Brand Names:

Generex Biotechnology Corp.
33 Harbour Square, Suite 202
Toronto, Ontario, Canada M5J 2G2
Internet: www.generex.com
Tel: +1 416.364.8288
Fax: +1 416.364.8782
Trademarks: Oralgen; Oralin; RapidMist
Technology: Spray device for aerosol administration to the oral cavity; oral insulin spray
Brand Names:

Glaxo Wellcome, Inc.
Five Moore Drive
Research Triangle Park, NC USA
Tel: +1 919.483.2100
Internet: www.glaxowellcome.com
Trademarks: Lamictal

Technology: Chewable tablets
Brand Name: Lamictal lamotrigine) Chewable Dispersible Tablets

Gum Base Co. SpA
Lainate (Milano), Italy
Tel: +011 39 02 931721
Fax: +011 39 02 93570533
Internet: www.gumbase.com
Trademarks: MedGum Base
Technology: Medicated chewing gum base
Brand Names:

Idea AG
Frankfurter Ring 193a
Munich
D-80807 Germany, E.U.
Tel: +011 49 89 324 6330
Fax: +011 49 89 324 1684
Internet: www.idea-ag.de
Trademarks: Transfersome
Technology: Enhanced transdermal/transmucosal delivery system
Brand Names:

Johnson & Johnson Merck
Consumer Pharmaceuticals Co.
Camp Hill Road
Fort Washington, PA 10034
Tel: +1 215.233.7000
Internet: www.pepcidac.com
Trademarks: Pepcid AC
Technology: Fast-dissolving chewable tablets
Brand Name: Pepcid AC (famotidine) Chewable Tablets

KV Pharmaceutical Company
2503 South Hanley Road
St. Louis, MO 63144 USA
Tel: +1 314.645.6600
Fax: +1 314.645.6732
Internet: www.ethex.com
Trademarks: Site Release; OraSert; OraSite; OraQuick; Trans-EP; MicroMask
Technology: Quick-dissolving technology; bioadhesive technologies
Brand Names:

Kyowa Hakko Kogyo Co., Ltd.
1-6-1 Otemachi, Chiyoda-Ku
Tokyo 100-8185, Japan
Tel: +011 81 3 3282 0052
Fax: +011 81 3 3282 0113
Internet: www.kyowa.co.jp
Trademarks: Solblet
Technology: Fast disintegration in the mouth tablets
Brand Names:

Lavipharm Laboratories, Inc.
69 Princeton-Hightstown Road
Hightstown, NJ 08520 USA
Tel: +1 609.448.3001
Fax: +1 609.371.6522
Internet: www.Lavipharm.com
Trademarks: Quick-Dis; Slow-Dis
Technology: Intra-oral delivery systems (IODS); quick and slow-dissolving films
Brand Names:

Lederle Consumer Health
Division of Whitehall-Robbins Healthcare
Five Giralda Farms
Madison, NJ 07940
Tel: +1 800.282.8805
Fax: +1
Internet: www.whitehall-robins.com
Trademarks: Caltrate, Centrum
Technology: Chewable vitamin supplements
Brand Names: Caltrate 600 Plus Chewables (calcium carbonate, calcium 600 mg); Centrum Kids Complete Shamu and His Crew Children's Chewable Vitamin/Mineral Formula

Lohmann Therapie Systeme GmbH
Lohmannstrasse 2
Andernach
D – 56626 Germany
Tel: +011 49 26 32 99 23 64
Fax: +011 49 26 32 99 25 15
Internet: www.ltslohmann.com
Trademarks:
Technology: Oral and buccal wafers – quick-dissolving film technology
Brand Names:

LTS Corporation
Lohmann Therapy-Systems
21 Henderson Drive
West Caldwell, NJ 07006 USA
Tel: +1 973.575.5170
Fax: +1 973.575.5174
Internet: www.ltslohmann.com
Trademarks:
Technology: Oral and buccal wafers – quick dissolving film technology
Brand Names:

Macrochem Corporation
110 Hartwell Avenue
Lexington, MA 02421 USA
Tel: +1 781.862.4003
Fax: +1 781.862.4338
Internet: www.macrochem.com
Trademarks: SEPA
Technology: Transdermal/mucosal permeation enhancer technology
Brand Names:

Mallinckrodt, Inc.
675 McDonnell Boulevard
P.O. Box 5840
Saint Louis, MO 63134 USA
Tel: +1 800.325.8888
Internet: www.Mallinckrodt.com
Trademarks:
Technology: Dispersible tablets
Brand Names: Methadone Dispersible Tablets

Mayor Pharmaceutical Laboratories
KareMore International
2401 S. 24th Street
Phoenix, AZ 85034 USA
Tel: +1 602.244.8899
Fax: +1 888.527.3329
Internet: www.karemor.com
Trademarks: Vitamist
Technology: Intra-oral vitamin sprays
Brand Names: Vitamist Intra-Oral Spray (dietary supplements)

McNeil Consumer Products Company
Fort Washington, PA 19034 USA
Tel: +1 215.233.7000
Internet: www.tylenol.com
Trademarks: Tylenol; Imodium; Motrin; Lactaid; Nicotrol
Technology: Chewable tablets; soft chews
Brand Names: Children's Tylenol Soft Chews (acetaminophen); Junior Strength Tylenol Chewable Tablets (acetaminophen 160 mg); Imodium Advanced (loperamide HCl 2 mg/simethicone) Chewable Tablets; Children's Motrin (ibuprofen) Chewable Tablets; Junior Strength Motrin (ibuprofen 100 mg) Chewable Tablets; Lactaid (lactose enzyme) Chewable Tablets; Nicotrol (nicotine inhalation system) Inhaler

Medi-Ject Corporation
161 Cheshire Lane, Suite 100
Minneapolis, MN 55441 USA
Tel: +1 612.475.7700
Fax: +1 612.476.1009
Internet: www.medi-ject.com
Trademarks: Medi-Jector Vision
Technology: Needle-free injection technology
Brand Names:

Merck & Co., Inc
P.O. Box 4
West Point, PA 19486-0004 USA
Tel: +1 800.672.6372
Fax: +1 800.637.2568
Internet: www.merck.com; www.pepcid.com
Trademarks: Pepcid RPD; Pepdine; Maxalt; Mintezol; Singulair
Technology: Quick-dissolve – Orally disintegrating tablets (innovator: RP Scherer)
Brand Names: Pepcid RPD Orally Disintegrating Tablets (Famotidine); Maxalt (rizatriptan benzoate) Orally Disintegrating Tablets; Mintezol (thiabendazole, USP) Chewable Tablets; Singulair (montelukast sodium) Chewable Tablets

NexMed, Inc.
350 Corporate Blvd.
Robbinsville, NJ 08691 USA
Tel: +1 609.208.9688
Fax: +1 609.208.1868
Internet: www.nexmed.com

Trademarks: NexACT
Technology: Transdermal/transmucosal permeation enhancer technology
Brand Names:

Novartis Consumer Health
560 Morris Avenue
Summit, NJ 07901-1312 USA
Tel: +1 800.452.0051
Fax: +1 800.635.2801
Internet: www.novartis-us-pharma.com
Trademarks: Triaminic; Maalox; Sunkist; Tegretol
Technology: Soft fast-dissolving chewable tablets (innovator: Cima Labs)
Brand Names: Triaminic Softchews (cold & cough); Triaminic Softchews (throat pain & cough); Maalox Quick Dissolve Antacid (calcium carbonate) Chewable Tablets; Tegretol (carbamazepine USP) Chewable Tablets; Sunkist Children's Chewable Multivitamins-Complete; Sunkist Vitamin C Citrus Complex Chewable Tablets

Noven Pharmaceuticals, Inc.
11960 Southwest 144th Street
Miami, FL 33186 USA
Tel: +1 305.253.5099
Fax: +1 305.253.0525
Internet: www.noven.com
Trademarks: DentiPatch
Technology: Oral transmucosal patch
Brand Names: DentiPatch Transmucosal Lidocaine Delivery System

Novopharm, USA, Inc.
165 East Commerce Drive
Schaumburg, IL 60173-5326 USA
Tel: +1 800.426.0769
Fax: +1 847.781.1154
Internet: www.novopharmbiotech.ca
Trademarks:
Technology: Chewable tablets
Brand Names: Amoxicillin Chewable Tablets

OraPharma, Inc.
732 Louis Drive
Warminster, PA 18974
Tel: +1 215.956.2000
Fax: +1
Internet: www.orapharma.com

Trademarks: Arestin
Technology: Microspheres for targeted delivery to periodontal tissue for treatment of periodontal disease, mucositis, and dental pain
Brand Names:

Park-Davis
Division of Warner Lambert
201 Tabor Road
Morris Plains, NJ 07950
Tel: +1 800.223.0432
Fax: +1 973.540.2248
Internet: www.warnerlambert.com
Trademarks: Nitrostat
Technology: Sublingual tablets
Brand Names: Nitrostat (nitroglycerin tablet, USP)

Penjet™ Corporation
P.O. Box 6911
Beverly Hills, CA 90212 USA
Tel: +1 310.201.0800
Fax: +1 310.201.0800
Internet: www.penjet.com
Trademarks: Penject
Technology: Needleless injection technology
Brand Names:

Pentech Pharmaceutical, Inc.
417 Harvester Court
Wheeling, IL 60090 USA
Tel: +1 847.459.9122
Fax: +1 847.459.5602
Internet: www.pentech-inc.com
Trademarks:
Technology: Buccal and sublingual tablets
Brand Names:

Perio Products, Ltd.
7 Hamarpeh 57
Har Hotzuim Industrial Area
PO Box 23950
Jerusalem
91237 Israel
Tel: +011 972 22589 8200
Fax: +011 972 2581 2722

Internet: www.ppl.co.il
Trademarks: SLP; Periochip; PerioSense
Technology: Novel buccal delivery system; periodontal drug delivery; liquid polymer technology for intraoral delivery
Brand Names:

Permatec Pharma AG
Hardstrasse 18
CH-4132 Muttenz (Basel) Switzerland
Tel: +011 41 61 465 92 92
Fax: +011 41 61 465 92 91
Internet: www.permatec.com
Trademarks: Permatec Easy-Tec
Technology: Fast-dissolving tablet technology
Brand Names:

Pfizer, Inc.
Consumer Health Care Group
235 E. 42nd Street
New York, NY 10017-5755 USA
Tel: +1 212.733.5656
Fax: +1 212.973.7437
Internet: www.pfizer.com
Tradename: Bonnie; Feldene Melt
Technology: Chewable tablets
Brand Name: Bonnie (meclizine hydrochloride) Chewable Tablets

Pharmacia & Upjohn
100 Route 206 North
Peapack, NJ 07977 USA
Tel: +1 908.901.8000
Fax: +1 908.901.8379
Internet: www.pnu.com
Tradename: Nicorette; Nicotrol; Dramamine
Technology: Chewing gum; oral inhalation device; nicotine inhalation system
Brand Name: Nicorette (nicotine polacrilex) Chewing Gum; Dramamine Chewable Tablets (dimenhydrinate USP 50 mg)

Pharma Innovation SA
Via Cantonale
Mezzovico Ticino
CH-6805 Switzerland
Internet: www.pharmainnovation.ch
Trademarks:

Technology: Chewable tablet technology
Brand Names:

Procter & Gamble Pharmaceuticals, Inc.
Health Care Research Center
8700 Mason Montgomery Road
Mason, OH 45040 USA
Tel: +1 800.448.4878
Fax: +1
Internet: www.Procter&gamble.com
Trademarks: Vicks; Chloraseptic
Technology: Lozenge; liquid sprays
Brand Names: Vicks Chloraseptic Sore Throat Lozenges; Vicks Chloraseptic Sore Throat Spray

PowderJect Pharmaceuticals Plc
Florey House
Robert Robinson AvenueTel: +011 41 91 935 9110
Fax: +011 41 91 935 9119

The Oxford Science Park
Oxford, OX4 4GA ENGLAND
Tel: +1 44 0 1865 332600
Fax: +1 44 0 1865 332601
Internet: www.powderject.com
Trademarks: Powderject; Smart Particle
Technology: Needle-free dry powder injection, oral lidocaine
Brand Names:

PowderJect Technologies, Inc.
6511 Dumbarton Circle
Fremont, CA 94555 USA
Tel: +1 510.742.9700
Fax: +1 510.742.9720
Internet: www.powderject.com
Trademarks: Powderject; Smart Particle
Technology: Needle-free dry powder injection, oral lidocaine
Brand Names:

Purepac Pharmaceutical Co.
200 Elmora Avenue
Elizabeth, NJ 07207 USA
Tel: +1 908.527.9100
Fax: +1 908.527.0649

Internet: www.purepac.com
Trademarks:
Technology: Chewable tablets, generic pharmaceuticals
Brand Names: Amoxicillin Chewable Tablets

Prographarm Group
29 rue Vernet
Paris 75008 France
Tel: +011 33 1 53 57 4646
Fax: +011 33 1 53 57 4647
Internet: www.prographarm.com
Trademarks: Flashtab
Technology: Fast-dissolving tablets
Brand Names:

Rhone-Poulenc Rorer Pharmaceuticals, Inc.
500 Arcola Road
Collegeville, PA 19426-0107 USA
Tel: +1 610.454.2332
Fax: +1 610. 454.8110
Internet: www.RPR.com
Trademarks: Nitrolingual
Technology: Lingual spray formulations
Brand Names: Nitrolingual Spray (nitroglycerin lingual aerosol)

R&D Laboratories, Inc.
4640 Admiralty Way, Suite 710
Marina Del Rey, CA 90292 USA
Tel: +1 800.338.9066
Fax: +1 310.305.9229
Trademarks: Calci-Chew
Technology: Chewable tablets
Brand Name: Calci-Chew (calcium carbonate USP) Medical Food

Ross Products Division
Abbott Laboratories
6480 Busch Boulevard
Columbus, OH 43215
Tel: +1 800.988.8441
Internet: www.pedialyte.com
Trademarks: Pedialyte
Technology: Freezer pops (innovator: The Jel Sert Company)
Brand Name: Pedialyte Freezer Pops (oral electrolyte maintenance solution)

RP Scherer Corporation
2075 West Big Beaver Road
P.O. Box 160
Troy, MI 48084 USA
Tel: +1 813.572.4000
Fax: +1 813.572.1607
Internet: www.rpscherer.com
Trademarks: Zydis
Technology: Freeze-dried fast-dissolving dosage forms
Brand Names:

Scherer DDS
Division of RP Scherer Corporation
Frankland Road, Blagrove
Swindon, Wiltshire
SN5 8RU UK
Tel: +011 44 1793 488 411
Fax: +011 44 1793 548201
Internet: www.rpscherer.com
Trademarks: Zydis
Technology: Freeze-dried fast-dissolving dosage forms
Brand Names:

Schering Corporation
Galloping Hill Road
Kenilworth, NJ 07033 USA
Tel: +1 800.222.4000
Fax: +1 908. 820.6400
Internet: www.schering.com
Trademarks: Claritin®
Technology: Freeze-dried fast-dissolving tablets (innovator: YSP Pharma)
Brand Names: Claritin Reditabs (loratadine) Rapidly-Disintegrating Tablets

SmithKline Beecham Consumer Healthcare, L.P.
SmithKline Beecham Inc.
P.O. Box 1467
Pittsburg, PA 15230 USA
Tel: +1 800.233.2426
Internet: www.sb.com
Trademarks: Nicorette; Augmentin; Os-Cal
Technology: Medicated gum (innovator: Pharmacia); chewable tablets
Brand Names: Nicorette (nicotine polacrilex 2 and 4 mg) Gum; Augmentin (amoxicillin clavulanate potassium) Chewable Tablets; Os-Cal Chewable Calcium Supplements (calcium 500 mg)

Strakan Pharmaceutical
3 East Gate House
East Street, Andover
SP10 1EP UK
Tel: +011 44 1264 323 748
Fax: +011 44 1264 362 108
Internet: www.strakan.com
Trademarks: Residerm
Technology: Localized drug delivery
Brand Names:

The Jel Sert Company
Highway 59 and Conde Street
P.O. Box 261
West Chicago, IL 60186-0261 USA
Tel: +1 630.876.4845
Fax: +1 630.231.7681
Internet: www.jelsert.com
Trademarks: Jel Sert
Technology: Manufacturer of freezer pops
Brand Names:

The Parthenon Co., Inc.
3311 W. 2400 South
Salt Lake, UT 84119 USA
Tel: +1 801.972.4734
Fax: +1 801.972.4734
Trademark: Devrom
Technology: Chewable tablets
Brand Name: Devrom Chewable Tablets (bismuth subgallate powder)

Therics, Inc.
115 Campus Drive
Princeton, NJ 08540 USA
Tel: +1 609.514.7200
Fax: +1 609.514.7219
Internet: www.therics.com
Trademarks: TheriForm; Theriflash
Technology: Quick-dissolve drug-delivery system
Brand Names:

Toyobo Company, Ltd.
17-9 Nihombashi Koamicho

chuo-ku, Tokyo
Japan
Tel: +011 81 03 3660-4800
Internet: www.toyobo.co.jp
Trademarks:
Technology: Transmucosal therapeutic system (TmTs); gingival mucoadhesive tablets
Brand Names:

Watson Foods Co., Inc.
301 Heffernan Drive
West Haven, CT 06516
Tel: +1 203.932.3000
Fax: +1 203.968.9415
Trademarks:
Techology: Oral transmucosal film techology; bioadhesive films rapid dissolving film strips
Brand Names: private label

Watson Pharmaceuticals
311 Bonnie Circle
Corona, CA 91720 USA
Tel: +1 909.270.1400
Fax: +1 909.270.1428
Internet: www.watson.com
Trademarks:
Technology: Oral transmucosal technology; bioadhesive lozenges
Brand Names:

Warner-Lambert Company
Consumer Health Products Group
201 Tabor Road
Morris Plains, NJ 07950 USA
Tel: +1 800.223.0182
Internet: www.warnerlambert.com
Trademark: Celestial Seasonings; Soothers; Halls; Mentho-Lyptus; Benadryl; Sudafed
Technology: drops; chewable tablets; fast-melt tablets (innovator: Yamanouchi Pharma Technologies)
Brand Name: Celestial Seasonings Soothers Herbal Throat Drops; Halls Juniors Sugar Free Cough Suppressant Drops (menthol 2.5 mg); Halls Mentho-Lyptus Cough Suppressant Drops (menthol 7 mg); Halls Sugar Free Mentho-Lyptus Cough Suppressant Drops (menthol 5 mg); Maximum Strength Halls Plus Cough Suppressant Drops (menthol 10 mg); Halls Vitamin C Supplement

Drops (Vitamin C 60 mg); Benadryl Allergy Chewables (diphenhydramine hydrochloride 12.5 mg); Benadryl Allergy & Cold Fastmelt Tablets; Children's Sudafed (pseudoephedrine) Nasal Decongestant Chewables

WE Pharmaceuticals, Inc.
P.O. Box 1142
Ramona, CA 92065
Tel: +1 760.788.9155
Brandnames: Ah-Chew
Technology: Chewable tablets
Brand Names: Ah-Chew (chlorpheniramine maleate, phenylephrine, methscopolamine nitrate) Chewable Tablets; Ah-Chew D (phenylephrine) Chewable Tablets

Whitehall-Robins Healthcare
American Home Products
Five Giralda Farms
Madison, NJ 07940 USA
Tel: +1 800.322.3129
Internet: www.whitehall-robins.com; www.ahp.com
Trademarks: Dimetapp; Dimetapp Get Better Bear
Technology: Chewable tablets; quick-dissolve tablets (innovator: RP Scherer); Freezer Pops (innovator: The Jel Sert Company)
Brand Names: Dimetapp (Cold & Allergy) Chewable Tablets, Quick Dissolve Tablets; Dimetapp Get Better Bear Freezer Pops For the Throat (demulcent/pectin)

Wyeth-Ayerst Pharmaceuticals
Division of American Home Products Corporation
P.O. Box 8299
Philadelphia, PA 19101 USA
Tel: +1 610.688.4400
Internet: www.ahp.com
Trademarks: Temesta Expidet; Seresta Expidet; Isordil
Technology: Lyophilized tablet (innovator: Scherer DDS)
Brand Names: Temesta Expidet (lorazapam) Lyophilized Tablets; Seresta Expidet (oxazepam) Lyophilized Tablets; Isordil (isosorbide dinitrate) Sublingual Tablets

Yamanouchi Shaklee Pharma (YSPharm), Inc.
Stanford Research Park
1050 Arastradero Road
Palo Alto, CA 94304 USA
Tel: +1 650.849.8630

Fax: +1 650.849.8565
Internet: www.yspharma.com
Trademarks: Wowtab
Technology: Without water quick-dissolving tablet technology
Brand Names:

Yongxin Trading Company, Ltd.
227-3, Second Floor
Zhong Xiao Road
Sanduan
Taibei Taiwan
Tel: +011 02 752 8502
Internet: en.faucet-china.com/comindex_E.JSP?cid=member0044600
Trademarks: The Pullulan
Technology: Quick-dissolving film technology; breath freshener
Brand Names: The Pullulan Breath Freshener (manufactured in Osaka, Japan)

Appendix 2

Quick Reference Guide to Books and Market Research Reports on Companies Developing Intraoral Drug Delivery Technologies and Products

William R. Pfister

This section provides an easy to use reference guide to books and market research reports containing information on companies developing intraoral drug delivery technologies and commercial products worldwide. Every effort is made to ensure that the details are correct, however, if details change, please either fax changes to 609.208.1868, or e-mail pfister@mindspring.com so we can update this guide for the next edition.

A. Books

1. Bioadhesive Drug Delivery Systems: Fundamentals, Novel Approaches, and Development, E. Mathiowitz, D.E. Chickering, and C.M. Lehr, Eds., Marcel Dekker, Inc., New York, 1999, 696 pp.
2. Oral Mucosal Drug Delivery, M.J. Rathbone, Ed., Marcel Dekker, Inc., New York, NY, 1998, 464 pp.

B. Market Research Reports

1. Orally Disintegrating Tablet and Film Technologies, Second Edition, Technology Catalysts International Corporation, Technology Catalysts International Corporation, 605 Park Avenue, Falls Church, VA 22046 USA, Tel: + 1 703.531.0256, Fax: + 1 703.237.0042, E-mail: dbingham@technology-catalysts.com; www.technology-catalysts.com, September 2004.
2. Novel Drug Delivery Systems Reports – NDDS, 22nd edition, January 2004, Oral Drug Delivery Systems and Materials, Delivery of Proteins and Peptides, Transdermal and Transmucosal Drug Delivery Systems, Drug Delivery Pipelines Online, Drug Delivery Deals Online, Technology Catalysts International Corporation, 605 Park Avenue, Falls Church, VA 22046 USA, Tel: + 1 703.237.9600, Fax: + 1 703.237.7967, E-mail: info@technology-catalysts.com; www.technology-catalysts.com.
3. Commercial Success in Drug Delivery, An Analysis of Key Technologies and Products, Reuters Healthcare Reports, Business Insights Reports, Charles House, 108-110 Finchley Road, London, NW3 5JJ, United Kingdom, Tel: + 1 44 20 7675 0990, Fax: + 1 44 20 7675 7533, www.reutersbusinessinsight.com, June 2003.
4. The Drug Delivery Companies Report 2003, Driving Drug Delivery Partnerships Worldwide, PharmaVentures, 2003.
5. Oral Fast-Dissolving Drug Delivery, August 2003, Technology Catalysts International Corporation, 605 Park Avenue, Falls Church, VA 22046 USA, Tel: + 1 703.531-0256, Fax: + 1 703.237-0042, E-mail: info@technology-catalysts.com; www.technology-catalysts.com.
6. Oral Drug Delivery Systems and Materials, Technology Assessment & Business Development, 20th ed., January 2002, Technology Catalysts International Corporation, 605 Park Avenue, Falls Church, VA 22046 USA, Tel: + 1 703.237.9600, Fax: + 1 703.237.7967, E-mail: info@technology-catalysts.com
7. Transdermal and Transmucosal Delivery Systems, Technology Assessment & Business Opportunities, 20th Ed., January 2002, Technology Catalysts International Corporation, 605 Park Avenue, Falls Church, VA 22046 USA, Tel: + 1 703.237.9600, Fax: + 1 703.237.7967, E-mail: info@technology-catalysts.com
8. The Drug Delivery Companies Report 2000, Commercial Opportunities within the Drug Delivery Industry, PharmaVentures Ltd., Strategic Intelligence Services, Magdalen Centre, Oxford Science Park, Oxford OX4 4GA, UK, Tel: + 1 44 0 1865 784177, Fax: + 1 44 0 1865 784178.
9. Advanced Drug Delivery Systems: New Developments, New Technologies, Business Communications Co., March 1, 2001.
10. Drug Delivery Technology: U.S. Markets & Developments, Theta Reports, June 1, 2001.
11. Drug Delivery Systems, Freedonia Group, July 1, 2001.
12. Drug Delivery 2000: A Key Company and Market Analysis, Datamonitor, January 1, 2001.
13. Drug Delivery Review 2001/01-Company and Market Analysis, Datamonitor, November 30, 2001.
14. Cremer, K., Drug Delivery Report: Orally Disintegrating Dosage Forms, Pharma Concepts GmbH & Co. KG, Box 58 04 13, D-10414, Germany, Fax: + 001 49 30 4473 4683, www.pharma-concepts.com, April 2001.

15. Drug Delivery Outlook 2001, Reuters Business Insight, July 1, 2001
16. Drug Delivery Technologies, AdvanceTech Monitor, September 1, 2001.
17. Novel Drug Delivery Systems, 16th edition, Delivery of Proteins and Peptides, Oral Drug Delivery Systems and Materials, Transdermal and Transmucosal Drug Delivery, Unlaunched Candidate Drugs for Controlled Delivery, Technology Catalysts International Corporation, 605 Park Avenue, Falls Church, VA 22046 USA, Tel: + 1 703.237.9600, Fax: + 1 703.237.7967, E-mail: info@technology-catalysts.com.

Appendix 3

Glossary: Abbreviations and Definitions

William R. Pfister

ε: Area fraction of the paracellular route.

μ: Process mean.

σ: Standard deviation.

6σ: Process range.

ACE: Angiotensin converting enzyme.

ADME: Adsorption, disposition, metabolism, and elimination.

ADS: Atrigel delivery system.

AIDS: Autoimmune deficiency disease.

AIDS: Auto immune deficiency syndrome.

ANC: Acid neutralization capacity.

ANDA: Abbreviated new drug application.

ATP: Adenosine triphosphate.

AUC: Area under the plasma concentration time curve.

$AUC_{0\text{-}\infty}$: Area under the plasma drug concentration time curve from time zero extrapolated to infinity.

AUC_0: Area under the drug plasma concentration time curve.

$AUC_{0\text{-}t}$: Area under the plasma drug concentration time curve from time zero to the time of last collection.

BA: Bioavailability. The rate and extent to which the active ingredient or active moiety is absorbed from the drug product and becomes available at the site of action.

BCS: Biopharmaceutics classification system.

BE: Bioequivalence. The absence of a significant difference in the rate and extent to which the active ingredient or active moiety in pharmaceutical equivalents or pharmaceutical alternatives becomes available at the site of drug action when administered at the same molar dose under similar conditions in an appropriately designed study. Where there is an intentional difference in rate (e.g., in certain controlled release dosage forms), certain pharmaceutical equivalents or alternatives may be considered bioequivalent if there is no significant difference in the extent to which the active ingredient from each product becomes available at the site of drug action. This applies only if the difference in the rate at which the active ingredient or moiety becomes available at the site of drug action is intentional and is reflected in the proposed labeling, is not essential to the attainment of effective body drug concentrations on chronic use, and is considered medically insignificant for the drug.

BS Batch size.

$C_{D:}$ Concentration of drug in the donor chamber.

CDER: Center for Drug Evaluation and Research.

CFR: Code of federal regulations.

cGMP: Current good manufacturing practices.

CGS 16617: Angiotensin converting enzyme inhibitor.

CHAPS: 3-3-Cholamidopropyl-dimethyl ammonio-1-propane sulfonate.

CLSM: Confocal laser scanning microscopy.

C_{max}: Maximum drug plasma concentration.

CM: Compendial Methods.Those described in the current edition of the United States Pharmacopeia (USP) and National Formulary (NF).

CMC: Chemistry, manufacturing, and controls.

CMC: Carboxymethylcellulose.

CNS: Central nervous system.

CP: Carbopol.

C_p: Potential capability.

CPB: Clinical pharmacology and biopharmaceutics.

C_{pk}: Performance index.

C_{pl} : Upper capability index.

C_{pu}: Lower capability index.

CR: Controlled oral.

CRIOD: Controlled release intraoral delivery device.

CRT: Cathode ray tube.

CT: Conventional tablet.

C_{trough}: Trough plasma concentration.

D: Diffusion coefficient.

d: Specification range.

D_c: Bottom plate diameter at commercial scale.

DDPC: Didecanoyl phosphatidyl choline.

DEC: Decomposition temperature.

DFC: Dilute formocresol.

DGAVP: Desglycinamide-arginine-vasopressin (a decapeptide).

DMSO: Dimethylsulfoxide.

DOE: Design of experiments.

D_p: Bottom plate diameter at pilot scale.

$D_{P:}$ Diffusion coefficient in the intercellular spaces.

$D_{T:}$ Diffusion coefficient in the lipophilic phase.

DTC: Direct-to-consumer.

EDTA: Ethylenediamine tetracetic acid.

EM: Electron microscopic.

Eu: Eudragit.

EVA: Ethylene vinyl acetate.

FD: Fluorescein isothiocyanate labeled dextrans.

FDA Food and Drug Administration (Agency).

FIP: International Pharmaceutical Federation.

FS: Ferric sulfate.

G.I: Gastrointestinal.

GC/MS: Gas chromatography/mass spectroscopy.

GDC: Glycodeoxycholate.

GDC: Sodium glycodeoxycholate.

GERD: Gastrointestinal reflux reflux disease.

GI: Gastrointestinal.

GIT: Gastrointestinal tract.

GMO: Glycerol monooleate.

GMP: Good Manufacturing Practice.

GRAS: Generally recognized as safe.

HC: Hydrocortisone.

HCT: Calcitonin.

HEC: Hydroxyethyl cellulose.

HLB: Hydrophilic-lipophilic balance.

h_P: Length of the paracellular route.

HPC: Hydroxypropylcellulose.

HPCD: Hydroxypropyl cyclodextrin.

HPMC: Hydroxypropylmethylcellulose.

h_T: Length of the transcellular route.

ICH: International Conference on Harmonization.

IM: Intramuscular injection.

IND: Investigational new drug.

IOD: Intraoral dosage form.

IODD: Intraoral drug delivery.

IODF: Intraoral dosage forms.

IODS: Intraoral delivery system.

IR: Immediate release.

IRS: Immediate release spray.

IS5MN: Isosorbide 5-mononitrate.

ISDN: Isosorbide dinitrate.

ISIS 3082: 20 mer with a sequence of 5′ - TGC ATC CCC CAG GCC ACC AT - 3′

IV: Intravenous.

J_P: Steady-state flux.

J_T: Transcellular pathway and the steady state flux.

k: Terminal elimination rate constant.

K: Non-centering correction factor.

KDa: KiloDaltons.

K_P: Partition coefficient between the lipophilic phase and the aqueous hydrophilic donor phase.

LC/MS/MS: Liquid chromatography/mass spectroscopy/mass spectroscopy.

LD50: Lethal dose resulting in 50 percent mortality.

LHRH: Leutinizing hormone releasing hormone.

LMDS: Liquid metered dose spray.

LO: Lozenge.

LOD: Loss on drying.

LPC: Lysophosphatidylcholine.

LSL: Lower specification limit.

MAD: Mucoadhesive bioerodable pre-formed disc.

MAO: Monoamine oxidase.

MCGs: Membrane coating granules.

MDT: Mouth dissolving tablet.

Melatonin: {N-[2-(5-Methoxy-1H-indo-3-yl)ethyl]acetamide} or N-acetyl-5-methoxytryptamine.

MR: Modified release.

MW: Molecular weight.

NaGC: Sodium glycocholate.

NCE: New chemical entity.

ND: Non-dissolving.

NDA: New drug application.

NME New molecular entity.

NMP: N-Methylpyrrolidone.

NO: Nitric oxide

NR: Nicotine replacement.

NRT: Nicotine replacement therapy.

NSAIDS: Nonsteroidal antiinflammatory drugs.

NTG: Nitroglycerin

NTS: Nicotine transdermal system.

ODT: Orally disintegrating tablet.

OGD: Office of Generic Drugs.

OND: Office of New Drugs.

OT: Oral transmucosal.

OTC: Over-the-counter.

OTT: Oral transmucosal tablet.

P: Permeability ($cm.sec^{-1}$).

PA: Pharmacological activity.

PA: Pharmaceutical alternatives. Drug products that contain the identical therapeutic moiety, or its precursor, but not necessarily in the same amount or dosage form or as the same salt or ester. Each such drug product individually meets either the identical or its own respective compendial or other applicable standard of identity, strength, quality, and purity, including potency and, where applicable, content uniformity, disintegration times and/or dissolution rates.

PAA: Polyacrylic acid.

PAT: Process analytical technology.

PakSolv®: Specially designed package and processing system.

PE: Pharmaceutical equivalents. Drug products in identical dosage forms that contain identical amounts of the identical active drug ingredient, i.e., the same salt or ester of the same therapeutic moiety, or, in the case of modified release dosage forms that require a reservoir or overage or such form as prefilled syringes where residual volume may vary, that deliver identical amounts of the active drug ingredient over the identical dosing period; do not necessarily contain the same inactive ingredients; and meet the identical compendial or other applicable standard of identity, strength, quality, and purity, including potency and, where applicable, content uniformity, disintegration times and/or dissolution rates.

PEG: Polyethylene glycol.

PHEMA: Polyhydroxyethyl methacrylic acid.

PIB: Polyisobutylene.

PK: Pharmacokinetics. The study of the time course of drug absorption, distribution, metabolism, and elimination.

pK_a: pH corresponding to an equal concentration of ionized and unionized drug.

PMA: Polymethacrylic acid.

PNP: p-Nitrophenol.

PQD: Product quality and design.

PVA: Polyvinyl alcohol.

PVC: Polyvinylchloride.

PVDC: Polyvinylchloride.

PVdC: Polyvinyldichloride.

PVP: Polyvinylpyrrolidone.

QC: Quality control.

QD: Quick dispersing oral drug delivery systems.

QD: Quick dissolving

QDCT: Quick dissolve chewable tablets.

QDF: Quick dissolving film.

QDOD: Quick dissolving oral dosage forms.

QDT: Quick dissolving tablet.

QDW: Quick dissolving wafer.

QOC: Quality of conformance.

QOD: Quality of design.

QOD: Quick dissolving intraoral dosage forms.

RCT: Root canal treatment.

RDD: Rapidly disintegrating/dissolving dosage form.

RDT: Rapidly disintegrating tablet.

RLD: Reference listed drug.

RM: Rapid melt.

RMT: Rapid melt tablet.

RSD: Relative standard deviation.

Rx: Prescription.

s: Sample standard deviation.

SCN: Suprachiasmatic nucleus.

SD: Slow dissolving.

SDST: Slow dissolving sublingual tablet.

SP: Suitability petition.

SPC: Statistical process capability.

SRP: Scaling and root planning.

ST: Sublingual table.

STDHF: Sodium taurodihydrofusidate.

$T_{1/2}$: Plasma half-life, or drug elimination half-life.

TDD: Transdermal.

TEER: Transepithelial electrical resistance.

T_g: Glass transition temperature.

TGF-: Transforming growth factor-.

T_{max}: Time to reach maximum plasma concentration.

TMD: Transmucosal drug delivery.

TMDS: Transmucosal delivery systems.

TRH: Thyrotropin releasing hormone (L-pyroglutamyl-L-histidyl-L-proline amide) (a tripeptide).

US: United States.

USL: Upper specification limit.

USP: United States Pharmacopoeia.

VAS: Visual analogue scale, median pain scale.

VPS: Verbal pain scale.

WOWTAB: Without water tablet.

x: Sample mean.

ZOE: Zinc oxide-eugenol paste.

δ: Thickness.

Index